AF608519

Error Inequalities in Polynomial Interpolation and Their Applications

Mathematics and Its Applications

Volume 262

Error Inequalities in Polynomial Interpolation and Their Applications

by

Ravi P. Agarwal
Department of Mathematics,
National University of Singapore,
Kent Ridge, Singapore

and

Patricia J. Y. Wong
Division of Mathematics,
Nanyang Technological University,
Singapore

SPRINGER SCIENCE+BUSINESS MEDIA, B.V.

Library of Congress Cataloging-in-Publication Data

Agarwal, Ravi P.
Error inequalities in polynomial interpolation and their applications / by Ravi P. Agarwal and Patricia J.Y. Wong.
p. cm. -- (Mathematics and its applications ; v. 262)
Includes bibliographical references and index.
ISBN 978-0-7923-2337-2 ISBN 978-94-011-2026-5 (eBook)
DOI 10.1007/978-94-011-2026-5
1. Interpolation. 2. Approximation theory. 3. Polynomials.
I. Wong, Patricia J. Y. II. Title. III. Series: Mathematics and its applications (Kluwer Academic Publishers) ; v. 262.
QA281.A33 1993
511'.4--dc20 93-15913

ISBN 978-0-7923-2337-2

Printed on acid-free paper

CONTENTS

Preface ix

Chapter 1 LIDSTONE INTERPOLATION

1.1 Introduction 1

1.2 Lidstone Polynomials 2

1.3 Interpolating Polynomial Representations 17

1.4 Error Representations 17

Peano's Representation 18

Cauchy's Representation 18

1.5 Error Estimates 19

1.6 Lidstone Boundary Value Problems 21

Existence and Uniqueness 21

Picard's and Approximate Picard's Iteration 29

Quasilinearization and Approximate Quasilinearization 43

References 59

Chapter 2 HERMITE INTERPOLATION

2.1 Introduction 62

2.2 Interpolating Polynomial Representations 63

Method of Lagrange 63

Method of Newton 67

2.3 Error Representations 71

Cauchy's Representation 71

Newton's Representation 72

Peano's Representation 73

A New Representation 77

2.4 Error Estimates 91

Error Estimates in Interpolation 91

Error Estimates for Derivatives 105

2.5 Some Applications 149

Generalized Maximum Principle 149

Hermite Boundary Value Problems 151

Generalized Liapunoff's Inequality 161

Generalized Hartman's Inequality 162

A Lower Bound for the Zeros of the Solutions 164

A Test for Disconjugacy 166

References 168

Chapter 3 ABEL - GONTSCHAROFF INTERPOLATION

3.1 Introduction 172

3.2 Interpolating Polynomial Representations 173

3.3 Error Representations 175

Repeated Integrals Representation 175

Peano's Representation 175

Cauchy's Representation 177

3.4 Error Estimates 178

3.5 Some Applications 186

References 189

Chapter 4 MISCELLANEOUS INTERPOLATION

4.1 Introduction 192

4.2 (n, p) and (p, n) Interpolation 192

4.3 $(0, 0; m, n-m)$ Interpolation 198

4.4 $(0; m, n-m)$ Interpolation 203

4.5 $(0, 2, 0; m, n-m)$ Interpolation 206

4.6 $(0 : l-1, l : l+j-1; m, n-m)$ Interpolation 208

4.7 $(0; \text{Lidstone})$ Interpolation 209

4.8 $(0,2,0;\text{Lidstone})$ Interpolation 210

4.9 $(1,3,0,1;\text{Lidstone})$ Interpolation 211

4.10 $(0:l-1,l:l+j-1;\text{Lidstone})$ Interpolation 213

4.11 $(0,2,1;\text{Lidstone})$ Interpolation 214

References 216

Chapter 5 PIECEWISE - POLYNOMIAL INTERPOLATION

5.1 Introduction 217

5.2 Preliminaries 218

5.3 Piecewise Hermite Interpolation 228

5.4 Piecewise Lidstone Interpolation 257

5.5 Two Variable Piecewise Hermite Interpolation 266

5.6 Two Variable Piecewise Lidstone Interpolation 274

References 278

Chapter 6 SPLINE INTERPOLATION

6.1 Introduction 281

6.2 Preliminaries 282

6.3 Cubic Spline Interpolation 285

6.4 Quintic Spline Interpolation : $\tau = 4$ 290

6.5 Approximated Quintic Splines : $\tau = 4$ 301

6.6 Quintic Spline Interpolation : $\tau = 3$ 309

6.7 Approximated Quintic Splines : $\tau = 3$ 315

6.8 Cubic Lidstone - Spline Interpolation 321

6.9 Quintic Lidstone - Spline Interpolation 323

6.10 L_2 - Error Bounds for Spline Interpolation 327

6.11 Two Variable Spline Interpolation 334

6.12 Two Variable Lidstone - Spline Interpolation 346

6.13 Some Applications 350

Linear Integral Equations 351

References 360

Name Index 363

Preface

Given a function $x(t) \in C^{(n)}[a,b]$, points $a = a_1 < a_2 < \cdots < a_r = b$ and subsets α_j of $\{0,1,\cdots,n-1\}$ with $\sum_{j=1}^r card(\alpha_j) = n$, the classical interpolation problem is to find a polynomial $P_{n-1}(t)$ of degree at most $(n-1)$ such that

$$P_{n-1}^{(i)}(a_j) = x^{(i)}(a_j) \;\; \text{for } i \in \alpha_j, \; j = 1,2,\cdots,r.$$

In the first four chapters of this monograph we shall consider respectively the cases: the Lidstone interpolation ($a = 0, b = 1, n = 2m, r = 2, \alpha_1 = \alpha_2 = \{0,2,\cdots,2m-2\}$), the Hermite interpolation ($\alpha_j = \{0,1,\cdots,k_j-1\}$), the Abel - Gontscharoff interpolation ($r = n, a_i \le a_{i+1}, \alpha_j = \{j-1\}$), and the several particular cases of the Birkhoff interpolation. For each of these problems we shall offer:

(1) explicit representations of the interpolating polynomial;

(2) explicit representations of the associated error function $e(t) = x(t) - P_{n-1}(t)$; and

(3) explicit optimal/sharp constants $C_{n,k}$ so that the inequalities

$$| e^{(k)}(t) | \le \; C_{n,k}(b-a)^{n-k} \max_{a \le t \le b} | x^{(n)}(t) |, \;\; 0 \le k \le n-1$$

are satisfied.

In addition, for the Hermite interpolation we shall provide explicit optimal/sharp constants $C(n,p,\nu)$ so that the inequality

$$\| e(t) \|_p \le \; C(n,p,\nu) \, \| x^{(n)}(t) \|_\nu, \;\; p,\nu \ge 1$$

holds.

Although these results are of fundamental importance in every aspect of numerical mathematics, we shall demonstrate their significance in the theory of ordinary differential equations such as maximum principles, boundary value problems, oscillation theory, disconjugacy and disfocality.

Polynomial interpolation often produces approximations that are wildly oscillatory. To overcome this difficulty we divide the interval $[a,b]$ into small intervals and in each subinterval consider polynomials of relatively low degree and finally 'piece together' these polynomials. This subject has steadily developed over the past fifty years, and at present there are thousands of research papers on piecewise - polynomial interpolation and their applications.

In the fifth chapter of this monograph we shall consider the piecewise Hermite and Lidstone interpolating problems and for each of these provide:

(i) explicit representations of the piecewise interpolating polynomial;

(ii) explicit error bounds for the derivatives of cubic and quintic piecewise interpolation in L_∞- norm; and

(iii) explicit error bounds for the derivatives of arbitrary order piecewise interpolation in L_2- norm.

In addition, these results are extended to two variable piecewise - polynomial interpolation.

Spline interpolation is an improvement over piecewise - polynomial interpolation. It uses less information of the given function, yet furnishes smoother interpolates.

In the final chapter of this monograph we shall consider the spline and Lidstone - spline interpolating problems and for each of these give:

(a) explicit representations of the interpolating spline; and

(b) explicit error bounds for the derivatives of cubic and quintic interpolating splines in L_∞- norm.

In addition, for the spline interpolation we shall obtain:

(α) explicit representations of the approximated splines for the quintic case, and precise error bounds for their derivatives in L_∞-norm; and

(β) explicit error bounds for the derivatives of arbitrary order interpolating spline in L_2- norm.

Finally, these results are generalized to two variable spline interpolation.

Throughout final chapter, we also give numerical illustrations of the importance as well as the sharpness of the results which are presented.

This work is a cumulation of the authors' research in this field, extending over a period of ten years. We hope reading this monograph will be a pleasure and rewarding as well.

Ravi P.Agarwal

Patricia J.Y.Wong

CHAPTER 1

LIDSTONE INTERPOLATION

1.1 INTRODUCTION

In the year 1929 Lidstone [15] introduced a generalization of Taylor's series, it approximates a given function in the neighborhood of two points instead of one. From the practical point of view such a development is very useful; and in terms of completely continuous functions it has been characterized in the work of Boas [9], Poritsky [19], Schoenberg [20], Whittaker [28, 29], Widder [30, 31], and others. In the field of approximation theory [12,27] the *Lidstone interpolating polynomial* $P_{(1.1.1)}(t)$ of degree $(2m-1)$ satisfies the *Lidstone conditions*

$$P^{(2i)}_{(1.1.1)}(0) = \alpha_i, \ P^{(2i)}_{(1.1.1)}(1) = \beta_i; \ 0 \le i \le m-1. \tag{1.1.1}$$

The plan of this chapter is as follows : In Section 1.2 we shall introduce Lidstone polynomials $\Lambda_n(t)$ of degree $(2n+1)$, provide their explicit representations and give their relations with Bernoulli and Euler's polynomials. Here, we shall also establish several equalities and inequalities involving the

Lidstone polynomials. While some of these inequalities will be needed later, others compare sharply with the several supporting results of Boas [8] and Widder [31] and hence are of independent interest. In Section 1.3 we shall use these representations of $\Lambda_n(t)$ to construct the Lidstone interpolating polynomial $P_{(1.1.1)}(t)$. In Section 1.4 we shall provide two different forms of the error function $e_{(1.1.1)}(t) = x(t) - P_{(1.1.1)}(t)$, where $x(t) \in C^{(2m)}[0,1]$ and $P_{(1.1.1)}(t)$ is the Lidstone interpolating polynomial of the function $x(t)$, i.e., it satisfies the conditions $P^{(2i)}_{(1.1.1)}(0) = x^{(2i)}(0)$, $P^{(2i)}_{(1.1.1)}(1) = x^{(2i)}(1)$, $0 \le i \le m-1$. Best possible pointwise as well as uniform bounds for $| e^{(i)}_{(1.1.1)}(t) |$, $0 \le i \le 2m-1$ are obtained in Section 1.5. Finally, in Section 1.6 we shall show that the results of Sections 1.2 - 1.5 are of fundamental importance in the study of Lidstone boundary value problems.

1.2 LIDSTONE POLYNOMIALS

Definition 1.2.1. The unique polynomial $\Lambda_n(t)$ of degree $(2n+1)$ defined by the relations

$$\begin{aligned} &\Lambda_0(t) = t \\ &\Lambda_n''(t) = \Lambda_{n-1}(t) \\ &\Lambda_n(0) = \Lambda_n(1) = 0, \ n \ge 1 \end{aligned} \tag{1.2.1}$$

is called *Lidstone polynomial.*

Lemma 1.2.1. The Lidstone polynomial $\Lambda_n(t)$ can be expressed as

$$\Lambda_n(t) = \int_0^1 g_n(t,s)s\, ds, \ n \ge 1 \tag{1.2.2}$$

where

$$g_1(t,s) = \begin{cases} (t-1)s, & s \le t \\ (s-1)t, & t \le s \end{cases} \tag{1.2.3}$$

$$g_n(t,s) = \int_0^1 g_1(t,t_1)g_{n-1}(t_1,s)\, dt_1, \ n \ge 2. \tag{1.2.4}$$

Proof. The proof uses a simple induction. From the theory of differential equations it is clear that the solution $\Lambda_1(t)$ of the boundary value problem

$$\Lambda_1''(t) = t$$

$$\Lambda_1(0) = \Lambda_1(1) = 0$$

can be written as

$$\Lambda_1(t) = \int_0^1 g_1(t,s)s\ ds.$$

Next, if (1.2.2) is true for $n \geq 1$, then the solution $\Lambda_{n+1}(t)$ of the boundary value problem

$$\Lambda_{n+1}''(t) = \int_0^1 g_n(t,s)s\ ds$$

$$\Lambda_{n+1}(0) = \Lambda_{n+1}(1) = 0$$

can be written as

$$\begin{aligned} \Lambda_{n+1}(t) &= \int_0^1 \left[g_1(t,t_1) \int_0^1 g_n(t_1,s)s\ ds \right] dt_1 \\ &= \int_0^1 \left[\int_0^1 g_1(t,t_1) g_n(t_1,s)\ dt_1 \right] s\ ds \\ &= \int_0^1 g_{n+1}(t,s)s\ ds. \quad \blacksquare \end{aligned}$$

Remark 1.2.1. $g_1(t,s) \leq 0,\ 0 \leq s,t \leq 1$ is obvious from (1.2.3). Thus, $0 \leq (-1)^n g_n(t,s) = |\ g_n(t,s)\ |,\ 0 \leq s,t \leq 1$ follows from (1.2.4). Hence, in view of (1.2.2) we have $(-1)^n \Lambda_n(t) \geq 0,\ 0 \leq t \leq 1$. ■

Lemma 1.2.2. The following equality holds

$$\int_0^1 g_n(t,s) \sin k\pi s\ ds = (-1)^n \frac{1}{(k\pi)^{2n}} \sin k\pi t,\ 0 \leq t \leq 1 \tag{1.2.5}$$

where k is a positive integer.

Proof. Since

$$\begin{aligned}\int_0^1 g_1(t,s)\sin k\pi s\,ds &= -(1-t)\int_0^t s\sin k\pi s\,ds - t\int_t^1(1-s)\sin k\pi s\,ds\\ &= -(1-t)\left[-\frac{1}{k\pi}t\cos k\pi t+\frac{1}{(k\pi)^2}\sin k\pi t\right]\\ &\quad -t\left[\frac{1}{k\pi}(1-t)\cos k\pi t+\frac{1}{(k\pi)^2}\sin k\pi t\right]\\ &= -\frac{1}{(k\pi)^2}\sin k\pi t\end{aligned}$$

the equality (1.2.5) easily follows from (1.2.4) by using an inductive argument. ∎

Corollary 1.2.3. The following equality holds

$$\int_0^1 |\ g_n(t,s)\ |\sin \pi s\,ds = \frac{1}{\pi^{2n}}\sin\pi t,\ \ 0\le t\le 1. \tag{1.2.6}$$ ∎

Lemma 1.2.4. The Lidstone polynomial $\Lambda_n(t)$ can be expressed as

$$\Lambda_n(t) = (-1)^n\frac{2}{\pi^{2n+1}}\sum_{k=1}^{\infty}\frac{(-1)^{k+1}}{k^{2n+1}}\sin k\pi t,\ \ n\ge 1. \tag{1.2.7}$$

Proof. Starting with the familiar Fourier series

$$s = \frac{2}{\pi}\sum_{k=1}^{\infty}\frac{(-1)^{k+1}}{k}\sin k\pi s,\ \ 0<s<1$$

we multiply the series by $g_n(t,s)$ and integrate with respect to s from 0 to 1. The relation (1.2.7) is then immediate from (1.2.2) and (1.2.5). ∎

Remark 1.2.2. It is well known (e.g. see Luke [16, p.23]) that the Bernoulli polynomial $B_{2n+1}(t)$ of degree $(2n+1)$ can be expressed as

$$B_{2n+1}(t) = (-1)^{n+1}\frac{2}{(2\pi)^{2n+1}}(2n+1)!\sum_{k=1}^{\infty}\frac{1}{k^{2n+1}}\sin 2k\pi t \tag{1.2.8}$$

and hence

$$\frac{2^{2n+1}}{(2n+1)!}B_{2n+1}\left(\frac{1+t}{2}\right) = (-1)^n\frac{2}{\pi^{2n+1}}\sum_{k=1}^{\infty}-\frac{1}{k^{2n+1}}\sin(k\pi+k\pi t)$$

i.e., it follows that

$$\Lambda_n(t) = \frac{2^{2n+1}}{(2n+1)!} B_{2n+1}\left(\frac{1+t}{2}\right). \tag{1.2.9}$$

This relation between Lidstone and Bernoulli polynomials is due to Whittaker [28]. ∎

Remark 1.2.3. Another explicit representation of Lidstone polynomial $\Lambda_n(t)$ is given by

$$\Lambda_n(t) = \frac{1}{6}\left[\frac{6t^{2n+1}}{(2n+1)!} - \frac{t^{2n-1}}{(2n-1)!}\right] - \sum_{k=0}^{n-2} \frac{2(2^{2k+3}-1)}{(2k+4)!} B_{2k+4} \times \frac{t^{2n-2k-3}}{(2n-2k-3)!}, \quad n \ge 1 \tag{1.2.10}$$

where B_{2k+4} is the $(2k+4)$th Bernoulli number. The proof of (1.2.10) is by induction. Indeed, for $n = 1$ the relation (1.2.2) gives

$$\Lambda_1(t) = \int_0^1 g_1(t,s)s\, ds = \int_0^t (t-1)s^2\, ds + \int_t^1 t(s^2 - s)\, ds = \frac{1}{6}t^3 - \frac{1}{6}t,$$

which is the same as (1.2.10) for $n = 1$. Now assuming (1.2.10) to be true for n, then from (1.2.7) it follows that

$$\frac{1}{6}\left[\frac{6t^{2n+1}}{(2n+1)!} - \frac{t^{2n-1}}{(2n-1)!}\right] - \sum_{k=0}^{n-2} \frac{2(2^{2k+3}-1)}{(2k+4)!} B_{2k+4} \frac{t^{2n-2k-3}}{(2n-2k-3)!} \tag{1.2.11}$$

$$= (-1)^n \frac{2}{\pi^{2n+1}} \sum_{k=1}^{\infty} \frac{(-1)^{k+1}}{k^{2n+1}} \sin k\pi t.$$

Integrating (1.2.11) twice from 0 to t, we obtain

$$\frac{1}{6}\left[\frac{6t^{2n+3}}{(2n+3)!} - \frac{t^{2n+1}}{(2n+1)!}\right] - \sum_{k=0}^{n-2} \frac{2(2^{2k+3}-1)}{(2k+4)!} B_{2k+4} \frac{t^{2n-2k-1}}{(2n-2k-1)!}$$

$$= (-1)^n \frac{2}{\pi^{2n+1}} \sum_{k=1}^{\infty} \frac{(-1)^{k+1}}{k^{2n+1}} \left[\frac{1}{k\pi}\left(t - \frac{1}{k\pi}\sin k\pi t\right)\right]$$

$$= (-1)^{n+1} \frac{2}{\pi^{2n+3}} \sum_{k=1}^{\infty} \frac{(-1)^{k+1}}{k^{2n+3}} \sin k\pi t + (-1)^n \frac{2}{\pi^{2n+2}} \sum_{k=1}^{\infty} \frac{(-1)^{k+1}}{k^{2n+2}} t$$

$$= \Lambda_{n+1}(t) - (-1)^{n+1} \frac{2}{\pi^{2n+2}} \frac{(2^{2n+1}-1)}{(2n+2)!} \pi^{2n+2} \mid B_{2n+2} \mid t, \tag{1.2.12}$$

where in (1.2.12) we have used another well known (Jordan [14, p. 244]) relation

$$\sum_{k=1}^{\infty}\frac{(-1)^{k+1}}{k^{2n+2}} = \frac{(2^{2n+1}-1)}{(2n+2)!}\pi^{2n+2}\mid B_{2n+2}\mid . \tag{1.2.13}$$

However, since $(-1)^n B_{2n+2} > 0$ from (1.2.12) the relation (1.2.10) for $(n+1)$ follows immediately. ∎

Remark 1.2.4. Equating (1.2.9) and (1.2.10), we get the polynomial expansion of $B_{2n+1}\left(\frac{1+t}{2}\right)$. Further, as a consequence of this equality it directly follows that $B_{2n+1}\left(\frac{1}{2}\right) = 0$. ∎

Lemma 1.2.5. The following expansion holds

$$\int_0^1 g_n(t,s)\,ds = (-1)^n\frac{4}{\pi^{2n+1}}\sum_{k=0}^{\infty}\frac{\sin(2k+1)\pi t}{(2k+1)^{2n+1}},\ n \geq 1. \tag{1.2.14}$$

Proof. The proof is similar to that of Lemma 1.2.4 except that it uses the series

$$1 = \frac{4}{\pi}\sum_{k=0}^{\infty}\frac{\sin(2k+1)\pi s}{(2k+1)},\ 0 < s < 1. \quad ∎$$

Remark 1.2.5. Lindelöf has shown that the Euler polynomials $E_{2n}(t)$ can be expanded in terms of Fourier series and the following equality holds (see Jordan [14, p. 294])

$$E_{2n}(t) = (-1)^n\frac{4}{\pi^{2n+1}}\sum_{k=0}^{\infty}\frac{\sin(2k+1)\pi t}{(2k+1)^{2n+1}},\ n \geq 1. \tag{1.2.15}$$

Thus, from (1.2.14) and (1.2.15) it follows that

$$\int_0^1 g_n(t,s)\,ds = E_{2n}(t),\ n \geq 1. \tag{1.2.16}$$

Further, in view of Remark 1.2.1 it is clear that $(-1)^n E_{2n}(t) \geq 0$. ∎

Remark 1.2.6. As in Remark 1.2.3, by induction we shall show that

$$\int_0^1 g_n(t,s)\,ds = \frac{1}{2}\left[\frac{2t^{2n}}{(2n)!} - \frac{t^{2n-1}}{(2n-1)!}\right] - \sum_{k=0}^{n-2}\frac{2(2^{2k+4}-1)}{(2k+4)!}B_{2k+4}\times \frac{t^{2n-2k-3}}{(2n-2k-3)!},\ n \geq 1. \tag{1.2.17}$$

For $n = 1$, by direct computation we have

$$\int_0^1 g_1(t,s)\, ds = \frac{1}{2}\left(t^2 - t\right),$$

which is the same as (1.2.17) for $n = 1$. Now assuming (1.2.17) to be true for n, then from (1.2.14) and (1.2.17) we have

$$\frac{1}{2}\left[\frac{2t^{2n}}{(2n)!} - \frac{t^{2n-1}}{(2n-1)!}\right] - \sum_{k=0}^{n-2} \frac{2(2^{2k+4}-1)}{(2k+4)!} B_{2k+4} \frac{t^{2n-2k-3}}{(2n-2k-3)!} \tag{1.2.18}$$

$$= (-1)^n \frac{4}{\pi^{2n+1}} \sum_{k=0}^{\infty} \frac{\sin(2k+1)\pi t}{(2k+1)^{2n+1}}.$$

Now integrating (1.2.18) twice from 0 to t, we get

$$\frac{1}{2}\left[\frac{2t^{2n+2}}{(2n+2)!} - \frac{t^{2n+1}}{(2n+1)!}\right] - \sum_{k=0}^{n-2} \frac{2(2^{2k+4}-1)}{(2k+4)!} B_{2k+4} \frac{t^{2n-2k-1}}{(2n-2k-1)!}$$

$$= (-1)^n \frac{4}{\pi^{2n+1}} \sum_{k=0}^{\infty} \frac{1}{(2k+1)^{2n+1}} \times$$

$$\left\{\frac{1}{(2k+1)\pi}\left[t - \frac{1}{(2k+1)\pi}\sin(2k+1)\pi t\right]\right\}$$

$$= (-1)^{n+1} \frac{4}{\pi^{2n+3}} \sum_{k=0}^{\infty} \frac{\sin(2k+1)\pi t}{(2k+1)^{2n+3}} + (-1)^n \frac{4}{\pi^{2n+2}} \sum_{k=0}^{\infty} \frac{1}{(2k+1)^{2n+2}} t$$

$$= \int_0^1 g_{n+1}(t,s)\, ds + \frac{2(2^{2n+2}-1)}{(2n+2)!} B_{2n+2} t, \tag{1.2.19}$$

where in (1.2.19) we have used the relation

$$\sum_{k=0}^{\infty} \frac{1}{(2k+1)^{2n+2}} = \frac{(2^{2n+2}-1)}{2(2n+2)!} (-1)^n \pi^{2n+2} B_{2n+2}, \tag{1.2.20}$$

which is a direct consequence of the known (Milne - Thomson [17, p. 138]) equality

$$\sum_{k=1}^{\infty} \frac{1}{k^{2n+2}} = (-1)^n \frac{(2\pi)^{2n+2}}{2(2n+2)!} B_{2n+2}. \tag{1.2.21}$$

Equation (1.2.19) is the same as (1.2.17) for $(n+1)$. ∎

Lemma 1.2.6. The following equality holds

$$\Lambda_n(1-t) = \int_0^1 g_n(t,s)(1-s)\,ds = (-1)^n \frac{2}{\pi^{2n+1}} \sum_{k=1}^{\infty} \frac{\sin k\pi t}{k^{2n+1}},\ n \geq 1. \tag{1.2.22}$$

Proof. From (1.2.2), (1.2.7) and (1.2.14), we have

$$\begin{aligned}
&\int_0^1 g_n(t,s)(1-s)\,ds \\
&= (-1)^n \frac{4}{\pi^{2n+1}} \sum_{k=0}^{\infty} \frac{\sin(2k+1)\pi t}{(2k+1)^{2n+1}} - (-1)^n \frac{2}{\pi^{2n+1}} \left\{ \sum_{k=0}^{\infty} \frac{\sin(2k+1)\pi t}{(2k+1)^{2n+1}} \right. \\
&\qquad \left. - \sum_{k=1}^{\infty} \frac{\sin 2k\pi t}{(2k)^{2n+1}} \right\} \\
&= (-1)^n \frac{2}{\pi^{2n+1}} \left\{ \sum_{k=0}^{\infty} \frac{\sin(2k+1)\pi t}{(2k+1)^{2n+1}} + \sum_{k=1}^{\infty} \frac{\sin 2k\pi t}{(2k)^{2n+1}} \right\} \\
&= (-1)^n \frac{2}{\pi^{2n+1}} \sum_{k=1}^{\infty} \frac{\sin k\pi t}{k^{2n+1}} \\
&= (-1)^n \frac{2}{\pi^{2n+1}} \sum_{k=1}^{\infty} \frac{(-1)^{k+1} \sin k\pi(1-t)}{k^{2n+1}} \\
&= \Lambda_n(1-t). \quad \blacksquare
\end{aligned}$$

Remark 1.2.7. From (1.2.9) it follows that

$$\Lambda_n(1-t) = \frac{2^{2n+1}}{(2n+1)!} B_{2n+1}\left(1 - \frac{t}{2}\right).$$

However, for the Bernoulli polynomials it is known (Luke [16, p. 20]) that $B_{2n+1}\left(1-\frac{t}{2}\right) = -B_{2n+1}\left(\frac{t}{2}\right)$, and hence

$$\begin{aligned}
\Lambda_n(1-t) &= -\frac{2^{2n+1}}{(2n+1)!} B_{2n+1}\left(\frac{t}{2}\right) \\
&= -\frac{2^{2n+1}}{(2n+1)!} \sum_{k=0}^{2n+1} \binom{2n+1}{k} \left(\frac{t}{2}\right)^k B_{2n+1-k}. \quad \blacksquare
\end{aligned} \tag{1.2.23}$$

Remark 1.2.8. Since

$$\Lambda_n(1-t) = \int_0^1 g_n(t,s)(1-s)\,ds = \int_0^1 g_n(t,s)\,ds - \int_0^1 g_n(t,s)s\,ds$$

from (1.2.2) and (1.2.16) it follows that

$$E_{2n}(t) = \Lambda_n(t) + \Lambda_n(1-t) \tag{1.2.24}$$

and hence from (1.2.9) and (1.2.23), we have

$$E_{2n}(t) = \frac{2^{2n+1}}{(2n+1)!}\left[B_{2n+1}\left(\frac{1+t}{2}\right) - B_{2n+1}\left(\frac{t}{2}\right)\right]. \tag{1.2.25}$$

Fort [13, p. 41] has defined Euler polynomials as we have here in (1.2.25). ∎

Lemma 1.2.7. The following holds

$$\int_0^1 (-1)^n g_n(t,s)\, ds = \int_0^1 |\, g_n(t,s)\,|\, ds = (-1)^n E_{2n}(t) \tag{1.2.26}$$

$$\leq (-1)^n E_{2n}\left(\frac{1}{2}\right) = \frac{(-1)^n E_{2n}}{2^{2n}(2n)!}, \tag{1.2.27}$$

where E_{2n} is the $2n$ th Euler number.

Proof. Equality (1.2.26) is clear from Remark 1.2.1 and the relation (1.2.16). Further, since the extrema of the function $E_{2n}(t)$ is at $t = \frac{1}{2}$ (Jordan [14, p. 293]) the inequality (1.2.27) is obvious. ∎

Lemma 1.2.8. The following holds

$$\int_0^1 |\, g'_n(t,s)\,|\, ds = (-1)^n \left[2E_{2n}(t) + (1-2t)E_{2n-1}(t)\right] \tag{1.2.28}$$

$$\leq (-1)^{n+1}\frac{2(2^{2n}-1)}{(2n)!}B_{2n}. \tag{1.2.29}$$

Proof. From (1.2.4) we have

$$\begin{aligned} \int_0^1 |\, g'_n(t,s)\,|\, ds &= \int_0^1\int_0^t t_1(-1)^{n-1}g_{n-1}(t_1,s)\, dt_1\, ds \\ &\quad + \int_0^1\int_t^1 (1-t_1)(-1)^{n-1}g_{n-1}(t_1,s)\, dt_1\, ds, \end{aligned}$$

which on changing the order of integration and using (1.2.26) gives

$$\int_0^1 |\, g_n'(t,s)\,|\, ds = (-1)^{n-1}\left[\int_0^t t_1 E_{2n-2}(t_1)\, dt_1 + \int_t^1 (1-t_1)E_{2n-2}(t_1)\, dt_1\right]. \tag{1.2.30}$$

Now in (1.2.30), we use the relation $E_k'(t) = E_{k-1}(t)$ (Jordan [14, p. 288]) and the conditions $E_{2n}(0) = E_{2n}(1) = 0$, to obtain

$$\begin{aligned}\int_0^1 |\, g_n'(t,s)\,|\, ds &= (-1)^{n-1}\left[tE_{2n-1}(t) - E_{2n}(t) + E_{2n}(0) - (1-t)E_{2n-1}(t) + E_{2n}(1) - E_{2n}(t)\right] \\ &= (-1)^n\left[2E_{2n}(t) + (1-2t)E_{2n-1}(t)\right].\end{aligned}$$

Next, since the derivative of the right side of (1.2.28) is $-(1-2t)(-1)^{n-1}\times E_{2n-2}(t)$ and $(-1)^{n-1}E_{2n-2}(t) \ge 0$ for all $t \in [0,1]$, it is immediate that

$$\int_0^1 |\, g_n'(t,s)\,|\, ds \le (-1)^n \max\{E_{2n-1}(0), -E_{2n-1}(1)\}.$$

However, since $E_{2n-1}(0) + E_{2n-1}(1) = 0$ (Jordan [14, p. 291]) it follows that

$$\int_0^1 |\, g_n'(t,s)\,|\, ds \le (-1)^n E_{2n-1}(0) = (-1)^n E_{2n}'(0). \tag{1.2.31}$$

Now, from (1.2.16) and (1.2.17), we have

$$E_{2n}'(t) = \frac{1}{2}\left[\frac{2t^{2n-1}}{(2n-1)!} - \frac{t^{2n-2}}{(2n-2)!}\right] - \sum_{k=0}^{n-2}\frac{2(2^{2k+4}-1)}{(2k+4)!}B_{2k+4}\frac{t^{2n-2k-4}}{(2n-2k-4)!}$$

and so

$$E_{2n}'(0) = -\frac{2(2^{2n}-1)}{(2n)!}B_{2n}.$$

Hence, the inequality (1.2.31) is the same as (1.2.29). ∎

Lemma 1.2.9. The following inequality holds

$$\frac{1}{\pi^{2n+1}}\sin \pi t \le (-1)^n \Lambda_n(t), \quad 0 \le t \le 1. \tag{1.2.32}$$

Proof. Since $\sin \pi t \le \pi t$, $0 \le t \le 1$ it follows that

$$(-1)^n \Lambda_n(t) = \int_0^1 |\, g_n(t,s)\,|\, s\, ds \ge \int_0^1 |\, g_n(t,s)\,|\, \frac{\sin \pi s}{\pi}\, ds$$

and now (1.2.32) follows from (1.2.6). ∎

Remark 1.2.9. Widder [31] has proved that

$$\frac{t^{n+1}(1-t)^{n+1}}{(n+2)!} \leq (-1)^n \Lambda_n(t),\ 0 \leq t \leq 1. \tag{1.2.33}$$

We note that (1.2.32) is sharper than (1.2.33). In fact, from the elementary inequality

$$t(1-t) \leq \frac{1}{\pi} \sin \pi t,\ 0 \leq t \leq 1 \tag{1.2.34}$$

we find that

$$\frac{t^{n+1}(1-t)^{n+1}}{(n+2)!} \leq \frac{1}{\pi^{n+1}} \frac{1}{(n+2)!} (\sin \pi t)^{n+1}.$$

Now, by an easy induction we note that

$$\frac{(\sin \pi t)^n}{(n+2)!} \leq \frac{1}{\pi^n}.$$

Hence, we obtain

$$\frac{t^{n+1}(1-t)^{n+1}}{(n+2)!} \leq \frac{1}{\pi^{2n+1}} \sin \pi t. \qquad \blacksquare$$

Remark 1.2.10. From (1.2.9) and (1.2.32) it follows that

$$(-1)^n B_{2n+1}\left(\frac{1+t}{2}\right) \geq \frac{(2n+1)!}{(2\pi)^{2n+1}} \sin \pi t,\ 0 \leq t \leq 1. \tag{1.2.35}$$ ■

Lemma 1.2.10. The following inequality holds

$$(-1)^n \Lambda_n(t) \leq \frac{1}{\pi^{2n}} \left(\frac{\pi}{3}\right) \sin \pi t,\ 0 \leq t \leq 1. \tag{1.2.36}$$

Proof. Since from (1.2.10)

$$(-1)\Lambda_1(t) = \frac{1}{6} t(1-t)(1+t) \leq \frac{1}{3} t(1-t)$$

inequality (1.2.34) gives

$$(-1)\Lambda_1(t) \leq \frac{1}{3\pi} \sin \pi t$$

and hence (1.2.36) is true for $n = 1$.

Next, since from (1.2.1)

$$(-1)^{k+1}\Lambda_{k+1}(t) = \int_0^1 | g_1(t,s) || \Lambda_k(s) | ds$$

if (1.2.36) is true for $k \geq 1$, then from (1.2.6) for $n = 1$ it follows that

$$(-1)^{k+1}\Lambda_{k+1}(t) \leq \int_0^1 | g_1(t,s) | \frac{1}{\pi^{2k}}\left(\frac{\pi}{3}\right)\sin \pi s\, ds = \frac{1}{\pi^{2k+2}}\left(\frac{\pi}{3}\right)\sin \pi t. \quad \blacksquare$$

Remark 1.2.11. Widder [31] has proved that there exists a constant M such that $(-1)^n\Lambda_n(t) \leq \frac{M}{\pi^{2n}}$. From (1.2.36) it is clear that $M \leq \frac{\pi}{3}$. This can also be proved directly from (1.2.7) as follows :

$$(1.2.37) \qquad (-1)^n\Lambda_n(t) \leq \frac{2}{\pi^{2n+1}}\sum_{k=1}^{\infty}\frac{1}{k^{2n+1}} \leq \frac{2}{\pi^{2n+1}}\sum_{k=1}^{\infty}\frac{1}{k^2} = \frac{1}{\pi^{2n}}\left(\frac{\pi}{3}\right).$$

Further, from (1.2.7) we also have

$$(1.2.38) \qquad | \Lambda_n'(t) | \leq \frac{2}{\pi^{2n}}\sum_{k=1}^{\infty}\frac{1}{k^{2n}} \leq \frac{2}{\pi^{2n}}\sum_{k=1}^{\infty}\frac{1}{k^2} = \frac{1}{\pi^{2n}}\left(\frac{\pi^2}{3}\right). \quad \blacksquare$$

Remark 1.2.12. From (1.2.9) and (1.2.36) it is immediate that

$$(-1)^n B_{2n+1}\left(\frac{1+t}{2}\right) \leq \frac{(2n+1)!}{(2\pi)^{2n}}\left(\frac{\pi}{6}\right)\sin \pi t, \; 0 \leq t \leq 1. \quad \blacksquare$$

Remark 1.2.13. From (1.2.32) and (1.2.36) it is obvious that

$$\frac{1}{\pi^{2n+1}}\sin \pi t \leq (-1)^n\Lambda_n(1-t) \leq \frac{1}{\pi^{2n}}\left(\frac{\pi}{3}\right)\sin \pi t, \; 0 \leq t \leq 1. \quad \blacksquare$$

Lemma 1.2.11. The following inequality holds

$$(1.2.39) \qquad \int_0^1 | g_n(t,s) | ds \leq \frac{1}{\pi^{2n-2}}\left(\frac{1}{2\pi}\right)\sin \pi t, \; 0 \leq t \leq 1.$$

Proof. Since from (1.2.17)

$$\int_0^1 | g_1(t,s) | ds = \frac{1}{2}t(1-t)$$

inequality (1.2.34) gives

$$\int_0^1 | g_1(t,s) | ds \leq \frac{1}{2\pi}\sin \pi t$$

and hence (1.2.39) is true for $n = 1$.

Next, if (1.2.39) is true for $k \geq 1$, then from (1.2.4) we have

$$\int_0^1 | g_{k+1}(t,s) | \, ds$$

$$\leq (1-t)\int_0^1 \int_0^t | g_k(t_1,s) | \, t_1 \, dt_1 \, ds + t\int_0^1 \int_t^1 | g_k(t_1,s) | \, (1-t_1) \, dt_1 \, ds$$

$$\leq \left[(1-t)\int_0^t t_1 \sin \pi t_1 \, dt_1 + t\int_t^1 (1-t_1)\sin \pi t_1 \, dt_1\right] \frac{1}{\pi^{2k-2}}\left(\frac{1}{2\pi}\right)$$

$$= \frac{1}{\pi^{2k}}\left(\frac{1}{2\pi}\right)\sin \pi t. \quad \blacksquare$$

Remark 1.2.14. From (1.2.26) and (1.2.39) it follows that

$$(1.2.40) \qquad | E_{2n}(t) | \leq \frac{1}{\pi^{2n-2}}\left(\frac{1}{2\pi}\right)\sin \pi t \leq \frac{1}{2\pi^{2n-1}}.$$

Jordan [14, p. 302] has proved that $| E_{2n}(t) | < \frac{2}{3\pi^{2n-1}}$, and hence (1.2.40) gives a better estimate. Further, from (1.2.40) we find that

$$\left| E_{2n}\left(\frac{1}{2}\right) \right| \leq \frac{1}{2\pi^{2n-1}}$$

and hence from the formula $E_{2n} = 2^{2n}(2n)! E_{2n}\left(\frac{1}{2}\right)$ it is immediate to have

$$(1.2.41) \qquad | E_{2n} | \leq \left(\frac{2}{\pi}\right)^{2n-1} (2n)!,$$

which is sharper than $| E_{2n} | < \frac{4}{3}\left(\frac{2}{\pi}\right)^{2n-1} (2n)!$ given by Jordan [14, p. 303]. $\blacksquare$

Lemma 1.2.12. If $(-1)^n x^{(2n)}(t)$ is nonnegative and concave in $[0,1]$, then

$$(1.2.42) \qquad \left| \int_0^1 g_n(t,s) x^{(2n)}(s) \, ds \right| \geq \frac{2M_{2n}}{\pi^{2n+2}} \sin \pi t, \ 0 \leq t \leq 1$$

where $M_{2n} = \max_{0 \leq t \leq 1} (-1)^n x^{(2n)}(t)$.

Proof. Suppose the nonnegative and concave function $(-1)^n x^{(2n)}(t)$ attains its maximum at $t = t_0$, then in $0 \leq t \leq t_0$ we have

$$\begin{aligned}
(-1)^n x^{(2n)}(t) &= (-1)^n x^{(2n)}\left(\frac{t_0 - t}{t_0} \cdot 0 + \frac{t}{t_0} \cdot t_0\right) \\
&\geq \frac{t_0 - t}{t_0}(-1)^n x^{(2n)}(0) + \frac{t}{t_0} M_{2n} \\
&\geq \frac{t}{t_0} M_{2n} \geq t M_{2n} \geq \left(t - t^2\right) M_{2n},
\end{aligned}$$

and similarly in $t_0 \le t \le 1$, we find

$$\begin{aligned}(-1)^n x^{(2n)}(t) &= (-1)^n x^{(2n)}\left(\frac{1-t}{1-t_0}\cdot t_0 + \frac{t-t_0}{1-t_0}\cdot 1\right)\\ &\ge \frac{1-t}{1-t_0}M_{2n} + \frac{t-t_0}{1-t_0}(-1)^n x^{(2n)}(1)\\ &\ge \frac{1-t}{1-t_0}M_{2n} \ge (1-t)M_{2n} \ge \left(t-t^2\right)M_{2n}.\end{aligned}$$

Therefore, for all $t \in [0,1]$ it follows that $(-1)^n x^{(2n)}(t) \ge t(1-t)M_{2n}$. Hence, from the elementary inequality

$$t(1-t) \ge \frac{2}{\pi^2}\sin \pi t \tag{1.2.43}$$

we find that $(-1)^n x^{(2n)}(t) \ge \frac{2}{\pi^2}\sin\pi t\ M_{2n}$. Now in view of (1.2.6), we obtain

$$\begin{aligned}\left|\int_0^1 g_n(t,s)x^{(2n)}(s)\,ds\right| &= \left|\int_0^1 \mid g_n(t,s)\mid (-1)^n x^{(2n)}(s)\,ds\right|\\ &\ge \frac{2}{\pi^2}M_{2n}\int_0^1 \mid g_n(t,s)\mid \sin\pi s\,ds\\ &= \frac{2M_{2n}}{\pi^{2n+2}}\sin\pi t. \qquad \blacksquare\end{aligned}$$

Remark 1.2.15. In particular if $t = \frac{1}{2}$ then the inequality (1.2.42) reduces to

$$\left|\int_0^1 g_n\left(\frac{1}{2},s\right)x^{(2n)}(s)\,ds\right| \ge \frac{2M_{2n}}{\pi^{2n+2}},$$

which is sharper than

$$\left|\int_0^1 g_n\left(\frac{1}{2},s\right)x^{(2n)}(s)\,ds\right| \ge \frac{4M_{2n}}{\pi^{2n+3}}\left(1-3^{-2n-1}\right)$$

obtained by Boas [8]. ∎

Lemma 1.2.13. The following inequality holds

$$\int_0^1 \mid g_n'(t,s)\mid ds \le \frac{1}{\pi^{2n-2}}\left(\frac{1}{2\pi}\right)[2\sin\pi t + \pi(1-2t)\cos\pi t],\ 0\le t\le 1. \tag{1.2.44}$$

Proof. From the inequality (1.2.39) for $n > 1$ and the relation (1.2.4) it is immediate to obtain that

$$\begin{aligned}\int_0^1 \mid g_n'(t,s)\mid ds &\le \frac{1}{\pi^{2n-4}}\left(\frac{1}{2\pi}\right)\left[\int_0^t t_1\sin\pi t_1\,dt_1 + \int_t^1 (1-t_1)\sin\pi t_1\,dt_1\right]\\ &= \frac{1}{\pi^{2n-2}}\left(\frac{1}{2\pi}\right)[2\sin\pi t + \pi(1-2t)\cos\pi t].\end{aligned}$$

Further, for $n = 1$ we have

$$\int_0^1 | g_1'(t,s) | \, ds = \frac{t^2 + (1-t)^2}{2}$$

however, since

$$t^2 + (1-t)^2 \le \frac{1}{\pi}[2\sin \pi t + \pi(1-2t)\cos \pi t], \ 0 \le t \le 1$$

the inequality (1.2.44) is true for $n = 1$ also. ∎

Lemma 1.2.14. The following inequality holds

$$(1.2.45) \quad \int_0^1 | g_n'(t,s) | \sin \pi s \, ds \le \frac{1}{\pi^{2n}}[2\sin \pi t + \pi(1-2t)\cos \pi t], \ 0 \le t \le 1.$$

Proof. For $n = 1$, a direct computation gives

$$\begin{aligned}
\int_0^1 | g_1'(t,s) | \sin \pi s \, ds &= \int_0^t s \sin \pi s \, ds + \int_t^1 (1-s)\sin \pi s \, ds \\
&= \frac{1}{\pi^2}[2\sin \pi t + \pi(1-2t)\cos \pi t].
\end{aligned}$$

For $n > 1$, we use (1.2.6) to get

$$\begin{aligned}
\int_0^1 | g_n'(t,s) | \sin \pi s \, ds &\le \int_0^t t_1 \left[\int_0^1 | g_{n-1}(t_1,s) | \sin \pi s \, ds\right] dt_1 \\
&\quad + \int_t^1 (1-t_1) \left[\int_0^1 | g_{n-1}(t_1,s) | \sin \pi s \, ds\right] dt_1 \\
&= \frac{1}{\pi^{2n-2}} \left[\int_0^t t_1 \sin \pi t_1 \, dt_1 + \int_t^1 (1-t_1)\sin \pi t_1 \, dt_1\right] \\
&= \frac{1}{\pi^{2n}}[2\sin \pi t + \pi(1-2t)\cos \pi t]. \qquad ∎
\end{aligned}$$

Lemma 1.2.15. The following inequality holds

$$(1.2.46) \quad \int_0^1 | g_n(t,s) | \, [2\sin \pi s + \pi(1-2s)\cos \pi s] \, ds \le \frac{1}{\pi^{2n-2}}\left(\frac{4}{\pi^2}\right)\sin \pi t,$$

$$0 \le t \le 1.$$

Proof. For $n = 1$, a direct computation gives (1.2.46). If (1.2.46) is true for $n = j \geq 1$, then as earlier we have

$$\int_0^1 |\, g_{j+1}(t,s) \,|\, [2\sin \pi s + \pi(1-2s)\cos \pi s]\, ds$$

$$\leq (1-t)\int_0^t t_1 \left\{ \int_0^1 |\, g_j(t_1,s) \,|\, [2\sin \pi s + \pi(1-2s)\cos \pi s]\, ds \right\} dt_1$$

$$+t\int_t^1 (1-t_1) \left\{ \int_0^1 |\, g_j(t_1,s) \,|\, [2\sin \pi s + \pi(1-2s)\cos \pi s]\, ds \right\} dt_1$$

$$\leq \frac{1}{\pi^{2j-2}}\left(\frac{4}{\pi^2}\right)\left[(1-t)\int_0^t t_1 \sin \pi t_1\, dt_1 + t\int_t^1 (1-t_1)\sin \pi t_1\, dt_1\right]$$

$$= \frac{1}{\pi^{2j}}\left(\frac{4}{\pi^2}\right)\sin \pi t. \qquad \blacksquare$$

Lemma 1.2.16. The following inequality holds

$$\int_0^1 |\, g_n'(t,s) \,|\, [2\sin \pi s + \pi(1-2s)\cos \pi s]\, ds \tag{1.2.47}$$

$$\leq \frac{1}{\pi^{2n-2}}\left(\frac{4}{\pi^2}\right)[2\sin \pi t + \pi(1-2t)\cos \pi t], \ 0 \leq t \leq 1.$$

Proof. For $n = 1$, the proof is by direct computation. If (1.2.47) is true for $n = j \geq 1$, then as an application of (1.2.46) we find that

$$\int_0^1 |\, g_{j+1}'(t,s) \,|\, [2\sin \pi s + \pi(1-2s)\cos \pi s]\, ds$$

$$\leq \int_0^t t_1 \left\{ \int_0^1 |\, g_j(t_1,s) \,|\, [2\sin \pi s + \pi(1-2s)\cos \pi s]\, ds \right\} dt_1$$

$$+\int_t^1 (1-t_1) \left\{ \int_0^1 |\, g_j(t_1,s) \,|\, [2\sin \pi s + \pi(1-2s)\cos \pi s]\, ds \right\} dt_1$$

$$\leq \frac{1}{\pi^{2j-2}}\left(\frac{4}{\pi^2}\right)\left[\int_0^t t_1 \sin \pi t_1\, dt_1 + \int_t^1 (1-t_1)\sin \pi t_1\, dt_1\right]$$

$$= \frac{1}{\pi^{2j}}\left(\frac{4}{\pi^2}\right)[2\sin \pi t + \pi(1-2t)\cos \pi t]. \qquad \blacksquare$$

1.3 INTERPOLATING POLYNOMIAL REPRESENTATIONS

Theorem 1.3.1. The Lidstone interpolating polynomial $P_{(1.1.1)}(t)$ can be expressed as

$$P_{(1.1.1)}(t) = \sum_{k=0}^{m-1} [\alpha_k \Lambda_k(1-t) + \beta_k \Lambda_k(t)]. \tag{1.3.1}$$

Proof. It is clear that $P_{(1.1.1)}(t)$ in (1.3.1) is a polynomial of degree at most $(2m-1)$. Further, since in view of (1.2.1)

$$\begin{aligned} P^{(2i)}_{(1.1.1)}(t) &= \sum_{k=i}^{m-1} \left[\alpha_k \Lambda_k^{(2i)}(1-t) + \beta_k \Lambda_k^{(2i)}(t)\right] \\ &= \sum_{k=i}^{m-1} [\alpha_k \Lambda_{k-i}(1-t) + \beta_k \Lambda_{k-i}(t)] \\ &= \sum_{k=0}^{m-i-1} [\alpha_{k+i} \Lambda_k(1-t) + \beta_{k+i} \Lambda_k(t)]\,,\ 0 \le i \le m-1 \end{aligned}$$

it follows that

$$P^{(2i)}_{(1.1.1)}(0) = \alpha_i \Lambda_0(1-t)\big|_{t=0} = \alpha_i,\ 0 \le i \le m-1$$

and

$$P^{(2i)}_{(1.1.1)}(1) = \beta_i \Lambda_0(t)\big|_{t=1} = \beta_i,\ 0 \le i \le m-1.$$

The uniqueness of $P_{(1.1.1)}(t)$ is obvious. ∎

Remark 1.3.1. Various explicit representations of $\Lambda_n(t)$ and $\Lambda_n(1-t)$ given in Section 1.2 can be used in (1.3.1) to obtain different forms of $P_{(1.1.1)}(t)$. ∎

1.4 ERROR REPRESENTATIONS

Let $\alpha_i = x^{(2i)}(0)$, $\beta_i = x^{(2i)}(1)$, $0 \le i \le m-1$ where the function $x(t)$ is assumed to be $2m$ times continuously differentiable on $[0,1]$. In such a case

$P_{(1.1.1)}(t)$ is called the Lidstone interpolating polynomial of the function $x(t)$. For the associated error $e_{(1.1.1)}(t) = x(t) - P_{(1.1.1)}(t)$ here we shall provide two different representations.

Peano's Representation

Theorem 1.4.1. If $x(t) \in C^{(2m)}[0,1]$, then

$$e_{(1.1.1)}(t) = \int_0^1 g_m(t,s)x^{(2m)}(s)\,ds. \tag{1.4.1}$$

Proof. Since $\frac{\partial^{2n} g_m(t,s)}{\partial t^{2n}} = g_{m-n}(t,s)$, and $g_{m-n}(0,s) = g_{m-n}(1,s) = 0,\ 0 \le n \le m-1$ it is clear that $e^{(2i)}_{(1.1.1)}(0) = e^{(2i)}_{(1.1.1)}(1) = 0,\ 0 \le i \le m-1$. Thus, it suffices to prove that $e^{(2m)}_{(1.1.1)}(t) = x^{(2m)}(t)$. But, this is immediate from

$$e^{(2m-2)}_{(1.1.1)}(t) = \int_0^1 g_1(t,s)x^{(2m)}(s)\,ds. \quad \blacksquare$$

Cauchy's Representation

Theorem 1.4.2. If $x(t) \in C^{(2m)}[0,1]$, then

$$e_{(1.1.1)}(t) = E_{2m}(t)x^{(2m)}(\xi), \tag{1.4.2}$$

where $\xi \in (0,1)$.

Proof. Since in view of Remark 1.2.1, $(-1)^m g_m(t,s) \ge 0,\ 0 \le s,t \le 1$, in (1.4.1) an application of the mean value theorem gives that

$$e_{(1.1.1)}(t) = x^{(2m)}(\xi)\int_0^1 g_m(t,s)\,ds. \tag{1.4.3}$$

The result now follows from (1.2.16). $\blacksquare$

Remark 1.4.1. Theorem 1.4.2 can also be proved directly. For this, we define an auxiliary function $\phi(\tau)$ as follows

$$\phi(\tau) = e_{(1.1.1)}(\tau) - \psi(t)E_{2n}(\tau),\ 0 < t < 1 \tag{1.4.4}$$

where $\psi(t) = e_{(1.1.1)}(t)/E_{2n}(t)$.

For this function $\phi(\tau)$ it is clear that $\phi^{(2i)}(0) = \phi^{(2i)}(1) = 0,\ 0 \le i \le m-1$ and $\phi(t) = 0$. Therefore, as an application of Rolle's theorem $\phi'(\tau)$ will have at least two zeros in $(0,1)$, and $\phi''(\tau)$ will have at least one zero in $(0,1)$. Now repeating the application of Rolle's theorem to the function $\phi''(\tau)$, and then continuing we find that $\phi^{(2m)}(\tau)$ vanishes at least once in $(0,1)$, i.e., there is a $\xi \in (0,1)$ such that $\phi^{(2m)}(\xi) = 0$. Thus, from (1.4.4) we find

$$\phi^{(2m)}(\xi) = x^{(2m)}(\xi) - \psi(t) = 0,$$

which is the same as (1.4.2). ∎

Remark 1.4.2. Two other representations of $e_{(1.1.1)}(t)$ can be obtained by using (1.2.14) and (1.2.17) in (1.4.3). ∎

1.5 ERROR ESTIMATES

The representation of the error function $e_{(1.1.1)}(t)$ obtained in Theorem 1.4.1 and some of the equalities and inequalities established in Section 1.2 will be used here to obtain best possible pointwise as well as uniform bounds for $|\ e^{(i)}_{(1.1.1)}(t)\ |,\ 0 \le i \le 2m-1$.

Theorem 1.5.1. If $x(t) \in C^{(2m)}[0,1]$, then the following hold

(1.5.1) $$|\ e^{(2i)}_{(1.1.1)}(t)\ |$$

$$= \left| x^{(2i)}(t) - \sum_{k=0}^{m-i-1} \left[x^{(2k+2i)}(0)\Lambda_k(1-t) + x^{(2k+2i)}(1)\Lambda_k(t) \right] \right|$$

$$\le (-1)^{m-i} E_{2m-2i}(t) M_{2m}$$

$$\le \frac{(-1)^{m-i} E_{2m-2i}}{2^{2m-2i}(2m-2i)!} M_{2m},\ 0 \le i \le m-1$$

and

$$(1.5.2)\quad |\, e^{(2i+1)}_{(1.1.1)}(t)\, |$$

$$= \left| x^{(2i+1)}(t) - \sum_{k=0}^{m-i-1} \left[x^{(2k+2i)}(0)\Lambda_k'(1-t) + x^{(2k+2i)}(1)\Lambda_k'(t) \right] \right|$$

$$\le (-1)^{m-i} \left[2E_{2m-2i}(t) + (1-2t)E_{2m-2i-1}(t) \right] M_{2m}$$

$$\le (-1)^{m-i+1} \frac{2(2^{2m-2i}-1)}{(2m-2i)!} B_{2m-2i} M_{2m},\ \ 0 \le i \le m-1$$

where $M_{2m} = \max_{0 \le t \le 1} |\, x^{(2m)}(t)\, |$.

Proof. Following as in Theorem 1.3.1, the relation (1.4.1) successively leads to

$$(1.5.3)\qquad x^{(2i)}(t) = \sum_{k=0}^{m-i-1} \left[x^{(2k+2i)}(0)\Lambda_k(1-t) + x^{(2k+2i)}(1)\Lambda_k(t) \right]$$

$$+ \int_0^1 g_{m-i}(t,s) x^{(2m)}(s)\, ds,\ \ 0 \le i \le m-1$$

from which it is clear that

$$|\, e^{(2i)}_{(1.1.1)}(t)\, | \le M_{2m} \int_0^1 |\, g_{m-i}(t,s)\, |\, ds,\ \ 0 \le i \le m-1.$$

Inequalities (1.5.1) are now immediate from Lemma 1.2.7.

From (1.5.3) it also follows that

$$|\, e^{(2i+1)}_{(1.1.1)}(t)\, | \le M_{2m} \int_0^1 |\, g'_{m-i}(t,s)\, |\, ds,\ \ 0 \le i \le m-1.$$

Inequalities (1.5.2) are now obvious from Lemma 1.2.8. ∎

Remark 1.5.1. Inequalities (1.5.1) and (1.5.2) are the best possible, as throughout the equality holds for the function $x(t) = E_{2m}(t)$ whose Lidstone interpolating polynomial $P_{(1.1.1)}(t) \equiv 0$, and only for this function up to a constant factor. ∎

1.6 LIDSTONE BOUNDARY VALUE PROBLEMS

The Lidstone boundary value problem consists of the general $2m$ th order nonlinear differential equation

$$(-1)^m x^{(2m)}(t) = f(t, \mathbf{x}(t)), \tag{1.6.1}$$

where $\mathbf{x}(t) = (x(t), x'(t), ..., x^{(q)}(t))$, $0 \le q \le 2m-1$ but fixed, f is continuous at least in the interior of the domain of interest; and the boundary conditions

$$x^{(2i)}(0) = \alpha_i, \;\; x^{(2i)}(1) = \beta_i, \;\; 0 \le i \le m-1. \tag{1.6.2}$$

If $f = 0$ then the problem (1.6.1), (1.6.2) obviously has a unique solution $x(t) = P_{(1.1.1)}(t)$; and if f is linear, i.e., $f = \sum_{i=0}^{q} a_i(t)x^{(i)}(t)$ then (1.6.1), (1.6.2) gives the possibility of interpolation by the solutions of the differential equation (1.6.1). As such in recent years this problem has attracted considerable attention (see [1-3,5-7,10,11,18, 21-26 and several references therein]) particularly because of its special cases frequently occur in engineering and other branches of physical sciences. For instance, the deflection of a uniformly loaded rectangular plate supported over the entire surface by an elastic foundation and rigidly supported along the edges leads to this type of problem with $m = 2$ and $f = a_0(t)x + b(t)$.

Existence and Uniqueness

The equalities and inequalities established in Sections 1.2 - 1.5 will be used here to provide necessary and sufficient conditions for the existence and uniqueness of the solutions of the Lidstone boundary value problem (1.6.1), (1.6.2).

Theorem 1.6.1. Suppose that

(i) $K_i > 0$, $0 \le i \le q$ are given real numbers and let Q be the maximum of $| f(t, x_0, x_1, ..., x_q) |$ on the compact set $[0,1] \times D_0$, where

$$D_0 = \{(x_0, x_1, ..., x_q) \ : \ | x_i | \le 2K_i, \ 0 \le i \le q\} ;$$

(ii) $Q \dfrac{(-1)^{m-i} E_{2m-2i}}{2^{2m-2i}(2m-2i)!} \le K_{2i}, \ 0 \le i \le \left[\dfrac{q}{2}\right];$

(iii) $Q \dfrac{(-1)^{m-i+1} 2 \,(2^{2m-2i}-1)\, B_{2m-2i}}{(2m-2i)!} \le K_{2i+1}, \ 0 \le i \le \left[\dfrac{q-1}{2}\right];$

(iv) $\max\{| \alpha_i |, | \beta_i |\} + \left[\sum_{k=1}^{m-i-1} \max\{| \alpha_{k+i} |, | \beta_{k+i} |\} \dfrac{(-1)^k E_{2k}}{2^{2k}(2k)!}\right]$

$$= C_{2i} \le K_{2i}, \ 0 \le i \le \left[\frac{q}{2}\right];$$

(v) $| \alpha_i - \beta_i | + \left[\sum_{k=1}^{m-i-1} \max\{| \alpha_{k+i} |, | \beta_{k+i} |\} \dfrac{(-1)^{k+1} 2 \left(2^{2k}-1\right) B_{2k}}{(2k)!}\right]$

$$= C_{2i+1} \le K_{2i+1}, \ 0 \le i \le \left[\frac{q-1}{2}\right].$$

Then, the boundary value problem (1.6.1), (1.6.2) has a solution in D_0.

Proof. From Theorem 1.4.1 it follows that the boundary value problem (1.6.1), (1.6.2) is equivalent to the following Fredholm type of integral equation

$$(1.6.3) \qquad x(t) = \sum_{k=0}^{m-1} [\alpha_k \Lambda_k(1-t) + \beta_k \Lambda_k(t)] + \int_0^1 | g_m(t,s) | \, f\,(s, \mathbf{x}(s))\, ds.$$

We define the set

$$B[0,1] = \left\{x(t) \in C^{(q)}[0,1] \ : \| x^{(i)} \| = \max_{0 \le t \le 1} | x^{(i)}(t) | \le 2K_i, \ 0 \le i \le q\right\}.$$

It is easy to verify that $B[0,1]$ is a closed convex subset of the Banach space $C^{(q)}[0,1]$. Consider an operator $T : C^{(q)}[0,1] \to C^{(2m)}[0,1]$ as follows

$$(1.6.4) \quad (Tx)(t) = \sum_{k=0}^{m-1} [\alpha_k \Lambda_k(1-t) + \beta_k \Lambda_k(t)] + \int_0^1 | g_m(t,s) | \, f\,(s, \mathbf{x}(s))\, ds.$$

Obviously, any fixed point of (1.6.4) is a solution of the boundary value problem (1.6.1), (1.6.2).

We shall show that T maps $B[0,1]$ into itself. For this, let $x(t) \in B[0,1]$ then from (1.6.4), (1.2.22), (1.2.27) and hypotheses (i), (ii), (iv) we find

$$\begin{aligned}
|(Tx)^{(2i)}(t)| &\le \sum_{k=0}^{m-i-1} [|\alpha_{k+i}||\Lambda_k(1-t)| + |\beta_{k+i}||\Lambda_k(t)|] \\
&\quad + Q\int_0^1 |g_{m-i}(t,s)|\,ds \\
&\le \max_{0\le t\le 1}[|\alpha_i|(1-t) + |\beta_i|t] + \sum_{k=1}^{m-i-1} \max\{|\alpha_{k+i}|, |\beta_{k+i}|\} \times \\
&\quad \int_0^1 |g_k(t,s)|(s+1-s)\,ds + Q\int_0^1 |g_{m-i}(t,s)|\,ds \\
&\le \max\{|\alpha_i|, |\beta_i|\} + \sum_{k=1}^{m-i-1} \max\{|\alpha_{k+i}|, |\beta_{k+i}|\} \frac{(-1)^k E_{2k}}{2^{2k}(2k)!} \\
&\quad + Q\frac{(-1)^{m-i} E_{2m-2i}}{2^{2m-2i}(2m-2i)!} \\
&\le K_{2i} + K_{2i} \\
&= 2K_{2i},\ 0 \le i \le \left[\frac{q}{2}\right]. \qquad (1.6.5)
\end{aligned}$$

Similarly, from (1.6.4), (1.2.22), (1.2.29) and hypotheses (i), (iii), (v) we get

$$\begin{aligned}
|(Tx)^{(2i+1)}(t)| &\le |\alpha_i \Lambda_0'(1-t) + \beta_i \Lambda_0'(t)| + \sum_{k=1}^{m-i-1} \max\{|\alpha_{k+i}|, |\beta_{k+i}|\} \times \\
&\quad [|\Lambda_k'(t)| + |\Lambda_k'(1-t)|] + Q\int_0^1 |g_{m-i}'(t,s)|\,ds \\
&\le |\alpha_i - \beta_i| + \sum_{k=1}^{m-i-1} \max\{|\alpha_{k+i}|, |\beta_{k+i}|\} \times \\
&\quad \int_0^1 |g_k'(t,s)|(s+1-s)\,ds + Q\int_0^1 |g_{m-i}'(t,s)|\,ds \\
&\le |\alpha_i - \beta_i|
\end{aligned}$$

$$+\sum_{k=1}^{m-i-1} \max\{|\alpha_{k+i}|, |\beta_{k+i}|\} \frac{(-1)^{k+1}2(2^{2k}-1)B_{2k}}{(2k)!}$$

$$+Q\frac{(-1)^{m-i+1}2(2^{2m-2i}-1)B_{2m-2i}}{(2m-2i)!}$$

$$\leq K_{2i+1} + K_{2i+1}$$

$$(1.6.6) \qquad = 2K_{2i+1}, \ 0 \leq i \leq \left[\frac{q-1}{2}\right].$$

This completes the proof of $TB[0,1] \subseteq B[0,1]$. The inequalities (1.6.5) and (1.6.6) imply that the sets $\left\{(Tx)^{(i)}(t) \ : \ x(t) \in B[0,1]\right\}$, $0 \leq i \leq q$ are uniformly bounded and equicontinuous in $[0,1]$. Hence $\overline{TB}[0,1]$ is compact follows from the Ascoli - Arzela theorem. The Schauder fixed point theorem is applicable and a fixed point of T in D_0 exists. ∎

Corollary 1.6.2. Assume that the function $f(t,x_0,x_1,...,x_q)$ on $[0,1]\times\Re^{q+1}$ satisfies the following condition

$$(1.6.7) \qquad |f(t,x_0,x_1,...,x_q)| \leq L + \sum_{i=0}^{q} L_i |x_i|^{\alpha_i},$$

where L, L_i, $0 \leq i \leq q$ are nonnegative constants, and $0 \leq \alpha_i < 1$, $0 \leq i \leq q$. Then, the boundary value problem (1.6.1), (1.6.2) has a solution.

Proof. For $x(t) \in B[0,1]$ the condition (1.6.7) implies that

$$|f(t,\mathbf{x}(t))| \leq L + \sum_{i=0}^{q} L_i(2K_i)^{\alpha_i} = Q_1, \text{ say.}$$

Now Corollary 1.6.2 follows immediately by observing that hypotheses of Theorem 1.6.1 are satisfied with Q replaced by Q_1 provided K_i, $0 \leq i \leq q$ are sufficiently large. ∎

Theorem 1.6.3. Suppose that the function $f(t,x_0,x_1,...,x_q)$ on $[0,1]\times D_1$ satisfies the following condition

$$(1.6.8) \qquad |f(t,x_0,x_1,...,x_q)| \leq L + \sum_{i=0}^{q} L_i |x_i|,$$

where

$$D_1 = \Big\{(x_0, x_1, \ldots, x_q) : | x_{2i} | \leq (1-\theta)^{-1} \frac{1}{\pi^{2m-2i-2}} \left(\frac{1}{2\pi}\right) C \sin \pi t + C_{2i}, \quad 0 \leq i \leq \left[\frac{q}{2}\right];$$

$$| x_{2i+1} | \leq (1-\theta)^{-1} \frac{1}{\pi^{2m-2i-2}} \left(\frac{1}{2\pi}\right) C[2 \sin \pi t + \pi(1-2t) \cos \pi t] + C_{2i+1}, \ 0 \leq i \leq \left[\frac{q-1}{2}\right]\Big\}$$

and

$$C = L + \sum_{i=0}^{[\frac{q}{2}]} L_{2i} C_{2i} + \sum_{i=0}^{[\frac{q-1}{2}]} L_{2i+1} C_{2i+1}, \tag{1.6.9}$$

$$\theta = \sum_{i=0}^{[\frac{q}{2}]} L_{2i} \frac{1}{\pi^{2m-2i}} + \sum_{i=0}^{[\frac{q-1}{2}]} L_{2i+1} \frac{1}{\pi^{2m-2i-2}} \left(\frac{4}{\pi^2}\right) < 1. \tag{1.6.10}$$

Then, the boundary value problem (1.6.1), (1.6.2) has a solution in D_1.

Proof. Let $y(t) = x(t) - P_{(1.1.1)}(t)$, so that the boundary value problem (1.6.1), (1.6.2) is equivalent to the following :

$$(-1)^m y^{(2m)}(t) = f\left(t,\ y(t) + P_{(1.1.1)}(t),\ y'(t) + P'_{(1.1.1)}(t),\ \ldots,\ y^{(q)}(t) + P^{(q)}_{(1.1.1)}(t)\right) \tag{1.6.11}$$

$$y^{(2i)}(0) = y^{(2i)}(1) = 0, \ 0 \leq i \leq m-1. \tag{1.6.12}$$

Define $M[0,1]$ as the space of q times continuously differentiable functions. If we introduce in $M[0,1]$ the finite norm

$$\| y \| = \max \Big\{ \sup_{0 \leq t \leq 1} \frac{\pi^{-2i} \, | y^{(2i)}(t) |}{\sin \pi t}, \ 0 \leq i \leq \left[\frac{q}{2}\right]; \quad \sup_{0 \leq t \leq 1} \frac{\pi^{-2i} \, | y^{(2i+1)}(t) |}{2 \sin \pi t + \pi(1-2t) \cos \pi t}, \ 0 \leq i \leq \left[\frac{q-1}{2}\right] \Big\} \tag{1.6.13}$$

then it becomes a Banach space. As in Theorem 1.6.1 we shall show that $T : M[0,1] \to M[0,1]$ defined by

(1.6.14) $(Ty)(t)$

$$= \int_0^1 \mid g_m(t,s) \mid f\left(s,\ y(s) + P_{(1.1.1)}(s),\ \dots,\ y^{(q)}(s) + P^{(q)}_{(1.1.1)}(s)\right) ds$$

maps the set

$$B_1[0,1] = \left\{ y(t) \in M[0,1]\ :\ \| y \| \le (1-\theta)^{-1}\frac{1}{\pi^{2m-2}}\left(\frac{1}{2\pi}\right) C \right\}$$

into itself. For this, if $y(t) \in B_1[0,1]$ then it immediately follows that

$$\mid y^{(2i)}(t) \mid \le (1-\theta)^{-1}\frac{1}{\pi^{2m-2i-2}}\left(\frac{1}{2\pi}\right) C \sin \pi t,\ 0 \le i \le \left[\frac{q}{2}\right]$$

$$\mid y^{(2i+1)}(t) \mid \le (1-\theta)^{-1}\frac{1}{\pi^{2m-2i-2}}\left(\frac{1}{2\pi}\right) C[2\sin \pi t + \pi(1-2t)\cos \pi t],$$

$$0 \le i \le \left[\frac{q-1}{2}\right].$$

Hence, $\left(y(t) + P_{(1.1.1)}(t),\ y'(t) + P'_{(1.1.1)}(t),\ \dots,\ y^{(q)}(t) + P^{(q)}_{(1.1.1)}(t)\right) \in D_1$. Thus, from (1.6.14), (1.6.8), (1.2.39), (1.2.6) and (1.2.46) we have

$$\begin{aligned}
\mid (Ty)^{(2k)}(t) \mid &\le \int_0^1 \mid g_{m-k}(t,s) \mid \left[L + \sum_{i=0}^{q} L_i \mid y^{(i)}(s) + P^{(i)}_{(1.1.1)}(s) \mid \right] ds \\
&\le \int_0^1 \mid g_{m-k}(t,s) \mid \left[L + \sum_{i=0}^{[\frac{q}{2}]} L_{2i} \mid P^{(2i)}_{(1.1.1)}(s) \mid \right. \\
&\quad + \sum_{i=0}^{[\frac{q-1}{2}]} L_{2i+1} \mid P^{(2i+1)}_{(1.1.1)}(s) \mid + \sum_{i=0}^{[\frac{q}{2}]} L_{2i} \frac{\sin \pi s}{\pi^{-2i}} \frac{\pi^{-2i} \mid y^{(2i)}(s) \mid}{\sin \pi s} \\
&\quad + \sum_{i=0}^{[\frac{q-1}{2}]} L_{2i+1} \frac{2\sin \pi s + \pi(1-2s)\cos \pi s}{\pi^{-2i}} \times \\
&\quad \left. \frac{\pi^{-2i} \mid y^{(2i+1)}(s) \mid}{2\sin \pi s + \pi(1-2s)\cos \pi s} \right] ds
\end{aligned}$$

$$\leq \int_0^1 | g_{m-k}(t,s) | \left[L + \sum_{i=0}^{[\frac{q}{2}]} L_{2i}C_{2i} + \sum_{i=0}^{[\frac{q-1}{2}]} L_{2i+1}C_{2i+1} \right.$$

$$+\sum_{i=0}^{[\frac{q}{2}]} L_{2i}\frac{\sin \pi s}{\pi^{-2i}} \| y \|$$

$$\left. + \sum_{i=0}^{[\frac{q-1}{2}]} L_{2i+1}\frac{2\sin\pi s + \pi(1-2s)\cos\pi s}{\pi^{-2i}} \| y \| \right] ds$$

$$\leq C\frac{1}{\pi^{2m-2k-2}}\left(\frac{1}{2\pi}\right)\sin\pi t + \sum_{i=0}^{[\frac{q}{2}]} L_{2i}\frac{1}{\pi^{2m-2k-2i}}\sin\pi t \| y \|$$

$$+ \sum_{i=0}^{[\frac{q-1}{2}]} L_{2i+1}\frac{1}{\pi^{2m-2k-2-2i}}\left(\frac{4}{\pi^2}\right)\sin\pi t \| y \|$$

$$(1.6.15) \qquad = \left[C\frac{1}{\pi^{2m-2}}\left(\frac{1}{2\pi}\right) + \theta \| y \|\right]\frac{\sin\pi t}{\pi^{-2k}}, \ 0 \leq k \leq \left[\frac{q}{2}\right].$$

Similarly, from (1.6.14), (1.6.8), (1.2.44), (1.2.45) and (1.2.47) we get

$$(1.6.16) \qquad | (Ty)^{(2k+1)}(t) | \leq \left[C\frac{1}{\pi^{2m-2}}\left(\frac{1}{2\pi}\right) + \theta \| y \|\right] \times$$

$$\frac{2\sin\pi t + \pi(1-2t)\cos\pi t}{\pi^{-2k}}, \ 0 \leq k \leq \left[\frac{q-1}{2}\right].$$

Now combining (1.6.15) and (1.6.16) we obtain

$$\begin{aligned} \| Ty \| &\leq C\frac{1}{\pi^{2m-2}}\left(\frac{1}{2\pi}\right) + \theta(1-\theta)^{-1}\frac{1}{\pi^{2m-2}}\left(\frac{1}{2\pi}\right)C \\ &= (1-\theta)^{-1}\frac{1}{\pi^{2m-2}}\left(\frac{1}{2\pi}\right)C. \quad \blacksquare \end{aligned}$$

Remark 1.6.1. In Theorem 1.6.3 the inequality (1.6.10) for $q = 0$ is the best possible, i.e., θ cannot be replaced by a smaller number. Indeed, in case of equality $L_0 = \pi^{2m}$ the boundary value problem $(-1)^m x^{(2m)} = L_0 x$; $x(0) = \epsilon(\neq 0)$, $x^{(2i)}(0) = 0$, $1 \leq i \leq m-1$; $x^{(2i)}(1) = 0$, $0 \leq i \leq m-1$ has no solution. ∎

Theorem 1.6.4. Suppose that the differential equation (1.6.1) together with the conditions

$$x^{(2i)}(0) = x^{(2i)}(1) = 0, \ 0 \le i \le m-1 \tag{1.6.17}$$

has a nontrivial solution $x(t)$ and the condition (1.6.8) with $L = 0$ on $[0,1] \times D_2$ is satisfied, where

$$\begin{aligned} D_2 \ = \ \Big\{ (x_0, x_1, ..., x_q) \ : \ & | \ x_{2i} \ |\le (-1)^{m-i} E_{2m-2i}(t) \max_{0 \le t \le 1} | \ x^{(2m)}(t) \ |, \\ & \qquad 0 \le i \le \left[\frac{q}{2}\right]; \\ & | \ x_{2i+1} \ |\le (-1)^{m-i} \left[2E_{2m-2i}(t) + (1-2t) E_{2m-2i-1}(t) \right] \times \\ & \qquad \max_{0 \le t \le 1} | \ x^{(2m)}(t) \ |, \ 0 \le i \le \left[\frac{q-1}{2}\right] \Big\}. \end{aligned}$$

Then, it is necessary that $\theta \ge 1$.

Proof. From (1.5.1) and (1.5.2) for any function $x(t) \in C^{(2m)}[0,1]$ satisfying (1.6.17) it follows that $\mathbf{x}(t) \in D_2$. Now since $x(t)$ is a nontrivial solution of (1.6.1), (1.6.17) we find that $\eta_i = \sup\limits_{0 \le t \le 1} \frac{\pi^{-2i} |x^{(2i)}(t)|}{\sin \pi t}$, $0 \le i \le \left[\frac{q}{2}\right]$; $\nu_i = \sup\limits_{0 \le t \le 1} \frac{\pi^{-2i} |x^{(2i+1)}(t)|}{2 \sin \pi t + \pi(1-2t) \cos \pi t}$, $0 \le i \le \left[\frac{q-1}{2}\right]$ exist and must be different from zero. Now as in Theorem 1.6.3 it is easy to obtain

$$\eta_j \le \theta \max \left\{ \eta_i, \ 0 \le i \le \left[\frac{q}{2}\right]; \ \nu_i, \ 0 \le i \le \left[\frac{q-1}{2}\right] \right\}, \ 0 \le j \le \left[\frac{q}{2}\right]$$

and

$$\nu_j \le \theta \max \left\{ \eta_i, \ 0 \le i \le \left[\frac{q}{2}\right]; \ \nu_i, \ 0 \le i \le \left[\frac{q-1}{2}\right] \right\}, \ 0 \le j \le \left[\frac{q-1}{2}\right].$$

Hence, it is necessary that $\theta \ge 1$. ∎

Remark 1.6.2. In Theorem 1.6.4 the inequality $\theta \ge 1$ for $q = 0$ is the best possible. Indeed, in case of equality $L_0 = \pi^{2m}$ the boundary value problem $(-1)^m x^{(2m)} = L_0 x$, (1.6.17) has nontrivial solutions $x(t) = c \ \sin \pi t$, where c is an arbitrary constant. ∎

Remark 1.6.3. If the condition (1.6.8) with $L = 0$ is satisfied, then obviously $x(t) \equiv 0$ is a solution of (1.6.1), (1.6.17); if $\theta < 1$ then Theorem 1.6.4 also guarantees its uniqueness in D_2. ∎

Theorem 1.6.5. Suppose that the function $f(t, x_0, x_1, ..., x_q)$ on $[0,1] \times D_1'$ satisfies the Lipschitz condition

$$(1.6.18) \qquad | f(t, x_0, x_1, ..., x_q) - f(t, \bar{x}_0, \bar{x}_1, ..., \bar{x}_q) | \leq \sum_{i=0}^{q} L_i | x_i - \bar{x}_i |$$

where D_1' is the same as D_1 with $L = \max_{0 \leq t \leq 1} | f(t, 0, ..., 0) |$. Then, the boundary value problem (1.6.1), (1.6.2) has a unique solution in D_1'.

Proof. Since the Lipschitz condition (1.6.18) implies (1.6.8), the existence of a solution follows from Theorem 1.6.3. To prove the uniqueness let $x(t)$ and $y(t)$ be two solutions of the boundary value problem (1.6.1), (1.6.2) in D_1'. Then, as in Theorem 1.6.3 we find that $\| x - y \| \leq \theta \| x - y \|$, and since $\theta < 1$ it follows that $\| x - y \| = 0$, i.e., $x(t) \equiv y(t)$. ∎

Remark 1.6.4. Once again in Theorem 1.6.5 the inequality $\theta < 1$ for $q = 0$ is the best possible. Indeed, in case of equality the boundary value problem $(-1)^m x^{(2m)} = L_0 x$, (1.6.17) has an infinite number of solutions. ∎

Picard's and Approximate Picard's Iteration

Picard's method of successive approximations has an important characteristic, namely that it is constructive; moreover, bounds of the difference between iterates and the solution are easily available. Here, we shall provide a priori as well as posteriori estimates on the Lipschitz constants so that Picard's iterative sequence $\{x_n(t)\}$ converges to the unique solution $x^*(t)$ of the Lidstone boundary value problem (1.6.1), (1.6.2). For the particular case $q = 0$, the obtained estimates are the best possible.

In actual computation of $\{x_n(t)\}$ only an approximate sequence $\{y_n(t)\}$ is constructed, which depends on replacing f by some simpler function. To find $y_{n+1}(t)$, we shall approximate f by f_n following relative and absolute error criteria, and obtain necessary and sufficient conditions for the convergence of $\{y_n(t)\}$ to $x^*(t)$.

It is well recognized that the method of upper and lower solutions, together with uniformly monotone convergent technique offers effective tools in proving and constructing multiple solutions of nonlinear problems. The upper and lower solutions generate an interval in a suitable partially ordered space, and serve as upper and lower bounds for solutions which can be improved by uniformly monotone convergent iterative procedures. Obviously, from the computational point of view monotone convergence has superiority over ordinary convergence. We shall discuss this fruitful technique for the boundary value problem (1.6.1), (1.6.2) with $q = 0$.

Lemma 1.6.6. [2] Let B be a Banach space and let $0 < r \in \Re$, $\bar{S}(x_0, r) = \{x \in B : \| x - x_0 \| \leq r\}$. Let T map $\bar{S}(x_0, r)$ into B and

(i) for all $x, y \in \bar{S}(x_0, r)$, $\| Tx - Ty \| \leq \alpha \| x - y \|$, where $0 < \alpha < 1$;

(ii) $r_0 = (1-\alpha)^{-1} \| Tx_0 - x_0 \| \leq r$.

Then, the following hold

(1) T has a fixed point x^* in $\bar{S}(x_0, r_0)$;

(2) x^* is the unique fixed point of T in $\bar{S}(x_0, r)$;

(3) the sequence $\{x_n\}$, where $x_{n+1} = Tx_n$; $n = 0,\ 1,\ \ldots$ converges to x^* with $\| x^* - x_n \| \leq \alpha^n r_0$, and $\| x^* - x_n \| \leq \alpha(1-\alpha)^{-1} \| x_n - x_{n-1} \|$;

(4) for any $x \in \bar{S}(x_0, r_0)$, $x^* = \lim_{n\to\infty} T^n x$. ∎

Lemma 1.6.7. [2] Let $(E, \leq)$ be a partially ordered space and $x_0 \leq y_0$ be two elements of E, and $[x_0, y_0]$ denotes the interval $\{x \in E : x_0 \leq x \leq y_0\}$. Further, let $T : [x_0, y_0] \to E$ be an isotone operator ($T(x) \leq T(y)$, whenever $x \leq y$) and let it possess the properties

(i) $x_0 \leq T(x_0)$;

(ii) the (nondecreasing) sequence $\{T^n(x_0)\}$ where $T^0(x_0) = x_0$, $T^{n+1}(x_0) = T\left[T^n(x_0)\right]$ for each $n = 0,\ 1,\ \ldots$ is well defined, i.e., $T^n(x_0) \leq y_0$ for each natural n;

(iii) the sequence $\{T^n(x_0)\}$ has $\sup x \in E$, i.e., $T^n(x_0) \uparrow x$;

(iv) $T^{n+1}(x_0) \uparrow T(x)$.

(i)′ $T(y_0) \leq y_0$;

(ii)′ the (nonincreasing) sequence $\{T^n(y_0)\}$ is well defined, i.e., $T^n(y_0) \geq x_0$ for each natural n;

(iii)′ the sequence $\{T^n(y_0)\}$ has $\inf y \in E$, i.e., $T^n(y_0) \downarrow y$;

(iv)′ $T^{n+1}(y_0) \downarrow T(y)$.

Then, $x = T(x)$ and for any other fixed point $z \in [x_0, y_0]$ of T, $x \leq z$ is true. (Then, $y = T(y)$ and for any other fixed point $z \in [x_0, y_0]$ of T, $z \leq y$ is valid.)

Moreover, if T possesses both properties (i) and (i)′, then the sequences $\{T^n(x_0)\}$, $\{T^n(y_0)\}$ are well defined and if, further, T has the properties (iii), (iii)′ and (iv), (iv)′ then

$$x_0 \leq T(x_0) \leq \cdots \leq T^n(x_0) \leq \cdots \leq x \leq y \leq \cdots \leq T^n(y_0) \leq \cdots \leq T(y_0) \leq y_0$$

and $x = T(x)$, $y = T(y)$; also any other fixed point $z \in [x_0, y_0]$ of T satisfies $x \leq z \leq y$. ∎

Definition 1.6.1. A function $\bar{x}(t) \in C^{(2m)}[0,1]$ is called an *approximate solution* of (1.6.1), (1.6.2) if there exist nonnegative constants δ and ϵ such that

$$\max_{0 \leq t \leq 1} \mid (-1)^m \bar{x}^{(2m)}(t) - f(t, \bar{\mathrm{x}}(t)) \mid \leq \delta \tag{1.6.19}$$

and

$$\mid P^{(2i)}_{(1.1.1)}(t) - \bar{P}^{(2i)}_{(1.6.22)}(t) \mid \leq \epsilon \, \frac{1}{\pi^{2m-2i-2}} \left(\frac{1}{2\pi}\right) \sin \pi t, \ 0 \leq i \leq \left[\frac{q}{2}\right] \tag{1.6.20}$$

$$\begin{aligned} &\mid P^{(2i+1)}_{(1.1.1)}(t) - \bar{P}^{(2i+1)}_{(1.6.22)}(t) \mid \\ &\quad \leq \epsilon \, \frac{1}{\pi^{2m-2i-2}} \left(\frac{1}{2\pi}\right) [2 \sin \pi t + \pi(1-2t) \cos \pi t], \ 0 \leq i \leq \left[\frac{q-1}{2}\right] \end{aligned} \tag{1.6.21}$$

where $P_{(1.1.1)}(t)$ and $\bar{P}_{(1.6.22)}(t)$ are polynomials of degree $(2m-1)$ satisfying (1.1.1), and

$$(1.6.22) \qquad \bar{P}^{(2i)}_{(1.6.22)}(0) = \bar{x}^{(2i)}(0),\ \bar{P}^{(2i)}_{(1.6.22)}(1) = \bar{x}^{(2i)}(1);\ 0 \le i \le m-1$$

respectively.

The inequality (1.6.19) means that there exists a continuous function $\eta(t)$ such that

$$(1.6.23) \qquad (-1)^m \bar{x}^{(2m)}(t) = f(t, \bar{\mathbf{x}}(t)) + \eta(t)$$

and

$$(1.6.24) \qquad \max_{0\le t\le 1} |\, \eta(t) \,| \le \delta.$$

Thus, from Theorem 1.4.1 the approximate solution $\bar{x}(t)$ can be expressed as

$$(1.6.25) \qquad \bar{x}(t) = \bar{P}_{(1.6.22)}(t) + \int_0^1 |\, g_m(t,s) \,|\, [f(s, \bar{\mathbf{x}}(s)) + \eta(s)]\, ds.$$

In the following results also we shall consider the Banach space $C^{(q)}[0,1]$ and for $y(t) \in C^{(q)}[0,1]$ the norm $\| y \|$ is defined in (1.6.13).

Theorem 1.6.8. With respect to the boundary value problem (1.6.1), (1.6.2) we assume that there exists an approximate solution $\bar{x}(t)$ and

(i) the function $f(t, x_0, x_1, ..., x_q)$ satisfies the Lipschitz condition (1.6.18) on $[0,1] \times D_3$, where

$$(1.6.26) \quad D_3 = \Big\{ (x_0, x_1, ..., x_q) \ : |\, x_{2i} - \bar{x}^{(2i)}(t) \,| \le \pi^{2i} N \ \sin \pi t,\ 0 \le i \le \left[\frac{q}{2}\right];$$
$$|\, x_{2i+1} - \bar{x}^{(2i+1)}(t) \,| \le \pi^{2i} N [2 \sin \pi t + \pi (1-2t) \cos \pi t],$$
$$0 \le i \le \left[\frac{q-1}{2}\right] \Big\};$$

(ii) $N_0 = (1-\theta)^{-1}(\epsilon + \delta) \dfrac{1}{\pi^{2m-2}} \left(\dfrac{1}{2\pi}\right) \le N.$

Then, the following hold

(1) there exists a solution $x^*(t)$ of (1.6.1), (1.6.2) in $\bar{S}(\bar{x}, N_0)$;

(2) $x^*(t)$ is the unique solution of (1.6.1), (1.6.2) in $\bar{S}(\bar{x}, N)$;

(3) the Picard iterative sequence $\{x_n(t)\}$, defined by

$$x_{n+1}(t) = P_{(1.1.1)}(t) + \int_0^1 \mid g_m(t,s) \mid f(s, \mathbf{x}_n(s))\, ds; \; n = 0,\ 1,\ \dots \tag{1.6.27}$$

where $x_0(t) = \bar{x}(t)$ converges to $x^*(t)$ with $\| x^* - x_n \| \le \theta^n N_0$, and

$$\| x^* - x_n \| \le \theta(1-\theta)^{-1} \| x_n - x_{n-1} \|;$$

(4) for any $x_0(t) = x(t) \in \bar{S}(\bar{x}, N_0)$, $x^*(t) = \lim_{n\to\infty} x_n(t)$.

Proof. Define an operator $T : C^{(q)}[0,1] \to C^{(q)}[0,1]$ as in (1.6.4). We shall show that T on $\bar{S}(\bar{x}, N)$ satisfies the conditions of Lemma 1.6.6. For this, if $x(t) \in \bar{S}(\bar{x}, N)$ then from the definition of norm in (1.6.13) it is clear that $\mathbf{x}(t) \in D_3$. Further, if $x(t), y(t) \in \bar{S}(\bar{x}, N)$ then as in Theorem 1.6.3 it follows that

$$\mid (Tx)^{(2i)}(t) - (Ty)^{(2i)}(t) \mid \le \pi^{2i}\theta \| x - y \| \sin \pi t, \; 0 \le i \le \left[\frac{q}{2}\right]$$

and

$$\mid (Tx)^{(2i+1)}(t) - (Ty)^{(2i+1)}(t) \mid \le \pi^{2i}\theta \| x - y \| [2 \sin \pi t + \pi(1-2t)\cos \pi t],$$
$$0 \le i \le \left[\frac{q-1}{2}\right].$$

The above inequalities imply that

$$\| Tx - Ty \| \le \theta \| x - y \| .$$

Next, from (1.6.4) and (1.6.25) we have

$$(T\bar{x})(t) - \bar{x}(t) = P_{(1.1.1)}(t) - \bar{P}_{(1.6.22)}(t) - \int_0^1 \mid g_m(t,s) \mid \eta(s)\, ds$$

and hence from (1.6.20), (1.6.21), (1.6.24) and the inequalities (1.2.39), (1.2.44) we easily find

$$\mid (T\bar{x})^{(2i)}(t) - \bar{x}^{(2i)}(t) \mid \le (\epsilon + \delta)\frac{1}{\pi^{2m-2i-2}}\left(\frac{1}{2\pi}\right) \sin \pi t, \; 0 \le i \le \left[\frac{q}{2}\right]$$

and

$$| (T\bar{x})^{(2i+1)}(t) - \bar{x}^{(2i+1)}(t) | \le (\epsilon+\delta)\frac{1}{\pi^{2m-2i-2}}\left(\frac{1}{2\pi}\right)[2\sin \pi t + \pi(1-2t)\cos \pi t],$$

$$0 \le i \le \left[\frac{q-1}{2}\right].$$

Combining the above inequalities, we obtain

$$\| T\bar{x} - \bar{x} \| \le (\epsilon + \delta)\frac{1}{\pi^{2m-2}}\left(\frac{1}{2\pi}\right)$$

and hence

$$(1-\theta)^{-1} \| T\bar{x} - \bar{x} \| \le (1-\theta)^{-1}(\epsilon + \delta)\frac{1}{\pi^{2m-2}}\left(\frac{1}{2\pi}\right) = N_0 \le N.$$

Thus, the conditions of Lemma 1.6.6 are satisfied and the conclusions (1) - (4) follow. ∎

In Theorem 1.6.8 the conclusion (3) ensures that the sequence $\{x_n(t)\}$ obtained from (1.6.27) converges to the solution $x^*(t)$ of the boundary value problem (1.6.1), (1.6.2). However, in practical evaluation this sequence is approximated by the computed sequence, say, $\{y_n(t)\}$. To find $y_{n+1}(t)$, the function f is approximated by f_n. Therefore, the computed sequence $\{y_n(t)\}$ satisfies the recurrence relation

$$(1.6.28) \qquad y_{n+1}(t) = P_{(1.1.1)}(t) + \int_0^1 | g_m(t,s) | f_n(s, \mathbf{y}_n(s))\, ds; \ n = 0,\ 1,\ ...$$

where $y_0(t) = x_0(t) = \bar{x}(t)$.

With respect to f_n, we shall assume the following condition :

Condition $\mathbf{C_1}$: For $\mathbf{y}_n(t)$ obtained from (1.6.28), the following inequality is satisfied

$$(1.6.29) \qquad | f(t, \mathbf{y}_n(t)) - f_n(t, \mathbf{y}_n(t)) | \le \Delta | f(t, \mathbf{y}_n(t)) |; \ n = 0,\ 1,\ ...$$

where Δ is a nonnegative constant.

Inequality (1.6.29) corresponds to the relative error in approximating the function f by f_n for the $(n+1)$th iteration.

Theorem 1.6.9. With respect to the boundary value problem (1.6.1), (1.6.2) we assume that there exists an approximate solution $\bar{x}(t)$, and the Condition C_1 is satisfied. Further, we assume that

(i) condition (i) of Theorem 1.6.8;

(ii) $\theta_1 = (1+\Delta)\theta < 1$;

(iii) $N_1 = (1-\theta_1)^{-1}(\epsilon+\delta+\Delta F)\frac{1}{\pi^{2m-2}}\left(\frac{1}{2\pi}\right) \le N$, where $F = \max_{0\le t\le 1} \mid f(t,\bar{\mathbf{x}}(t)) \mid$.

Then,

(1) all the conclusions (1) - (4) of Theorem 1.6.8 hold;

(2) the sequence $\{y_n(t)\}$ obtained from (1.6.28) remains in $\bar{S}(\bar{x}, N_1)$;

(3) the sequence $\{y_n(t)\}$ converges to $x^*(t)$, the solution of (1.6.1), (1.6.2) if and only if $\lim_{n\to\infty} a_n = 0$, where

$$a_n = \| y_{n+1}(t) - P_{(1.1.1)}(t) - \int_0^1 \mid g_m(t,s) \mid f(s,\mathbf{y}_n(s))\, ds \| \tag{1.6.30}$$

and the following error estimate holds

$$\| x^* - y_{n+1} \| \le (1-\theta)^{-1}\left[\theta \| y_{n+1} - y_n \| + \frac{1}{\pi^{2m-2}}\left(\frac{1}{2\pi}\right)\Delta \times \max_{0\le t\le 1} \mid f(t,\mathbf{y}_n(t)) \mid\right]. \tag{1.6.31}$$

Proof. Since $\theta_1 < 1$ implies $\theta < 1$ and obviously $N_0 \le N_1$, the conditions of Theorem 1.6.8 are satisfied and the conclusion (1) follows.

To prove (2), we note that $\bar{x}(t) \in \bar{S}(\bar{x}, N_1)$, and from (1.6.25) and (1.6.28) we find

$$\begin{aligned} y_1(t) - \bar{x}(t) &= P_{(1.1.1)}(t) - \bar{P}_{(1.6.22)}(t) \\ &\quad + \int_0^1 \mid g_m(t,s) \mid [f_0(s,\bar{\mathbf{x}}(s)) - f(s,\bar{\mathbf{x}}(s)) - \eta(s)]\, ds. \end{aligned}$$

Thus, as earlier in Theorem 1.6.8 it follows that

$$\mid y_1^{(2i)}(t) - \bar{x}^{(2i)}(t) \mid \le (\epsilon+\delta+\Delta F)\frac{1}{\pi^{2m-2i-2}}\left(\frac{1}{2\pi}\right)\sin \pi t,\ 0 \le i \le \left[\frac{q}{2}\right] \tag{1.6.32}$$

and

$$(1.6.33)\qquad |\, y_1^{(2i+1)}(t) - \bar{x}^{(2i+1)}(t) \,| \le (\epsilon + \delta + \Delta F)\frac{1}{\pi^{2m-2i-2}}\left(\frac{1}{2\pi}\right)[2\sin\pi t + \pi(1-2t)\cos\pi t],\ 0 \le i \le \left[\frac{q-1}{2}\right].$$

From (1.6.32) and (1.6.33), we have

$$\|\, y_1 - \bar{x} \,\| \le (\epsilon + \delta + \Delta F)\frac{1}{\pi^{2m-2}}\left(\frac{1}{2\pi}\right) \le N_1,$$

i.e., $y_1(t) \in \bar{S}(\bar{x}, N_1)$.

Now we shall assume that $y_n(t) \in \bar{S}(\bar{x}, N_1)$ and show that $y_{n+1}(t) \in \bar{S}(\bar{x}, N_1)$. For this, from (1.6.25) and (1.6.28) we have

$$y_{n+1}(t) - \bar{x}(t) = P_{(1.1.1)}(t) - \bar{P}_{(1.6.22)}(t) + \int_0^1 |\, g_m(t,s) \,| \,[f_n(s, \mathbf{y}_n(s)) - f(s, \bar{\mathbf{x}}(s)) - \eta(s)]\, ds$$

and hence it follows that

$$\begin{aligned} |\, y_{n+1}^{(2i)}(t) - \bar{x}^{(2i)}(t) \,| & \\ \le (\epsilon + \delta)\frac{1}{\pi^{2m-2i-2}}\left(\frac{1}{2\pi}\right)\sin\pi t & \\ + \int_0^1 |\, g_{m-i}(t,s) \,| \,[|\, f_n(s, \mathbf{y}_n(s)) - f(s, \mathbf{y}_n(s)) \,| & \\ + |\, f(s, \mathbf{y}_n(s)) - f(s, \bar{\mathbf{x}}(s)) \,|]\, ds & \\ \le (\epsilon + \delta)\frac{1}{\pi^{2m-2i-2}}\left(\frac{1}{2\pi}\right)\sin\pi t & \\ + \int_0^1 |\, g_{m-i}(t,s) \,| \left[\Delta F + (1+\Delta)\sum_{i=0}^{q} L_i \,|\, y_n^{(i)}(s) - \bar{x}^{(i)}(s) \,|\right] ds & \end{aligned}$$

$$(1.6.34)\qquad \le (\epsilon + \delta + \Delta F)\frac{1}{\pi^{2m-2i-2}}\left(\frac{1}{2\pi}\right)\sin\pi t + \theta_1 \,\|\, y_n - \bar{x} \,\|\, \frac{\sin\pi t}{\pi^{-2i}},\quad 0 \le i \le \left[\frac{q}{2}\right]$$

and, similarly

$$\begin{aligned} (1.6.35)\qquad |\, y_{n+1}^{(2i+1)}(t) - \bar{x}^{(2i+1)}(t) \,| & \\ \le (\epsilon + \delta + \Delta F)\frac{1}{\pi^{2m-2i-2}}\left(\frac{1}{2\pi}\right)[2\sin\pi t + \pi(1-2t)\cos\pi t] & \\ + \theta_1 \,\|\, y_n - \bar{x} \,\|\, \frac{2\sin\pi t + \pi(1-2t)\cos\pi t}{\pi^{-2i}},\quad 0 \le i \le \left[\frac{q-1}{2}\right]. & \end{aligned}$$

Combining (1.6.34) and (1.6.35) and using the fact that $y_n(t) \in \bar{S}(\bar{x}, N_1)$, it follows that

$$\| y_{n+1} - \bar{x} \|$$

$$\leq (\epsilon + \delta + \Delta F)\frac{1}{\pi^{2m-2}}\left(\frac{1}{2\pi}\right) + \theta_1(1-\theta_1)^{-1}(\epsilon + \delta + \Delta F)\frac{1}{\pi^{2m-2}}\left(\frac{1}{2\pi}\right)$$

$$= (1-\theta_1)^{-1}(\epsilon + \delta + \Delta F)\frac{1}{\pi^{2m-2}}\left(\frac{1}{2\pi}\right) = N_1$$

and hence $y_{n+1}(t) \in \bar{S}(\bar{x}, N_1)$. This completes the proof of (2).

Next, from (1.6.27) and (1.6.28), we have

$$\begin{aligned} x_{n+1}(t) - y_{n+1}(t) \quad &= \quad P_{(1.1.1)}(t) + \int_0^1 \mid g_m(t,s) \mid f(s, \mathbf{y}_n(s))\, ds - y_{n+1}(t) \\ &\quad + \int_0^1 \mid g_m(t,s) \mid [f(s, \mathbf{x}_n(s)) - f(s, \mathbf{y}_n(s))]\, ds \end{aligned}$$

and hence, as earlier it is easy to obtain

$$\| x_{n+1} - y_{n+1} \|$$

$$\leq \| y_{n+1}(t) - P_{(1.1.1)}(t) - \int_0^1 \mid g_m(t,s) \mid f(s, \mathbf{y}_n(s))\, ds \| + \theta \| x_n - y_n \|$$

$$= a_n + \theta \| x_n - y_n \| .$$

Since $x_0(t) = y_0(t) = \bar{x}(t)$, the above inequality gives

$$\| x_{n+1} - y_{n+1} \| \leq \sum_{i=0}^{n} \theta^{n-i} a_i. \tag{1.6.36}$$

Now using (1.6.36) in the triangle inequality, we get

$$\| x^* - y_{n+1} \| \leq \sum_{i=0}^{n} \theta^{n-i} a_i + \| x_{n+1} - x^* \| . \tag{1.6.37}$$

In (1.6.37), Theorem 1.6.8 ensures that $\lim_{n\to\infty} \| x_{n+1} - x^* \| = 0$. Thus, $\lim_{n\to\infty} a_n = 0$ is a necessary and sufficient condition for the convergence of the sequence $\{y_n(t)\}$ to $x^*(t)$ follows from Toeplitz lemma "for any $0 \leq \alpha < 1$, let $s_n = \sum_{i=0}^{n} \alpha^{n-i} d_i$; $n = 0, 1, \ldots$, then $\lim_{n\to\infty} s_n = 0$ if and only if $\lim_{n\to\infty} d_n = 0$".

Finally, to prove (1.6.31) we note that

$$x^*(t) - y_{n+1}(t) = \int_0^1 | g_m(t,s) | [f(s, \mathbf{x}^*(s)) - f(s, \mathbf{y}_n(s)) + f(s, \mathbf{y}_n(s)) - f_n(s, \mathbf{y}_n(s))] \, ds$$

and as earlier we find

$$(1.6.38) \quad \| x^* - y_{n+1} \| \le \theta \| x^* - y_n \| + \frac{1}{\pi^{2m-2}} \left(\frac{1}{2\pi} \right) \Delta \max_{0 \le t \le 1} | f(t, \mathbf{y}_n(t)) | .$$

From (1.6.38) the inequality (1.6.31) is immediate. ∎

In our next result we shall assume that

Condition $\mathbf{C_2}$: For $\mathbf{y}_n(t)$ obtained from (1.6.28), the following inequality is satisfied

$$(1.6.39) \qquad | f(t, \mathbf{y}_n(t)) - f_n(t, \mathbf{y}_n(t)) | \le \nabla; \; n = 0, \, 1, \, \ldots$$

where ∇ is a nonnegative constant.

Inequality (1.6.39) corresponds to the absolute error in approximating the function f by f_n for the $(n+1)$th iteration.

Theorem 1.6.10. With respect to the boundary value problem (1.6.1), (1.6.2) we assume that there exists an approximate solution $\bar{x}(t)$, and the Condition C_2 is satisfied. Further, we assume that

(i) condition (i) of Theorem 1.6.8;

(ii) $N_2 = (1-\theta)^{-1}(\epsilon + \delta + \nabla) \dfrac{1}{\pi^{2m-2}} \left(\dfrac{1}{2\pi} \right) \le N.$

Then,

(1) all the conclusions (1) - (4) of Theorem 1.6.8 hold;

(2) the sequence $\{y_n(t)\}$ obtained from (1.6.28) remains in $\bar{S}(\bar{x}, N_2)$;

(3) the sequence $\{y_n(t)\}$ converges to $x^*(t)$, the solution of (1.6.1), (1.6.2) if and only if $\lim_{n\to\infty} a_n = 0$, and the following error estimate holds

$$\| x^* - y_{n+1} \| \le (1-\theta)^{-1}\left[\theta \| y_{n+1} - y_n \| + \frac{1}{\pi^{2m-2}}\left(\frac{1}{2\pi}\right)\nabla\right].$$

Proof. The proof is similar to that of Theorem 1.6.9. ∎

Definition 1.6.2. A function $\mu(t) \in C^{(2m)}[0,1]$ is said to be a *lower solution* of (1.6.1), (1.6.2) with $q = 0$ provided

$$(-1)^m \mu^{(2m)}(t) \le f(t, \mu(t)),\ t \in [0,1]$$

$$(-1)^i \left[\mu^{(2i)}(0) - \alpha_i\right] \le 0,\ (-1)^i \left[\mu^{(2i)}(1) - \beta_i\right] \le 0;\ 0 \le i \le m-1.$$

Similarly, a function $\nu(t) \in C^{(2m)}[0,1]$ is called an *upper solution* of (1.6.1), (1.6.2) with $q = 0$ if

$$(-1)^m \nu^{(2m)}(t) \ge f(t, \nu(t)),\ t \in [0,1]$$

$$(-1)^i \left[\nu^{(2i)}(0) - \alpha_i\right] \ge 0,\ (-1)^i \left[\nu^{(2i)}(1) - \beta_i\right] \ge 0;\ 0 \le i \le m-1.$$

Lemma 1.6.11. Let $\mu(t)$ and $\nu(t)$ be lower and upper solutions of (1.6.1), (1.6.2) with $q = 0$, and $P_\mu(t)$ and $P_\nu(t)$ be the polynomials of degree $(2m-1)$ satisfying

$$P_\mu^{(2i)}(0) = \mu^{(2i)}(0),\ P_\mu^{(2i)}(1) = \mu^{(2i)}(1);\ 0 \le i \le m-1$$

and

$$P_\nu^{(2i)}(0) = \nu^{(2i)}(0),\ P_\nu^{(2i)}(1) = \nu^{(2i)}(1);\ 0 \le i \le m-1$$

respectively. Then, for all $t \in [0,1]$

$$P_\mu(t) \le P_{(1.1.1)}(t) \le P_\nu(t).$$

Proof. Explicitly $P_\mu(t)$ can be written as

$$\begin{aligned} P_\mu(t) &= \sum_{i=0}^{m-1} \left[\mu^{(2i)}(0)\Lambda_i(1-t) + \mu^{(2i)}(1)\Lambda_i(t)\right] \\ &= \sum_{i=0}^{m-1} \left[(-1)^i \mu^{(2i)}(0)(-1)^i \Lambda_i(1-t) + (-1)^i \mu^{(2i)}(1)(-1)^i \Lambda_i(t)\right]. \end{aligned}$$

Now since $(-1)^i\Lambda_i(t)$ and $(-1)^i\Lambda_i(1-t)$ are nonnegative for all $t \in [0,1]$, from the definition of $\mu(t)$ it follows that

$$\begin{aligned} P_\mu(t) &\le \sum_{i=0}^{m-1}\left[(-1)^i\alpha_i(-1)^i\Lambda_i(1-t)+(-1)^i\beta_i(-1)^i\Lambda_i(t)\right] \\ &= \sum_{i=0}^{m-1}\left[\alpha_i\Lambda_i(1-t)+\beta_i\Lambda_i(t)\right] = P_{(1.1.1)}(t). \end{aligned}$$

The proof of the inequality $P_{(1.1.1)}(t) \le P_\nu(t)$ is similar. ∎

In the following result for $x(t) \in C[0,1]$ we shall consider the norm $\| x \| = \max_{0\le t\le 1} | x(t) |$, and introduce a partial ordering as follows : for $x,\ y \in C[0,1]$ we say that $x \le y$ if and only if $x(t) \le y(t)$ for all $t \in [0,1]$.

Theorem 1.6.12. With respect to the boundary value problem (1.6.1), (1.6.2) with $q = 0$ we assume that $f(t, u_0)$ is nondecreasing in u_0. Further, let there exist lower and upper solutions $x_0(t)$, $y_0(t)$ such that $x_0 \le y_0$. Then, the sequences $\{x_n(t)\}$, $\{y_n(t)\}$ where $x_n(t)$ and $y_n(t)$ are defined by the iterative schemes

$$x_{n+1}(t) = P_{(1.1.1)}(t) + \int_0^1 | g_m(t,s) | f(s, x_n(s))\, ds$$

$$y_{n+1}(t) = P_{(1.1.1)}(t) + \int_0^1 | g_m(t,s) | f(s, y_n(s))\, ds; \;\; n = 0,\ 1,\ \ldots$$

are well defined, and $\{x_n(t)\}$ converges to an element $x(t) \in C[0,1]$, $\{y_n(t)\}$ converges to an element $y(t) \in C[0,1]$ (the convergence being in the norm of $C[0,1]$). Further, $x_0 \le x_1 \le \cdots \le x_n \le \cdots \le x \le y \le \cdots \le y_n \le \cdots \le y_1 \le y_0$; $x(t)$, $y(t)$ are solutions of (1.6.1), (1.6.2) with $q = 0$, and each solution $z(t)$ of this problem which is such that $z \in [x_0, y_0]$ satisfies $x \le z \le y$.

Proof. First, we shall show that the operator $T : C[0,1] \to C[0,1]$ defined by

$$(Tx)(t) = P_{(1.1.1)}(t) + \int_0^1 | g_m(t,s) | f(s, x(s))\, ds$$

is isotone. For this, let $x(t)$, $y(t) \in C[0,1]$ and $x \le y$, then from the partial ordering it follows that $x(s) \le y(s)$ for all $s \in [0,1]$, and hence from the monotone property of f, we have $f(s,x(s)) \leq f(s,y(s))$, $s \in [0,1]$. Next, since $(-1)^m g_m(t,s) = | g_m(t,s) |$ it follows that

$$| g_m(t,s) | f(s,x(s)) \leq | g_m(t,s) | f(s,y(s)); \quad s,t \in [0,1].$$

From this the inequality $T(x) \leq T(y)$ is obvious, and this completes the proof of T being isotone.

Next, since $x_0(t)$ is a lower solution, Lemma 1.6.11 gives that

$$\begin{aligned} x_0(t) &= P_{x_0}(t) + \int_0^1 | g_m(t,s) | (-1)^m x_0^{(2m)}(s)\, ds \\ &\leq P_{(1.1.1)}(t) + \int_0^1 | g_m(t,s) | f(s,x_0(s))\, ds = (Tx_0)(t), \end{aligned}$$

i.e., $x_0 \leq T(x_0)$. The inequality $T(y_0) \leq y_0$ can be proved analogously. Thus, the conditions (i) and (i)$'$ of Lemma 1.6.7 hold and in conclusion the sequences $\{T^n(x_0)\}$, $\{T^n(y_0)\}$ are well defined.

Since $T^n(x_0) = T\left[T^{n-1}(x_0)\right]$, we have $T^n(x_0) = x_n$ and $T^n(y_0) = y_n$. The sequence $\{x_n(t)\}$ is nondecreasing and bounded from above by $y_0(t)$, $t \in [0,1]$. Similarly, the sequence $\{y_n(t)\}$ is nonincreasing and bounded from below by $x_0(t)$, $t \in [0,1]$. Hence, in conclusion the sequences $\{x_n(t)\}$, $\{y_n(t)\}$ are uniformly bounded in $[0,1]$. Further, since the functions $x_n(t)$, $y_n(t)$ are the solutions of appropriate boundary value problems, these sequences are equicontinuous also. Thus, Arzela - Ascoli theorem is applicable and there exist subsequences of $\{x_n(t)\}$, $\{y_n(t)\}$ which converge uniformly in $[0,1]$. However, since these sequences are monotonic, we conclude that the whole sequences $\{x_n(t)\}$, $\{y_n(t)\}$ converge uniformly to some $x(t)$, $y(t)$, and $x \leq y$ in the partial ordering of $C[0,1]$. Summarizing these arguments, we find that $T^n x_0 \uparrow x$ and $T^n y_0 \downarrow y$.

Finally, the continuity of the operator T implies that $T^{n+1}x_0 = T\left[T^n x_0\right] \uparrow Tx$ and $T^{n+1}y_0 = T\left[T^n y_0\right] \downarrow Ty$.

Hence the conditions of Lemma 1.6.7 are satisfied and the conclusions of Theorem 1.6.12 follow. ∎

Remark 1.6.5. If $f(t,u_0)$ is nondecreasing in u_0, then uniqueness of the solutions of (1.6.1), (1.6.2) with $q = 0$ is not guaranteed, e.g., the boundary value problem $(-1)^m x^{(2m)} = \pi^{2m} x$, (1.6.17) has an infinite number of solutions $x(t) = c\ \sin\pi t$, where c is an arbitrary constant. However, if $f(t, u_0)$ is nonincreasing then the problem (1.6.1), (1.6.2) with $q = 0$ has at most one solution. To prove this, we assume that $x(t)$ and $y(t)$ both are solutions of (1.6.1), (1.6.2) so that

$$(-1)^m[x^{(2m)}(t) - y^{(2m)}(t)] = f(t, x(t)) - f(t, y(t))$$

and hence

$$\begin{aligned}(-1)^m[x(t) - y(t)][x^{(2m)}(t) - y^{(2m)}(t)] &= [x(t) - y(t)][f(t, x(t)) - f(t, y(t))] \\ &\le 0,\end{aligned}$$

where the inequality follows as a consequence of the nonincreasing nature of $f(t, u_0)$ in u_0. Now an integration by parts gives that

$$\begin{aligned}&(-1)^m \int_0^1 [x(t) - y(t)][x^{(2m)}(t) - y^{(2m)}(t)]\, dt \\ &\qquad = (-1)^{2m} \int_0^1 [x^{(m)}(t) - y^{(m)}(t)]^2\, dt \le 0,\end{aligned}$$

which is possible only when $x(t) \equiv y(t)$. ∎

Theorem 1.6.13. The linear differential equation

$$(1.6.40) \qquad (-1)^m x^{(2m)} = f(t)x + g(t)$$

together with the boundary conditions (1.6.2) has a unique solution if $\max_{0\le t\le 1} f(t) \le 0$.

Proof. For the linear problem (1.6.40), (1.6.2) obviously the uniqueness implies the existence. Hence, if $f(t) \le 0$ for all $t \in [0,1]$ our Remark 1.6.5 is applicable and the result follows. ∎

Remark 1.6.6. From Theorem 1.6.13 the differential equation

$$(-1)^m x^{(2m)} = -\pi^{2m} x + 2\pi^{2m} \sin\pi t$$

together with the boundary conditions (1.6.17) has a unique solution $x(t) = \sin\pi t$. However, the iterative procedure given by

$$(-1)^m x_{n+1}^{(2m)} = -\pi^{2m} x_n + 2\pi^{2m} \sin\pi t$$

$$x_{n+1}^{(2i)}(0) = x_{n+1}^{(2i)}(1) = 0,\ 0 \le i \le m-1$$

with $x_0(t) = 0$ oscillates ($x_{2n+1}(t) = 2\sin\pi t$, $x_{2n}(t) = 0$), and hence Theorem 1.6.13 does not imply the convergence of Picard's iterative scheme. ∎

Quasilinearization and Approximate Quasilinearization

Newton's method when applied to higher order boundary value problems has been labeled as quasilinearization. Here, we shall employ this powerful technique to generate the iterative sequence $\{x_n(t)\}$ for the Lidstone boundary value problem (1.6.1), (1.6.2). Then, as further applications of the equalities and inequalities obtained in Sections 1.2 - 1.5 we shall provide a priori conditions for the linear as well as quadratic convergence of this sequence $\{x_n(t)\}$ to the unique solution $x^*(t)$ of (1.6.1), (1.6.2). Computational aspects of this sequence $\{x_n(t)\}$ similar to those of Picard's method are also discussed.

The quasilinear iterative scheme for (1.6.1), (1.6.2) is defined as

$$(1.6.41)\quad (-1)^m x_{n+1}^{(2m)}(t) = f(t, \mathbf{x}_n(t)) + \sum_{i=0}^{q}\left(x_{n+1}^{(i)}(t) - x_n^{(i)}(t)\right)\frac{\partial}{\partial x_n^{(i)}(t)} f(t, \mathbf{x}_n(t))$$

$$(1.6.42)\qquad x_{n+1}^{(2i)}(0) = \alpha_i,\ x_{n+1}^{(2i)}(1) = \beta_i;\ 0 \le i \le m-1,\ n = 0,\ 1,\ \ldots$$

where $x_0(t) = \bar{x}(t)$ is an approximate solution of (1.6.1), (1.6.2).

In the following results once again we shall consider the Banach space $C^{(q)}[0,1]$ and for $y(t) \in C^{(q)}[0,1]$ the norm $\| y \|$ is defined in (1.6.13).

Theorem 1.6.14. With respect to the boundary value problem (1.6.1), (1.6.2) we assume that there exists an approximate solution $\bar{x}(t)$, and

(i) the function $f(t, x_0, x_1, \ldots, x_q)$ is continuously differentiable with respect to all x_i, $0 \le i \le q$ on $[0,1] \times D_3$;

(ii) there exist nonnegative constants L_i, $0 \le i \le q$ such that for all $(t, x_0, x_1, \ldots, x_q) \in [0,1] \times D_3$

$$\left|\frac{\partial}{\partial x_i} f(t, x_0, x_1, \ldots, x_q)\right| \le L_i;$$

(iii) $3\theta < 1$;

(iv) $N_3 = (1 - 3\theta)^{-1}(\epsilon + \delta)\frac{1}{\pi^{2m-2}}\left(\frac{1}{2\pi}\right) \le N.$

Then, the following hold

(1) the sequence $\{x_n(t)\}$ generated by the iterative scheme (1.6.41), (1.6.42) remains in $\bar{S}(\bar{x}, N_3)$;

(2) the sequence $\{x_n(t)\}$ converges to the unique solution $x^*(t)$ of the boundary value problem (1.6.1), (1.6.2);

(3) a bound on the error is given by

$$\| x_n - x^* \|$$

$$\le \left(\frac{2\theta}{1-\theta}\right)^n \left(1 - \frac{2\theta}{1-\theta}\right)^{-1} \| x_1 - \bar{x} \| \tag{1.6.43}$$

$$\le \left(\frac{2\theta}{1-\theta}\right)^n \left(1 - \frac{2\theta}{1-\theta}\right)^{-1} (1-\theta)^{-1}(\epsilon + \delta)\frac{1}{\pi^{2m-2}}\left(\frac{1}{2\pi}\right). \tag{1.6.44}$$

Proof. First, we shall show that the sequence $\{x_n(t)\}$ remains in $\bar{S}(\bar{x}, N_3)$. For this, we define an operator T as

$$(Tx)(t) = P_{(1.1.1)}(t) + \int_0^1 | g_m(t,s) | \Big[f(s, \mathbf{x}(s)) + \sum_{i=0}^{q} \left((Tx)^{(i)}(s) - x^{(i)}(s) \right) \frac{\partial}{\partial x^{(i)}(s)} f(s, \mathbf{x}(s)) \Big] ds. \tag{1.6.45}$$

Since $\bar{x}(t) \in \bar{S}(\bar{x}, N_3)$ it is sufficient to show that if $x(t) \in \bar{S}(\bar{x}, N_3)$ then $(Tx)(t) \in \bar{S}(\bar{x}, N_3)$. For this, from (1.6.25) and (1.6.45) we have

$$(Tx)(t) - \bar{x}(t) = P_{(1.1.1)}(t) - \bar{P}_{(1.6.22)}(t) + \int_0^1 | g_m(t,s) | \Big[f(s, \mathbf{x}(s)) - f(s, \bar{\mathbf{x}}(s)) + \sum_{i=0}^{q} \left((Tx)^{(i)}(s) - x^{(i)}(s) \right) \frac{\partial}{\partial x^{(i)}(s)} f(s, \mathbf{x}(s)) - \eta(s) \Big] ds. \tag{1.6.46}$$

Now since $x(t) \in \bar{S}(\bar{x}, N_3)$ implies that $\mathbf{x}(t) \in D_3$, we find

$$| f(s,\mathbf{x}(s)) - f(s,\bar{\mathbf{x}}(s)) |$$

$$\leq \sum_{i=0}^{q} L_i \mid x^{(i)}(s) - \bar{x}^{(i)}(s) \mid$$

$$\leq \sum_{i=0}^{[\frac{q}{2}]} L_{2i} \mid x^{(2i)}(s) - \bar{x}^{(2i)}(s) \mid + \sum_{i=0}^{[\frac{q-1}{2}]} L_{2i+1} \mid x^{(2i+1)}(s) - \bar{x}^{(2i+1)}(s) \mid$$

$$(1.6.47) \qquad \leq \phi \sin \pi s \parallel x - \bar{x} \parallel + \psi[2 \sin \pi s + \pi(1-2s) \cos \pi s] \parallel x - \bar{x} \parallel,$$

where $\phi = \sum_{i=0}^{[\frac{q}{2}]} \pi^{2i} L_{2i}$ and $\psi = \sum_{i=0}^{[\frac{q-1}{2}]} \pi^{2i} L_{2i+1}$.

Further, we have

$$\sum_{i=0}^{q} | (Tx)^{(i)}(s) - x^{(i)}(s) | \left| \frac{\partial}{\partial x^{(i)}(s)} f(s,\mathbf{x}(s)) \right|$$

$$\leq \sum_{i=0}^{q} L_i \mid (Tx)^{(i)}(s) - x^{(i)}(s) \mid$$

$$(1.6.48) \qquad \leq \phi \sin \pi s \,(\parallel Tx - \bar{x} \parallel + \parallel x - \bar{x} \parallel)$$

$$+\psi[2 \sin \pi s + \pi(1-2s) \cos \pi s] \,(\parallel Tx - \bar{x} \parallel + \parallel x - \bar{x} \parallel)\,.$$

For $0 \leq k \leq \left[\frac{q}{2}\right]$, using (1.6.20), (1.6.24), (1.6.47) and (1.6.48) in (1.6.46) gives

$$| (Tx)^{(2k)}(t) - \bar{x}^{(2k)}(t) |$$

$$\leq \epsilon \, \frac{1}{\pi^{2m-2k-2}} \left(\frac{1}{2\pi}\right) \sin \pi t + \int_0^1 | g_{m-k}(t,s) | \, [(2\phi \parallel x - \bar{x} \parallel + \phi \parallel Tx - \bar{x} \parallel) \times$$

$$\sin \pi s + (2\psi \parallel x - \bar{x} \parallel + \psi \parallel Tx - \bar{x} \parallel) \, (2 \sin \pi s + \pi(1-2s) \cos \pi s) + \delta] \, ds,$$

which on applying (1.2.6), (1.2.39) and (1.2.46) leads to

$$| (Tx)^{(2k)}(t) - \bar{x}^{(2k)}(t) |$$

$$\leq \epsilon \, \frac{1}{\pi^{2m-2k-2}} \left(\frac{1}{2\pi}\right) \sin \pi t + (2\phi \parallel x - \bar{x} \parallel + \phi \parallel Tx - \bar{x} \parallel) \frac{1}{\pi^{2m-2k}} \sin \pi t$$

$$+(2\psi \parallel x-\bar{x} \parallel +\psi \parallel Tx-\bar{x} \parallel)\frac{1}{\pi^{2m-2k-2}}\left(\frac{4}{\pi^2}\right)\sin \pi t$$

$$+\delta \frac{1}{\pi^{2m-2k-2}}\left(\frac{1}{2\pi}\right)\sin \pi t$$

$$(1.6.49) \quad = (\epsilon+\delta)\frac{1}{\pi^{2m-2k-2}}\left(\frac{1}{2\pi}\right)\sin \pi t + (2\phi+8\psi)\frac{1}{\pi^{2m-2k}} \parallel x-\bar{x} \parallel \sin \pi t$$

$$+(\phi+4\psi)\frac{1}{\pi^{2m-2k}} \parallel Tx-\bar{x} \parallel \sin \pi t.$$

Similarly, for $0 \leq k \leq \left[\frac{q-1}{2}\right]$ it follows that

$$(1.6.50) \quad |\,(Tx)^{(2k+1)}(t) - \bar{x}^{(2k+1)}(t)\,|$$

$$\leq (\epsilon+\delta)\frac{1}{\pi^{2m-2k-2}}\left(\frac{1}{2\pi}\right)[2\sin \pi t + \pi(1-2t)\cos \pi t]$$

$$+(2\phi+8\psi)\frac{1}{\pi^{2m-2k}} \parallel x-\bar{x} \parallel [2\sin \pi t + \pi(1-2t)\cos \pi t]$$

$$+(\phi+4\psi)\frac{1}{\pi^{2m-2k}} \parallel Tx-\bar{x} \parallel [2\sin \pi t + \pi(1-2t)\cos \pi t].$$

Combining (1.6.49) and (1.6.50), we obtain

$$\parallel Tx-\bar{x} \parallel \;\leq\; (\epsilon+\delta)\frac{1}{\pi^{2m-2}}\left(\frac{1}{2\pi}\right) + (2\phi+8\psi)\frac{1}{\pi^{2m}} \parallel x-\bar{x} \parallel$$

$$+(\phi+4\psi)\frac{1}{\pi^{2m}} \parallel Tx-\bar{x} \parallel,$$

which is the same as

$$\parallel Tx-\bar{x} \parallel \leq (1-\theta)^{-1}\left[(\epsilon+\delta)\frac{1}{\pi^{2m-2}}\left(\frac{1}{2\pi}\right) + 2\theta N_3\right] = N_3.$$

Next, we shall show the convergence of the sequence $\{x_n(t)\}$. From (1.6.41), (1.6.42) we have

$$(1.6.51)\quad x_{n+1}(t) - x_n(t) = \int_0^1 \mid g_m(t,s) \mid \Bigg\{ f(s, \mathbf{x}_n(s)) - f(s, \mathbf{x}_{n-1}(s)) + \sum_{i=0}^{q} \Bigg[\left(x_{n+1}^{(i)}(s) - x_n^{(i)}(s)\right) \frac{\partial}{\partial x_n^{(i)}(s)} f(s, \mathbf{x}_n(s)) - \left(x_n^{(i)}(s) - x_{n-1}^{(i)}(s)\right) \frac{\partial}{\partial x_{n-1}^{(i)}(s)} f(s, \mathbf{x}_{n-1}(s)) \Bigg] \Bigg\} ds.$$

Thus, following as above and using the fact that $\{x_n(t)\} \subseteq \bar{S}(\bar{x}, N_3)$, we find

$$\begin{aligned}
&\mid x_{n+1}^{(2k)}(t) - x_n^{(2k)}(t) \mid \\
&\leq \int_0^1 \mid g_{m-k}(t,s) \mid \{ \phi \sin \pi s \parallel x_n - x_{n-1} \parallel \\
&+ \psi[2 \sin \pi s + \pi(1-2s) \cos \pi s] \parallel x_n - x_{n-1} \parallel \\
&+ \phi \sin \pi s \parallel x_{n+1} - x_n \parallel \\
&+ \psi[2 \sin \pi s + \pi(1-2s) \cos \pi s] \parallel x_{n+1} - x_n \parallel \\
&+ \phi \sin \pi s \parallel x_n - x_{n-1} \parallel \\
&+ \psi[2 \sin \pi s + \pi(1-2s) \cos \pi s] \parallel x_n - x_{n-1} \parallel \} \, ds
\end{aligned}$$

$$(1.6.52)\quad \begin{aligned} &\leq (2\phi + 8\psi) \frac{1}{\pi^{2m-2k}} \parallel x_n - x_{n-1} \parallel \sin \pi t \\ &+ (\phi + 4\psi) \frac{1}{\pi^{2m-2k}} \parallel x_{n+1} - x_n \parallel \sin \pi t, \ 0 \leq k \leq \left[\frac{q}{2}\right] \end{aligned}$$

and

$$(1.6.53)\quad \begin{aligned} &\mid x_{n+1}^{(2k+1)}(t) - x_n^{(2k+1)}(t) \mid \\ &\leq (2\phi + 8\psi) \frac{1}{\pi^{2m-2k}} \parallel x_n - x_{n-1} \parallel [2 \sin \pi t + \pi(1-2t) \cos \pi t] \\ &+ (\phi + 4\psi) \frac{1}{\pi^{2m-2k}} \parallel x_{n+1} - x_n \parallel [2 \sin \pi t + \pi(1-2t) \cos \pi t], \\ &\qquad 0 \leq k \leq \left[\frac{q-1}{2}\right]. \end{aligned}$$

From (1.6.52) and (1.6.53), we deduce that

$$\parallel x_{n+1} - x_n \parallel \leq 2\theta \parallel x_n - x_{n-1} \parallel + \theta \parallel x_{n+1} - x_n \parallel,$$

which is the same as

$$\| x_{n+1} - x_n \| \le \frac{2\theta}{1-\theta} \| x_n - x_{n-1} \| .$$

Thus, an easy induction leads to

$$(1.6.54) \qquad \| x_{n+1} - x_n \| \le \left(\frac{2\theta}{1-\theta}\right)^n \| x_1 - x_0 \| .$$

Since $3\theta < 1$ means that $\frac{2\theta}{1-\theta} < 1$, inequality (1.6.54) implies that $\{x_n(t)\}$ is a Cauchy sequence and hence converges to some $x^*(t) \in \bar{S}(\bar{x}, N_3)$. This $x^*(t)$ is the unique solution of (1.6.1), (1.6.2) and can easily be verified.

The error bound (1.6.43) follows from (1.6.54) and the triangle inequality

$$\begin{aligned}
&\| x_{n+p} - x_n \| \\
&\le \| x_{n+p} - x_{n+p-1} \| + \| x_{n+p-1} - x_{n+p-2} \| + \cdots + \| x_{n+1} - x_n \| \\
&\le \left[\left(\frac{2\theta}{1-\theta}\right)^{n+p-1} + \left(\frac{2\theta}{1-\theta}\right)^{n+p-2} + \cdots + \left(\frac{2\theta}{1-\theta}\right)^n\right] \| x_1 - \bar{x} \| \\
&\le \left(\frac{2\theta}{1-\theta}\right)^n \left(1 - \frac{2\theta}{1-\theta}\right)^{-1} \| x_1 - \bar{x} \|
\end{aligned}$$

and now taking $p \to \infty$.

Finally, to prove (1.6.44), from (1.6.25) and (1.6.41), (1.6.42) we have

$$\begin{aligned}
&x_1(t) - x_0(t) \\
&= P_{(1.1.1)}(t) - \bar{P}_{(1.6.22)}(t) + \int_0^1 | g_m(t,s) | \left[\sum_{i=0}^{q} \left(x_1^{(i)}(s) - x_0^{(i)}(s)\right) \times \right. \\
&\qquad \left. \frac{\partial}{\partial x_0^{(i)}(s)} f(s, \mathbf{x}_0(s)) - \eta(s)\right] ds
\end{aligned}$$

and as before we find

$$\begin{aligned}
&| x_1^{(2k)}(t) - x_0^{(2k)}(t) | \\
&\le \epsilon \frac{1}{\pi^{2m-2k-2}} \left(\frac{1}{2\pi}\right) \sin \pi t + \int_0^1 | g_{m-k}(t,s) | \, [\phi \sin \pi s \, \| x_1 - x_0 \| \\
&\quad + \psi(2 \sin \pi s + \pi(1-2s) \cos \pi s) \, \| x_1 - x_0 \| + \delta] \, ds
\end{aligned}$$

$$\leq (\epsilon+\delta)\frac{1}{\pi^{2m-2k-2}}\left(\frac{1}{2\pi}\right)\sin \pi t \tag{1.6.55}$$

$$+(\phi+4\psi)\frac{1}{\pi^{2m-2k}} \parallel x_1 - x_0 \parallel \sin \pi t, \; 0 \leq k \leq \left[\frac{q}{2}\right]$$

and

$$| x_1^{(2k+1)}(t) - x_0^{(2k+1)}(t) | \tag{1.6.56}$$

$$\leq (\epsilon+\delta)\frac{1}{\pi^{2m-2k-2}}\left(\frac{1}{2\pi}\right)[2\sin \pi t + \pi(1-2t)\cos \pi t]$$

$$+(\phi+4\psi)\frac{1}{\pi^{2m-2k}} \parallel x_1 - x_0 \parallel [2\sin \pi t + \pi(1-2t)\cos \pi t],$$

$$0 \leq k \leq \left[\frac{q-1}{2}\right].$$

Inequalities (1.6.55) and (1.6.56) yield

$$\parallel x_1 - x_0 \parallel \leq (\epsilon+\delta)\frac{1}{\pi^{2m-2}}\left(\frac{1}{2\pi}\right) + \theta \parallel x_1 - x_0 \parallel,$$

which is the same as

$$\parallel x_1 - x_0 \parallel \leq (1-\theta)^{-1}(\epsilon+\delta)\frac{1}{\pi^{2m-2}}\left(\frac{1}{2\pi}\right). \tag{1.6.57}$$

Using the above inequality in (1.6.43) we get the required estimate (1.6.44). ∎

Theorem 1.6.15. Let the conditions of Theorem 1.6.14 be satisfied. Further, let $f(t, x_0, x_1, ..., x_q)$ be twice continuously differentiable with respect to all x_i, $0 \leq i \leq q$ on $[0,1] \times D_3$, and

$$\left| \frac{\partial^2}{\partial x_i \partial x_j} f(t, x_0, x_1, ..., x_q) \right| \leq L_i L_j K, \; 0 \leq i, j \leq q.$$

Then,

$$\| x_{n+1} - x_n \| \le \alpha \| x_n - x_{n-1} \|^2 \le \frac{1}{\alpha}(\alpha \| x_1 - x_0 \|)^{2^n} \tag{1.6.58}$$

$$\le \frac{1}{\alpha}\left[\frac{(\epsilon+\delta)}{\pi^{2m-1}}\frac{\theta}{(1-\theta)^2}\frac{K}{4}\max\{\pi\psi, \phi+2\psi\}\right]^{2^n},$$

where $\alpha = \frac{\theta}{1-\theta}\frac{K}{2}\max\{\pi\psi, \phi+2\psi\}$. Thus, the convergence is quadratic if

$$\frac{(\epsilon+\delta)}{\pi^{2m-1}}\frac{\theta}{(1-\theta)^2}\frac{K}{4}\max\{\pi\psi, \phi+2\psi\} < 1.$$

Proof. The sequence $\{x_n(t)\} \subseteq \bar{S}(\bar{x}, N_3)$ implies that $\mathbf{x}_n(t) \in D_3$ for all n. Further, since f is twice continuously differentiable, we have

(1.6.59) $f(t, \mathbf{x}_n(t))$

$$= f(t, \mathbf{x}_{n-1}(t)) + \sum_{i=0}^{q}\left(x_n^{(i)}(t) - x_{n-1}^{(i)}(t)\right)\frac{\partial}{\partial x_{n-1}^{(i)}(t)} f(t, \mathbf{x}_{n-1}(t))$$

$$+\frac{1}{2}\left[\sum_{i=0}^{q}\left(x_n^{(i)}(t) - x_{n-1}^{(i)}(t)\right)\frac{\partial}{\partial p_i(t)}\right]^2 f(t, p_0(t), p_1(t), \ldots, p_q(t)),$$

where $p_i(t)$ lies between $x_{n-1}^{(i)}(t)$ and $x_n^{(i)}(t)$, $0 \le i \le q$.

Using (1.6.59) in (1.6.51), we obtain

$$x_{n+1}(t) - x_n(t)$$

$$= \int_0^1 | g_m(t,s) | \left\{ \sum_{i=0}^{q}\left(x_{n+1}^{(i)}(s) - x_n^{(i)}(s)\right)\frac{\partial}{\partial x_n^{(i)}(s)} f(s, \mathbf{x}_n(s)) \right.$$

$$\left. +\frac{1}{2}\left[\sum_{i=0}^{q}\left(x_n^{(i)}(s) - x_{n-1}^{(i)}(s)\right)\frac{\partial}{\partial p_i(s)}\right]^2 f(s, p_0(s), p_1(s), \ldots, p_q(s)) \right\} ds.$$

Thus, for $0 \le k \le \left[\frac{q}{2}\right]$ we find

$$
\begin{aligned}
&| x_{n+1}^{(2k)}(t) - x_n^{(2k)}(t) | \\
&\quad \leq \int_0^1 | g_{m-k}(t,s) | \Big\{ \phi \sin \pi s \parallel x_{n+1} - x_n \parallel \\
&\quad +\psi(2\sin \pi s + \pi(1-2s)\cos \pi s) \parallel x_{n+1} - x_n \parallel \\
&\quad +\frac{1}{2}[\phi \sin \pi s + \psi(2\sin \pi s + \pi(1-2s)\cos \pi s)]^2 K \parallel x_n - x_{n-1} \parallel^2 \Big\} ds \\
&\quad \leq \phi \parallel x_{n+1} - x_n \parallel \frac{1}{\pi^{2m-2k}} \sin \pi t \\
&\quad +\psi \parallel x_{n+1} - x_n \parallel \frac{1}{\pi^{2m-2k-2}} \left(\frac{4}{\pi^2}\right) \sin \pi t \\
&\quad +\frac{1}{2} K \parallel x_n - x_{n-1} \parallel^2 \max_{0\leq t\leq 1} [\phi \sin \pi t + \psi(2\sin \pi t + \pi(1-2t)\cos \pi t)] \times \\
&\qquad\qquad \int_0^1 | g_{m-k}(t,s) | [\phi \sin \pi s + \psi(2\sin \pi s + \pi(1-2s)\cos \pi s)] \, ds
\end{aligned}
$$

$$
\begin{aligned}
(1.6.60) \leq\ & (\phi + 4\psi)\frac{1}{\pi^{2m-2k}} \parallel x_{n+1} - x_n \parallel \sin \pi t \\
&+\frac{1}{2} K \parallel x_n - x_{n-1} \parallel^2 \frac{1}{\pi^{2m-2k}} \max\{\pi\psi, \phi + 2\psi\} \cdot (\phi + 4\psi) \sin \pi t
\end{aligned}
$$

and, similarly for $0 \leq k \leq \left[\frac{q-1}{2}\right]$

$$
\begin{aligned}
(1.6.61) \quad & | x_{n+1}^{(2k+1)}(t) - x_n^{(2k+1)}(t) | \\
&\leq (\phi + 4\psi)\frac{1}{\pi^{2m-2k}} \parallel x_{n+1} - x_n \parallel [2\sin \pi t + \pi(1-2t)\cos \pi t] \\
&\quad +\frac{1}{2} K \parallel x_n - x_{n-1} \parallel^2 \frac{1}{\pi^{2m-2k}} \max\{\pi\psi, \phi + 2\psi\} \times \\
&\qquad\qquad (\phi + 4\psi)[2\sin \pi t + \pi(1-2t)\cos \pi t].
\end{aligned}
$$

Combining (1.6.60) and (1.6.61), we get

$$
\parallel x_{n+1} - x_n \parallel \leq \theta \parallel x_{n+1} - x_n \parallel + \frac{1}{2} K\theta \max\{\pi\psi, \phi + 2\psi\} \parallel x_n - x_{n-1} \parallel^2,
$$

which is the same as the first part of (1.6.58). The second part of (1.6.58) follows by an easy induction. Finally, the last part is an application of (1.6.57). ∎

The conclusion (3) of Theorem 1.6.14 ensures that the sequence $\{x_n(t)\}$ generated from the scheme (1.6.41), (1.6.42) converges linearly to the unique solution $x^*(t)$ of the boundary value problem (1.6.1), (1.6.2). Theorem 1.6.15 provides sufficient conditions for its quadratic convergence. However, in practical evaluation this sequence is approximated by the computed sequence, say, $\{y_n(t)\}$ which satisfies the recurrence relation

$$(1.6.62)\quad (-1)^m y_{n+1}^{(2m)}(t) = f_n(t, \mathbf{y}_n(t)) + \sum_{i=0}^{q} \left(y_{n+1}^{(i)}(t) - y_n^{(i)}(t) \right) \frac{\partial}{\partial y_n^{(i)}(t)} f_n(t, \mathbf{y}_n(t))$$

$$(1.6.63)\qquad y_{n+1}^{(2i)}(0) = \alpha_i,\ y_{n+1}^{(2i)}(1) = \beta_i;\ 0 \le i \le m-1,\ n = 0,\ 1,\ \ldots$$

where $y_0(t) = x_0(t) = \bar{x}(t)$.

With respect to f_n we shall assume the following condition :

Condition $\mathbf{C_3}$: $f_n(t, x_0, x_1, \ldots, x_q)$ is continuously differentiable with respect to all x_i, $0 \le i \le q$ on $[0,1] \times D_3$ with

$$\left| \frac{\partial}{\partial x_i} f_n(t, x_0, x_1, \ldots, x_q) \right| \le L_i$$

and

$$(1.6.64)\quad |\ f(t, x_0, x_1, \ldots, x_q) - f_n(t, x_0, x_1, \ldots, x_q)\ | \le \lambda_n\ |\ f(t, x_0, x_1, \ldots, x_q)\ |,$$

where λ_n, $n = 0,\ 1,\ \ldots$ are nonnegative constants and $\lambda_n \le \Delta$.

The inequality (1.6.64) corresponds to the relative error in approximating the function f by f_n for the $(n+1)$th iteration.

Theorem 1.6.16. With respect to the boundary value problem (1.6.1), (1.6.2) we assume that there exists an approximate solution $\bar{x}(t)$, and the Condition C_3 is satisfied. Further, we assume that

(i) conditions (i) and (ii) of Theorem 1.6.14;

(ii) $\theta_2 = (3 + \Delta)\theta < 1$;

(iii) $N_4 = (1-\theta_2)^{-1}(\epsilon + \delta + \Delta F)\frac{1}{\pi^{2m-2}}\left(\frac{1}{2\pi}\right) \le N$, where $F = \max_{0 \le t \le 1} |\ f(t, \bar{\mathbf{x}}(t))\ |$.

Then,

(1) all the conclusions (1) - (3) of Theorem 1.6.14 hold;

(2) the sequence $\{y_n(t)\}$ generated by the iterative scheme (1.6.62), (1.6.63) remains in $\bar{S}(\bar{x}, N_4)$;

(3) the sequence $\{y_n(t)\}$ converges to $x^*(t)$, the unique solution of (1.6.1), (1.6.2) if and only if $\lim_{n\to\infty} a_n = 0$, and the following error estimate holds

$$\| x^* - y_{n+1} \| \le (1-\theta)^{-1}\Big[2\theta \| y_{n+1} - y_n \| + \frac{1}{\pi^{2m-2}}\left(\frac{1}{2\pi}\right)\Delta \times \max_{0\le t\le 1} | f(t, \mathbf{y}_n(t)) | \Big]. \tag{1.6.65}$$

Proof. Since $\theta_2 < 1$ implies that $3\theta < 1$ and clearly $N_3 \le N_4$, the conditions of Theorem 1.6.14 are fulfilled and part (1) follows.

To prove part (2), we note that $\bar{x}(t) \in \bar{S}(\bar{x}, N_4)$ and from (1.6.25), (1.6.62), (1.6.63) we have

$$y_1(t) - \bar{x}(t)$$

$$= P_{(1.1.1)}(t) - \bar{P}_{(1.6.22)}(t) + \int_0^1 | g_m(t,s) | \Big[f_0(s, \mathbf{y}_0(s)) - f(s, \mathbf{y}_0(s))$$

$$+ \sum_{i=0}^{q} \left(y_1^{(i)}(s) - y_0^{(i)}(s)\right) \frac{\partial}{\partial y_0^{(i)}(s)} f_0(s, \mathbf{y}_0(s)) - \eta(s)\Big] \, ds.$$

Thus, for $0 \le k \le \left[\frac{q}{2}\right]$, as before we get

$$| y_1^{(2k)}(t) - \bar{x}^{(2k)}(t) |$$

$$\le \epsilon \frac{1}{\pi^{2m-2k-2}}\left(\frac{1}{2\pi}\right) \sin \pi t$$

$$+ \int_0^1 | g_{m-k}(t,s) | \,[\Delta F + \phi \sin \pi s \| y_1 - y_0 \|$$

$$+ \psi(2\sin \pi s + \pi(1-2s)\cos \pi s) \| y_1 - y_0 \| + \delta] \, ds$$

$$\le (\epsilon + \delta + \Delta F) \frac{1}{\pi^{2m-2k-2}}\left(\frac{1}{2\pi}\right) \sin \pi t \tag{1.6.66}$$

$$+ (\phi + 4\psi) \frac{1}{\pi^{2m-2k}} \| y_1 - y_0 \| \sin \pi t.$$

Similarly, for $0 \le k \le \left[\frac{q-1}{2}\right]$ we find

$$(1.6.67)\quad |\ y_1^{(2k+1)}(t) - \bar{x}^{(2k+1)}(t)\ |$$

$$\le (\epsilon + \delta + \Delta F)\frac{1}{\pi^{2m-2k-2}}\left(\frac{1}{2\pi}\right)[2\sin\pi t + \pi(1-2t)\cos\pi t]$$

$$+(\phi + 4\psi)\frac{1}{\pi^{2m-2k}}\ \|\ y_1 - y_0\ \|\ [2\sin\pi t + \pi(1-2t)\cos\pi t].$$

Combining (1.6.66) and (1.6.67), we get

$$\|\ y_1 - \bar{x}\ \| \le (\epsilon + \delta + \Delta F)\frac{1}{\pi^{2m-2}}\left(\frac{1}{2\pi}\right) + \theta\ \|\ y_1 - y_0\ \|,$$

which is the same as

$$(1.6.68)\qquad \|\ y_1 - \bar{x}\ \| \le (1-\theta)^{-1}(\epsilon + \delta + \Delta F)\frac{1}{\pi^{2m-2}}\left(\frac{1}{2\pi}\right) \le N_4.$$

Therefore, $y_1(t) \in \bar{S}(\bar{x}, N_4)$. Next, we assume that $y_n(t) \in \bar{S}(\bar{x}, N_4)$ and show that $y_{n+1}(t) \in \bar{S}(\bar{x}, N_4)$. For this, again from (1.6.25), (1.6.62), (1.6.63) we have

$$y_{n+1}(t) - \bar{x}(t)$$

$$= P_{(1.1.1)}(t) - \bar{P}_{(1.6.22)}(t) + \int_0^1 |\ g_m(t,s)\ | \Bigg[f_n(s, \mathbf{y}_n(s))$$

$$+ \sum_{i=0}^{q}\left(y_{n+1}^{(i)}(s) - y_n^{(i)}(s)\right)\frac{\partial}{\partial y_n^{(i)}(s)} f_n(s, \mathbf{y}_n(s)) - f(s, \mathbf{y}_0(s)) - \eta(s)\Bigg] ds.$$

Thus, for $0 \le k \le \left[\frac{q}{2}\right]$ as before, we obtain

$$|\ y_{n+1}^{(2k)}(t) - \bar{x}^{(2k)}(t)\ |$$

$$\le \epsilon\ \frac{1}{\pi^{2m-2k-2}}\left(\frac{1}{2\pi}\right)\sin\pi t + \int_0^1 |\ g_{m-k}(t,s)\ | \times$$

$$\Bigg[\sum_{i=0}^{q} |\ y_{n+1}^{(i)}(s) - y_n^{(i)}(s)\ | \left|\frac{\partial}{\partial y_n^{(i)}(s)} f_n(s, \mathbf{y}_n(s))\right|$$

$$+(1+\Delta)\ |\ f(s, \mathbf{y}_n(s)) - f(s, \mathbf{y}_0(s))\ | + \Delta\ |\ f(s, \mathbf{y}_0(s))\ | + \delta\Bigg] ds$$

$$\le \epsilon \frac{1}{\pi^{2m-2k-2}}\left(\frac{1}{2\pi}\right)\sin \pi t + \int_0^1 \mid g_{m-k}(t,s) \mid \{\phi \sin \pi s \parallel y_{n+1} - y_n \parallel$$

$$+\psi(2\sin \pi s + \pi(1-2s)\cos \pi s) \parallel y_{n+1} - y_n \parallel$$

$$+(1+\Delta)[\phi \sin \pi s \parallel y_n - y_0 \parallel$$

$$+\psi(2\sin \pi s + \pi(1-2s)\cos \pi s) \parallel y_n - y_0 \parallel] + \Delta F + \delta\}\ ds$$

$$(1.6.69)\quad \le (\epsilon + \delta + \Delta F)\frac{1}{\pi^{2m-2k-2}}\left(\frac{1}{2\pi}\right)\sin \pi t$$

$$+(\phi + 4\psi)\frac{1}{\pi^{2m-2k}} \parallel y_{n+1} - y_n \parallel \sin \pi t$$

$$+(1+\Delta)(\phi + 4\psi)\frac{1}{\pi^{2m-2k}} \parallel y_n - y_0 \parallel \sin \pi t.$$

Likewise, for $0 \le k \le \left[\frac{q-1}{2}\right]$, we find

$$(1.6.70)\quad \mid y_{n+1}^{(2k+1)}(t) - \bar{x}^{(2k+1)}(t) \mid$$

$$\le (\epsilon + \delta + \Delta F)\frac{1}{\pi^{2m-2k-2}}\left(\frac{1}{2\pi}\right)[2\sin \pi t + \pi(1-2t)\cos \pi t]$$

$$+(\phi + 4\psi)\frac{1}{\pi^{2m-2k}} \parallel y_{n+1} - y_n \parallel [2\sin \pi t + \pi(1-2t)\cos \pi t]$$

$$+(1+\Delta)(\phi + 4\psi)\frac{1}{\pi^{2m-2k}} \parallel y_n - y_0 \parallel [2\sin \pi t + \pi(1-2t)\cos \pi t].$$

Inequalities (1.6.69) and (1.6.70) imply that

$$\parallel y_{n+1} - \bar{x} \parallel$$

$$\le (\epsilon + \delta + \Delta F)\frac{1}{\pi^{2m-2}}\left(\frac{1}{2\pi}\right) + \theta \parallel y_{n+1} - y_n \parallel + (1+\Delta)\theta \parallel y_n - y_0 \parallel$$

$$\le (\epsilon + \delta + \Delta F)\frac{1}{\pi^{2m-2}}\left(\frac{1}{2\pi}\right) + (2+\Delta)\theta \parallel y_n - y_0 \parallel + \theta \parallel y_{n+1} - y_0 \parallel$$

and since $y_n(t) \in \bar{S}(\bar{x}, N_4)$, we find

$$\parallel y_{n+1} - \bar{x} \parallel \le (1-\theta)^{-1}\left[(\epsilon + \delta + \Delta F)\frac{1}{\pi^{2m-2}}\left(\frac{1}{2\pi}\right) + (2+\Delta)\theta N_4\right] = N_4.$$

Next, to prove part (3), from the definition of $x_{n+1}(t)$, we have

$$\begin{aligned} x_{n+1}(t) - y_{n+1}(t) &= P_{(1.1.1)}(t) + \int_0^1 \mid g_m(t,s) \mid f(s, \mathbf{y}_n(s))ds - y_{n+1}(t) \\ &+ \int_0^1 \mid g_m(t,s) \mid \Bigg[f(s, \mathbf{x}_n(s)) - f(s, \mathbf{y}_n(s)) \\ &+ \sum_{i=0}^{q} \left(x_{n+1}^{(i)}(s) - x_n^{(i)}(s) \right) \frac{\partial}{\partial x_n^{(i)}(s)} f(s, \mathbf{x}_n(s)) \Bigg] ds \end{aligned}$$

and hence as earlier, we get

$$\begin{aligned} (1.6.71) \quad & \mid x_{n+1}^{(2k)}(t) - y_{n+1}^{(2k)}(t) \mid \\ & \leq \frac{1}{\pi^{-2k}} a_n \sin \pi t + (\phi + 4\psi) \frac{1}{\pi^{2m-2k}} \parallel x_n - y_n \parallel \sin \pi t \\ & + (\phi + 4\psi) \frac{1}{\pi^{2m-2k}} \parallel x_{n+1} - x_n \parallel \sin \pi t, \ 0 \leq k \leq \left[\frac{q}{2} \right] \end{aligned}$$

and

$$\begin{aligned} (1.6.72) \quad & \mid x_{n+1}^{(2k+1)}(t) - y_{n+1}^{(2k+1)}(t) \mid \\ & \leq \frac{1}{\pi^{-2k}} a_n [2 \sin \pi t + \pi(1-2t) \cos \pi t] \\ & + (\phi + 4\psi) \frac{1}{\pi^{2m-2k}} \parallel x_n - y_n \parallel [2 \sin \pi t + \pi(1-2t) \cos \pi t] \\ & + (\phi + 4\psi) \frac{1}{\pi^{2m-2k}} \parallel x_{n+1} - x_n \parallel [2 \sin \pi t + \pi(1-2t) \cos \pi t], \\ & \qquad 0 \leq k \leq \left[\frac{q-1}{2} \right]. \end{aligned}$$

Combining (1.6.71) and (1.6.72), we obtain

$$\parallel x_{n+1} - y_{n+1} \parallel \leq a_n + \theta \parallel x_n - y_n \parallel + \theta \parallel x_{n+1} - x_n \parallel .$$

Thus, from (1.6.54) it follows that

$$\parallel x_{n+1} - y_{n+1} \parallel \leq a_n + \theta \parallel x_n - y_n \parallel + \theta \left(\frac{2\theta}{1-\theta} \right)^n \parallel x_1 - x_0 \parallel .$$

Since $x_0(t) = y_0(t) = \bar{x}(t)$, the above inequality gives

$$(1.6.73) \qquad \parallel x_{n+1} - y_{n+1} \parallel \leq \sum_{i=0}^{n} \theta^{n-i} \left[a_i + \theta \left(\frac{2\theta}{1-\theta} \right)^i \parallel x_1 - \bar{x} \parallel \right].$$

Now coupling (1.6.73) with the triangle inequality, we obtain

(1.6.74) $$\| y_{n+1}-x^* \| \le \| x_{n+1}-x^* \| + \sum_{i=0}^{n} \theta^{n-i} \left[a_i + \theta \left(\frac{2\theta}{1-\theta} \right)^i \| x_1 - \bar{x} \| \right].$$

In (1.6.74), Theorem 1.6.14 ensures that $\lim_{n\to\infty} \| x_{n+1} - x^* \| = 0$. Thus, from the Toeplitz lemma we find that $\lim_{n\to\infty} \| y_{n+1} - x^* \| = 0$ if and only if $\lim_{n\to\infty} \left[a_n + \theta \left(\frac{2\theta}{1-\theta}\right)^n \| x_1 - \bar{x} \| \right] = 0$. However, since $\lim_{n\to\infty} \left(\frac{2\theta}{1-\theta}\right)^n = 0$, this condition is the same as $\lim_{n\to\infty} a_n = 0$.

Finally, to prove (1.6.65), we note that

$$x^*(t) - y_{n+1}(t)$$

$$= \int_0^1 | g_m(t,s) | \Big[f(s, \mathbf{x}^*(s)) - f(s, \mathbf{y}_n(s)) + f(s, \mathbf{y}_n(s)) - f_n(s, \mathbf{y}_n(s))$$

$$- \sum_{i=0}^{q} \left(y_{n+1}^{(i)}(s) - y_n^{(i)}(s) \right) \frac{\partial}{\partial y_n^{(i)}(s)} f_n(s, \mathbf{y}_n(s)) \Big] \, ds$$

and as earlier, we find

(1.6.75) $$| x^{*(2k)}(t) - y_{n+1}^{(2k)}(t) |$$

$$\le (\phi + 4\psi) \frac{1}{\pi^{2m-2k}} [\| x^* - y_n \| + \| y_{n+1} - y_n \|] \sin \pi t$$

$$+ \Delta \left\{ \max_{0\le t\le 1} | f(t, \mathbf{y}_n(t)) | \right\} \frac{1}{\pi^{2m-2k-2}} \left(\frac{1}{2\pi} \right) \sin \pi t, \quad 0 \le k \le \left[\frac{q}{2}\right]$$

and

(1.6.76) $$| x^{*(2k+1)}(t) - y_{n+1}^{(2k+1)}(t) |$$

$$\le (\phi + 4\psi) \frac{1}{\pi^{2m-2k}} [\| x^* - y_n \| + \| y_{n+1} - y_n \|] \times$$

$$[2 \sin \pi t + \pi(1-2t) \cos \pi t]$$

$$+ \Delta \left\{ \max_{0\le t\le 1} | f(t, \mathbf{y}_n(t)) | \right\} \frac{1}{\pi^{2m-2k-2}} \left(\frac{1}{2\pi} \right) \times$$

$$[2 \sin \pi t + \pi(1-2t) \cos \pi t], \quad 0 \le k \le \left[\frac{q-1}{2}\right].$$

Inequalities (1.6.75) and (1.6.76) can be combined, to get

$$\begin{aligned}
&\| x^* - y_{n+1} \| \\
&\quad \le \theta\,[\| x^* - y_n \| + \| y_{n+1} - y_n \|] + \Delta \max_{0\le t\le 1} | f(t, \mathbf{y}_n(t)) | \frac{1}{\pi^{2m-2}}\left(\frac{1}{2\pi}\right) \\
&\quad \le \theta\,[\| x^* - y_{n+1} \| + 2 \| y_{n+1} - y_n \|] + \Delta \max_{0\le t\le 1} | f(t, \mathbf{y}_n(t)) | \frac{1}{\pi^{2m-2}}\left(\frac{1}{2\pi}\right),
\end{aligned}$$

which is the same as (1.6.65). ∎

Theorem 1.6.17. Let the conditions of Theorem 1.6.16 be satisfied. Further, let $f_n = f_0$ for all $n = 1, 2, \ldots$ and $f_0(t, x_0, x_1, \ldots, x_q)$ be twice continuously differentiable with respect to all x_i, $0 \le i \le q$ on $[0,1] \times D_3$, and

$$\left| \frac{\partial^2}{\partial x_i \partial x_j} f_0(t, x_0, x_1, \ldots, x_q) \right| \le L_i L_j K, \quad 0 \le i, j \le q.$$

Then,

$$\begin{aligned}
(1.6.77) \qquad \| y_{n+1} - y_n \| &\le \alpha \| y_n - y_{n-1} \|^2 \le \frac{1}{\alpha} (\alpha \| y_1 - y_0 \|)^{2^n} \\
&\le \frac{1}{\alpha} \left[\frac{(\epsilon + \delta + \Delta F)}{\pi^{2m-1}} \frac{\theta}{(1-\theta)^2} \frac{K}{4} \max\{\pi\psi, \phi + 2\psi\} \right]^{2^n},
\end{aligned}$$

where α is the same as in Theorem 1.6.15.

Proof. As in Theorem 1.6.15, we have

$$\begin{aligned}
&y_{n+1}(t) - y_n(t) \\
&\quad = \int_0^1 | g_m(t,s) | \left\{ \sum_{i=0}^{q} \left(y_{n+1}^{(i)}(s) - y_n^{(i)}(s) \right) \frac{\partial}{\partial y_n^{(i)}(s)} f_0(s, \mathbf{y}_n(s)) \right. \\
&\qquad \left. + \frac{1}{2} \left[\sum_{i=0}^{q} \left(y_n^{(i)}(s) - y_{n-1}^{(i)}(s) \right) \frac{\partial}{\partial p_i(s)} \right]^2 f_0(s, p_0(s), p_1(s), \ldots, p_q(s)) \right\} ds,
\end{aligned}$$

where $p_i(t)$ lies between $y_{n-1}^{(i)}(t)$ and $y_n^{(i)}(t)$, $0 \le i \le q$.

Thus, as earlier we get

$$\| y_{n+1} - y_n \| \le \theta \| y_{n+1} - y_n \| + (1-\theta)\alpha \| y_n - y_{n-1} \|^2,$$

which is the same as the first part of (1.6.77). The last part of (1.6.77) follows from (1.6.68). ∎

REFERENCES

1. **R.P.Agarwal** and **G.Akrivis,** Boundary value problems occurring in plate deflection theory, J. Comp. Appl. Math. 8(1982), 145-154.

2. **R.P.Agarwal,** Boundary Value Problems for Higher Order Differential Equations, World Scientific, Singapore • Philadelphia, 1986.

3. **R.P.Agarwal** and **P.J.Y.Wong,** Lidstone polynomials and boundary value problems, Computers Math. Applic. 17(1989), 1397-1421.

4. **R.P.Agarwal,** Sharp inequalities in polynomial interpolation, in General Inequalities 6, Ed. W.Walter, International Series of Numerical Mathematics, Birkhäuser Verlag 103(1992), 73-92.

5. **R.P.Agarwal** and **P.J.Y.Wong,** Quasilinearization and approximate quasilinearization for Lidstone boundary value problems, Intern. J. Computer Math. 42(1992), 99-116.

6. **P.Baldwin,** Asymptotic estimates of the eigenvalues of a sixth - order boundary - value problem obtained by using global phase - integral method, Phil. Trans. R. Soc. London A322(1987), 281-305.

7. **P.Baldwin,** Localized instability in a Bénard layer, Applicable Analysis 24(1987), 117-156.

8. **R.P.Boas,** A note on functions of exponential type, Bull. Amer. Math. Soc. 47(1941), 750-754.

9. **R.P.Boas,** Representation of functions by Lidstone series, Duke Math. J. 10(1943), 239-245.

10. **A.Boutayeb** and **E.H.Twizell,** Finite - difference methods for twelfth - order boundary value problems, J. Comp. Appl. Math. 35(1991), 133-138.

11. **M.M.Chawla** and **C.P.Katti,** Finite difference methods for two - point boundary value problems involving higher order differential equations, BIT 19(1979), 27-33.

12. **P.J.Davis,** Interpolation and Approximation, Blaisdell Publishing Co., Boston, 1961.

13. **T.Fort,** Finite Differences and Difference Equations in the Real Domain, Oxford University Press, London, 1948.

14. **C.Jordan,** Calculus of Finite Differences, Chelsea Pub. Co., New York, 1960.

15. **G.J.Lidstone,** Notes on the extension of Aitken's theorem (for polynomial interpolation) to the Everett types, Proc. Edinburgh Math. Soc. (2), 2(1929), 16-19.

16. **Y.L.Luke,** The Special Functions and their Approximations, Academic Press, New York, 1969.

17. **L.M.Milne - Thomson,** The Calculus of Finite Differences, MacMillan, London, 1960.

18. **Muhammad Aslam Noor** and **S.I.Tirmizi,** Numerical methods for unilateral problems, J. Comp. Appl. Math. 16(1986), 387-395.

19. **H.Poritsky,** On certain polynomial and other approximations to analytic functions, Trans. Amer. Math. Soc. 34(1932), 274-331.

20. **I.J.Schoenberg,** On certain two - point expansions of integral functions of exponential type, Bull. Amer. Math. Soc. 42(1936), 284-288.

21. **J.Toomre, J.R.Jahn, J.Latour** and **E.A.Spiegel,** Stellar convection theory II : single - mode study of the second convection zone in an A - type star, Astrophys. J. 207(1976), 545-563.

22. **E.H.Twizell** and **S.I.A.Tirmizi,** A sixth order multiderivative method for two beam problems, Int. J. Numer. Methods Engg. 23(1986), 2089-2102.

23. **E.H.Twizell,** Numerical methods for sixth - order boundary value problems, International Series of Numerical Mathematics 86(1988), 495-506.

24. **E.H.Twizell** and **S.I.A.Tirmizi,** Multiderivative methods for nonlinear beam problems, Comm. Appl. Numer. Methods 4(1988), 43-50.

25. **E.H.Twizell** and **A.Boutayeb,** Numerical methods for the solution of special and general sixth - order boundary value problems, with applications to Bénard layer eigenvalue problems, Proc. R. Soc. London A431(1990), 433-450.

26. **R.A.Usmani,** Solving boundary value problems in plate deflection theory, Simulation, December(1981), 195-206.

27. **A.K.Varma** and **G.Howell,** Best error bounds for derivatives in two point Birkhoff interpolation problem, J. Approximation Theory 38(1983), 258-268.

28. **J.M.Whittaker,** On Lidstone's series and two-point expansions of analytic functions, Proc. London Math. Soc. (2), 36(1933-34), 451-459.

29. **J.M.Whittaker,** Interpolatory Function Theory, Cambridge, 1935.

30. **D.V.Widder,** Functions whose even derivatives have a prescribed sign, Proc. National Acad. Sciences 26(1940), 657-659.

31. **D.V.Widder,** Completely convex functions and Lidstone series, Trans. Amer. Math. Soc. 51(1942), 387-398.

CHAPTER 2

HERMITE INTERPOLATION

2.1 INTRODUCTION

Let $-\infty < a < b < \infty$, and $a \le a_1 < a_2 < \cdots < a_r \le b$ $(r \ge 2)$ be the given points. It is well known that the *Hermite interpolating polynomial* $P_{(2.1.1)}(t)$ of degree $(n-1)$ $(n \ge r)$ satisfying the *Hermite conditions*

$$(2.1.1) \qquad P^{(i)}_{(2.1.1)}(a_j) = A_{i,j};\ 0 \le i \le k_j,\ 1 \le j \le r,\ \sum_{j=1}^{r} k_j + r = n$$

exists uniquely [10,22]. It is of interest to note that (2.1.1) include in particular the

(i) *Simple Hermite* or *Osculatory conditions* : $n = 2m,\ r = m$

$$(2.1.2) \qquad P_{(2.1.2)}(a_j) = A_j,\ P'_{(2.1.2)}(a_j) = B_j,\ 1 \le j \le m$$

(ii) *Lagrange conditions* : $r = n$

$$(2.1.3) \qquad P_{(2.1.3)}(a_j) = A_j,\ 1 \le j \le n$$

(iii) *$(m, n-m)$ conditions* : $r = 2,\ 1 \le m \le n-1$ but fixed

$$P^{(i)}_{(2.1.4)}(a_1) = A_i,\ 0 \le i \le m-1$$

(2.1.4)

$$P^{(i)}_{(2.1.4)}(a_2) = B_i,\ 0 \le i \le n-m-1$$

(iv) *Two - point Taylor conditions* : $n = 2m,\ r = 2$

(2.1.5) $$P^{(i)}_{(2.1.5)}(a_1) = A_i,\ P^{(i)}_{(2.1.5)}(a_2) = B_i,\ 0 \le i \le m-1.$$

The plan of this chapter is as follows : In Section 2.2 we shall discuss the methods of Lagrange and Newton for constructing the Hermite interpolating polynomial $P_{(2.1.1)}(t)$. In Section 2.3 we shall provide four different representations of the error function $e_{(2.1.1)}(t) = x(t) - P_{(2.1.1)}(t)$, where $x(t) \in C^{(n)}[a,b]$ and $P_{(2.1.1)}(t)$ is the Hermite interpolating polynomial of the function $x(t)$, i.e., it satisfies the conditions $P^{(i)}_{(2.1.1)}(a_j) = x^{(i)}(a_j);\ 0 \le i \le k_j,\ 1 \le j \le r,\ \sum_{j=1}^{r} k_j + r = n$. These representations are used in Section 2.4 to obtain estimates for $\| e_{(2.1.1)} \|_p$ $(p \ge 1)$, and the pointwise as well as uniform bounds for $| e^{(k)}_{(2.1.1)}(t) |,\ 0 \le k \le n-1$. While some of the obtained inequalities here are proved to be the best possible, others compare sharply with the known results. A variety of applications which dwell upon the importance of the results of Sections 2.2 - 2.4 are illustrated in Section 2.5.

2.2 INTERPOLATING POLYNOMIAL REPRESENTATIONS

The methods of Lagrange and Newton are classical for computing an interpolating polynomial. In fact, these methods are exceptional for their remarkable simplicity.

Method of Lagrange

The success of the Lagrange method for the construction of an interpolating polynomial heavily depends on the following :

Lemma 2.2.1. The unique polynomial $H_{ij}(t)$; $0 \leq i \leq k_j$, $1 \leq j \leq r$ of degree at most $(n-1)$ satisfying

$$(2.2.1)\qquad \begin{aligned} H_{ij}^{(m)}(a_\ell) &= 0;\ 1 \leq \ell \leq r,\ \ell \neq j,\ 0 \leq m \leq k_\ell \\ H_{ij}^{(m)}(a_j) &= \delta_{im},\ 0 \leq m \leq k_j \end{aligned}$$

can be explicitly expressed as

$$(2.2.2)\qquad H_{ij}(t) = \frac{1}{i!}\frac{\omega(t)}{(t-a_j)^{k_j+1-i}}\sum_{k=0}^{k_j-i}\frac{1}{k!}\left[\frac{(t-a_j)^{k_j+1}}{\omega(t)}\right]_{t=a_j}^{(k)}(t-a_j)^k,$$

where

$$(2.2.3)\qquad \omega(t) = \prod_{j=1}^{r}(t-a_j)^{k_j+1}.$$

Proof. Conditions (2.2.1) imply that $H_{ij}(t)$ has zeros at a_ℓ, $\ell \neq j$ of multiplicities $(k_\ell+1)$, and a zero at a_j of multiplicity i. Thus, we can write

$$(2.2.4)\qquad H_{ij}(t) = \prod_{\substack{\ell=1 \\ \ell\neq j}}^{r}(t-a_\ell)^{k_\ell+1}(t-a_j)^i\bar{H}_{ij}(t) = \frac{\omega(t)}{(t-a_j)^{k_j+1-i}}\bar{H}_{ij}(t),$$

where $\bar{H}_{ij}(t)$ is a polynomial of degree (k_j-i) which does not vanish at $t=a_j$. Let

$$(2.2.5)\qquad \bar{H}_{ij}(t) = \sum_{k=0}^{k_j-i}A_k(t-a_j)^k.$$

Then, from (2.2.4) it follows that

$$(2.2.6)\qquad \sum_{k=0}^{k_j-i}A_k(t-a_j)^k = \frac{(t-a_j)^{k_j+1}}{\omega(t)}\frac{H_{ij}(t)}{(t-a_j)^i}.$$

To determine A_k, we differentiate (2.2.6), k times and put $t=a_j$ to get

$$(2.2.7)\qquad A_k = \frac{1}{k!}\lim_{t\to a_j}\left[\frac{(t-a_j)^{k_j+1}}{\omega(t)}\frac{H_{ij}(t)}{(t-a_j)^i}\right]^{(k)},\ 0 \leq k \leq k_j-i.$$

Now applying the Leibnitz rule, to find

$$\left[\frac{(t-a_j)^{k_j+1}}{\omega(t)}\frac{H_{ij}(t)}{(t-a_j)^i}\right]^{(k)} \tag{2.2.8}$$

$$= \sum_{s=0}^{k}\binom{k}{s}\left[\frac{(t-a_j)^{k_j+1}}{\omega(t)}\right]^{(s)}\left[\frac{H_{ij}(t)}{(t-a_j)^i}\right]^{(k-s)}.$$

Since the derivatives $\left[\frac{(t-a_j)^{k_j+1}}{\omega(t)}\right]^{(s)}$ are continuous at $t = a_j$, we have

$$\lim_{t\to a_j}\left[\frac{(t-a_j)^{k_j+1}}{\omega(t)}\right]^{(s)} = \left[\frac{(t-a_j)^{k_j+1}}{\omega(t)}\right]^{(s)}_{t=a_j}. \tag{2.2.9}$$

To determine $\lim\limits_{t\to a_j}\left[\frac{H_{ij}(t)}{(t-a_j)^i}\right]^{(k-s)}$, we write

$$\frac{H_{ij}(t)}{(t-a_j)^i} = \sum_{\mu=0}^{n-1-i} B_\mu (t-a_j)^\mu \tag{2.2.10}$$

and hence

$$\lim_{t\to a_j}\left[\frac{H_{ij}(t)}{(t-a_j)^i}\right]^{(k-s)} = (k-s)!B_{k-s}. \tag{2.2.11}$$

From (2.2.10), we note that

$$H_{ij}(t) = \sum_{\mu=0}^{n-1-i} B_\mu (t-a_j)^{\mu+i}$$

from which it follows that

$$B_{k-s} = \frac{H_{ij}^{(k-s+i)}(a_j)}{(k-s+i)!}. \tag{2.2.12}$$

Clearly,

$$k-s+i \le k+i \le k_j - i + i = k_j$$

and therefore $H_{ij}^{(k-s+i)}(a_j)$ is nonzero only when $s = k$. In this case, from (2.2.12), we have

$$B_0 = \frac{1}{i!}. \tag{2.2.13}$$

Combining (2.2.8), (2.2.9), (2.2.11), (2.2.12) and (2.2.13), the expression (2.2.7) becomes

$$A_k = \frac{1}{i!k!}\left[\frac{(t-a_j)^{k_j+1}}{\omega(t)}\right]^{(k)}_{t=a_j},$$

which on substituting into (2.2.6) leads to (2.2.2). ∎

Theorem 2.2.2. The Hermite interpolating polynomial $P_{(2.1.1)}(t)$ can be expressed as

$$(2.2.14)\quad P_{(2.1.1)}(t) = \sum_{j=1}^{r} \sum_{i=0}^{k_j} H_{ij}(t) A_{i,j}$$

$$= \sum_{j=1}^{r} \sum_{i=0}^{k_j} \sum_{k=0}^{k_j-i} \frac{1}{i!k!} \left[\frac{(t-a_j)^{k_j+1}}{\omega(t)} \right]^{(k)}_{t=a_j} \frac{\omega(t)}{(t-a_j)^{k_j+1-i-k}} A_{i,j}. \quad ∎$$

Corollary 2.2.3. The osculatory interpolating polynomial $P_{(2.1.2)}(t)$ can be written as

$$(2.2.15)\qquad P_{(2.1.2)}(t) = \sum_{j=1}^{m} \left[1 - \frac{\pi''_m(a_j)}{\pi'_m(a_j)}(t-a_j) \right] L^2_{m,j}(t) A_j + \sum_{j=1}^{m} (t-a_j) L^2_{m,j}(t) B_j,$$

where

$$(2.2.16)\qquad \pi_m(t) = \prod_{j=1}^{m} (t-a_j) \quad \text{and} \quad L_{m,j}(t) = \frac{\pi_m(t)}{(t-a_j)\pi'_m(a_j)}. \quad ∎$$

Corollary 2.2.4. The Lagrange interpolating polynomial $P_{(2.1.3)}(t)$ appears as

$$(2.2.17)\qquad P_{(2.1.3)}(t) = \sum_{j=1}^{n} L_{n,j}(t) A_j = \sum_{j=1}^{n} \prod_{\substack{k=1 \\ k \neq j}}^{n} \left(\frac{t-a_k}{a_j-a_k} \right) A_j. \quad ∎$$

Corollary 2.2.5. The $(m, n-m)$ interpolating polynomial $P_{(2.1.4)}(t)$ can be written as

$$(2.2.18)\qquad P_{(2.1.4)}(t) = \sum_{i=0}^{m-1} \tau_i(t) A_i + \sum_{i=0}^{n-m-1} \eta_i(t) B_i,$$

where

$$(2.2.19)\quad \tau_i(t) = \frac{1}{i!}(t-a_1)^i \left(\frac{t-a_2}{a_1-a_2} \right)^{n-m} \sum_{k=0}^{m-1-i} \binom{n-m+k-1}{k} \left(\frac{t-a_1}{a_2-a_1} \right)^k$$

and

$$(2.2.20)\quad \eta_i(t) = \frac{1}{i!}(t-a_2)^i \left(\frac{t-a_1}{a_2-a_1} \right)^{m} \sum_{k=0}^{n-m-1-i} \binom{m+k-1}{k} \left(\frac{t-a_2}{a_1-a_2} \right)^k. \quad ∎$$

Corollary 2.2.6. The two - point Taylor interpolating polynomial $P_{(2.1.5)}(t)$ can be expressed as

(2.2.21) $P_{(2.1.5)}(t)$

$$= \sum_{i=0}^{m-1} \sum_{k=0}^{m-1-i} \binom{m+k-1}{k} \left[\frac{(t-a_1)^i}{i!} \left(\frac{t-a_2}{a_1-a_2}\right)^m \left(\frac{t-a_1}{a_2-a_1}\right)^k A_i \right.$$
$$\left. + \frac{(t-a_2)^i}{i!} \left(\frac{t-a_1}{a_2-a_1}\right)^m \left(\frac{t-a_2}{a_1-a_2}\right)^k B_i \right]. \quad \blacksquare$$

Method of Newton

For this method it is convenient to assume that $A_j = f(a_j)$, $1 \le j \le n$ where the function $f(t)$ is defined at the distinct (but not necessarily increasing) points $a_1, \ldots, a_n$.

Definition 2.2.1. The *divided differences* of orders $0, 1, \ldots, k$ $(\le n-1)$ are defined recursively by the relations

$$f[a_1] = f(a_1), \quad f[a_1, a_2] = \frac{f[a_1] - f[a_2]}{a_1 - a_2}, \ \ldots,$$
(2.2.22)
$$f[a_1, \ldots, a_{k+1}] = \frac{f[a_1, \ldots, a_k] - f[a_2, \ldots, a_{k+1}]}{a_1 - a_{k+1}}.$$

On the basis of various applications of the divided differences, several representations have been obtained. However, the following is very useful.

(2.2.23) $f[a_1, \ldots, a_{k+1}]$

$$= \begin{vmatrix} 1 & a_1 & a_1^2 & \ldots & a_1^{k-1} & f(a_1) \\ 1 & a_2 & a_2^2 & \ldots & a_2^{k-1} & f(a_2) \\ \ldots & \ldots & \ldots & \ldots & \ldots & \ldots \\ 1 & a_{k+1} & a_{k+1}^2 & \ldots & a_{k+1}^{k-1} & f(a_{k+1}) \end{vmatrix} \div \begin{vmatrix} 1 & a_1 & \ldots & a_1^k \\ 1 & a_2 & \ldots & a_2^k \\ \ldots & \ldots & \ldots & \ldots \\ 1 & a_{k+1} & \ldots & a_{k+1}^k \end{vmatrix}.$$

If $f(t)$ is k times continuously differentiable in the interval $\min\{a_1, \ldots, a_{k+1}\} \le t \le \max\{a_1, \ldots, a_{k+1}\}$, then the divided differences can be represented in terms of repeated integrals. Some such well known representations are :

Hermite's Integral Representation

$$(2.2.24)\quad f[a_1,\dots,a_{k+1}] = \int_0^1 ds_1 \int_0^{s_1} ds_2 \cdots \int_0^{s_{k-1}} ds_k$$

$$\times f^{(k)}\left(s_k[a_{k+1}-a_k] + \cdots + s_1[a_2-a_1] + a_1\right),$$

where $k \geq 1$ and $s_0 = 1$.

Genocchi's First Integral Representation

$$(2.2.25)\quad f[a_1,\dots,a_{k+1}] = \underbrace{\int\int\cdots\int}_{k \text{ times}} f^{(k)}\left(s_1 a_1 + \cdots + s_{k+1}a_{k+1}\right) ds_2 \cdots ds_{k+1},$$

where $0 \leq s_2 + \cdots + s_{k+1} \leq 1,\ 0 \leq s_i \leq 1,\ s_1 = 1 - s_2 - \cdots - s_{k+1}$.

Genocchi's Second Integral Representation

$$(2.2.26)\quad f[a_1,\dots,a_{k+1}] = \underbrace{\int_0^1\int_0^1\cdots\int_0^1}_{k \text{ times}} s_1^{k-1}s_2^{k-2}\cdots s_{k-1} f^{(k)}\big(a_1 + (a_2-a_1)s_1$$

$$+ (a_3-a_2)s_1s_2 + \cdots + (a_{k+1}-a_k)s_1\cdots s_k\big)\, ds_1\cdots ds_k.$$

The following properties of the divided differences are fundamental.

Symmetry Property. $f[a_1,\dots,a_{k+1}]$ is independent of the order of the points $a_1,\dots,a_{k+1}$. In fact,

$$(2.2.27)\qquad f[a_1,\dots,a_{k+1}] = \sum_{j=1}^{k+1} f(a_j) \Big/ \prod_{\substack{i=1\\ i\neq j}}^{k+1} (a_j - a_i).$$

Linearity Property. If $f = \alpha g + \beta h$, where α and β are constants, then

$$f[a_1,\dots,a_{k+1}] = \alpha g[a_1,\dots,a_{k+1}] + \beta h[a_1,\dots,a_{k+1}].$$

To develop the interpolating polynomial in terms of divided differences we assume that the function f is also defined at a point $t \neq a_j,\ 1 \leq j \leq n$. Then, from the Definition 2.2.1, we have

$$f[t,a_1] = \frac{f(t)-f(a_1)}{t-a_1}$$

and hence

$$f(t) = f(a_1) + (t - a_1)f[t, a_1]. \tag{2.2.28}$$

Next, since

$$f[t, a_1, a_2] = \frac{f[t, a_1] - f[a_1, a_2]}{t - a_2}$$

we find

$$f[t, a_1] = f[a_1, a_2] + (t - a_2)f[t, a_1, a_2]. \tag{2.2.29}$$

Substituting (2.2.29) into (2.2.28), to obtain

$$f(t) = f(a_1) + (t - a_1)f[a_1, a_2] + (t - a_1)(t - a_2)f[t, a_1, a_2].$$

Continuing in this way, we get

$$f(t) = P(t) + R(t), \tag{2.2.30}$$

where

$$P(t) = f(a_1) + (t - a_1)f[a_1, a_2] + (t - a_1)(t - a_2)f[a_1, a_2, a_3] + \cdots + (t - a_1)\cdots(t - a_{n-1})f[a_1, ..., a_n] \tag{2.2.31}$$

and

$$R(t) = (t - a_1)\cdots(t - a_n)f[t, a_1, ..., a_n]. \tag{2.2.32}$$

We claim that $P(t) \equiv P_{(2.1.3)}(t)$. For this, it suffices to note that $P(t)$ is a polynomial of degree at most $(n - 1)$, and in view of (2.2.30), (2.2.32) and (2.2.27), we have

$$\begin{aligned} P(a_k) &= f(a_k) - R(t)\big|_{t=a_k} \\ &= f(a_k) - \prod_{i=1}^{n}(t - a_i)\left[\sum_{j=1}^{n} \frac{f(a_j)}{\prod\limits_{\substack{i=1 \\ i \neq j}}^{n} (a_j - a_i)(a_j - t)} + \frac{f(t)}{\prod\limits_{i=1}^{n}(t - a_i)}\right]\Bigg|_{t=a_k} \\ &= f(a_k), \ 1 \le k \le n. \end{aligned}$$

If some of the arguments coincide, then Definition 2.2.1 is not suitable. Since in this case ratios of the form $\frac{0}{0}$ appear. In such a case it is natural to define

$$f[a_i,a_i] = \lim_{a_i^{(1)}\to a_i} \frac{f[a_i^{(1)}]-f[a_i]}{a_i^{(1)}-a_i} = f'(a_i) \tag{2.2.33}$$

provided $f'(a_i)$ exists. To give a meaning to $f[a_i,a_i,a_i]$ we use (2.2.23), to obtain

$$\begin{aligned} f[a_i,a_i,a_i] &= \lim_{a_i^{(1)}\to a_i}\ \lim_{a_i^{(2)}\to a_i} \begin{vmatrix} 1 & a_i & f(a_i) \\ 1 & a_i^{(1)} & f(a_i^{(1)}) \\ 1 & a_i^{(2)} & f(a_i^{(2)}) \end{vmatrix} \div \begin{vmatrix} 1 & a_i & a_i^2 \\ 1 & a_i^{(1)} & (a_i^{(1)})^2 \\ 1 & a_i^{(2)} & (a_i^{(2)})^2 \end{vmatrix} \\ &= \begin{vmatrix} 1 & a_i & f(a_i) \\ 0 & 1 & f'(a_i) \\ 0 & 0 & f''(a_i) \end{vmatrix} \div \begin{vmatrix} 1 & a_i & a_i^2 \\ 0 & 1 & 2a_i \\ 0 & 0 & 2 \end{vmatrix} \\ &= \frac{1}{2!} f''(a_i) \end{aligned}$$

provided $f''(a_i)$ exists. Similarly, if $f^{(r)}(a_i)$ exists then

$$\begin{aligned} f[\underbrace{a_i,\cdots,a_i}_{(r+1)\text{ times}}] &= \lim_{a_i^{(1)}\to a_i} \cdots \lim_{a_i^{(r)}\to a_i} f[a_i,a_i^{(1)},\ldots,a_i^{(r)}] \\ &= \frac{1}{r!} f^{(r)}(a_i). \end{aligned} \tag{2.2.34}$$

Finally, we note that if the required derivatives of $f(t)$ exist, and all the lower order differences have been calculated, then

$$\begin{aligned} & f[\underbrace{a_1,\cdots,a_1}_{(k_1+1)\text{ times}};\cdots;\underbrace{a_i,\cdots,a_i}_{(k_i+1)\text{ times}};\underbrace{a_{i+1},\cdots,a_{i+1}}_{(r+1)\text{ times}}] \\ &= \frac{1}{a_1-a_{i+1}}\{f[\underbrace{a_1,\cdots,a_1}_{(k_1+1)\text{ times}};\cdots;\underbrace{a_i,\cdots,a_i}_{(k_i+1)\text{ times}};\underbrace{a_{i+1},\cdots,a_{i+1}}_{r\text{ times}}] \\ &\qquad - f[\underbrace{a_1,\cdots,a_1}_{k_1\text{ times}};\cdots;\underbrace{a_i,\cdots,a_i}_{(k_i+1)\text{ times}};\underbrace{a_{i+1},\cdots,a_{i+1}}_{(r+1)\text{ times}}]\}. \end{aligned} \tag{2.2.35}$$

From the above considerations we find that if $A_{i,j} = f^{(i)}(a_j)$, where for the function $f(t)$ the required derivatives exist, then the interpolating polynomial

$P(t) \equiv P_{(2.1.3)}(t)$ defined in (2.2.31) becomes

$$P_{(2.1.1)}(t) = f(a_1) + (t-a_1)f[a_1,a_1] + \cdots + (t-a_1)^{k_1} f[\underbrace{a_1,\cdots,a_1}_{(k_1+1)\text{ times}}] \tag{2.2.36}$$

$$+(t-a_1)^{k_1+1} f[\underbrace{a_1,\cdots,a_1}_{(k_1+1)\text{ times}}; a_2]$$

$$+(t-a_1)^{k_1+1}(t-a_2) f[\underbrace{a_1,\cdots,a_1}_{(k_1+1)\text{ times}}; a_2, a_2]$$

$$+\cdots + (t-a_1)^{k_1+1}\ldots(t-a_{r-1})^{k_{r-1}+1}(t-a_r)^{k_r}\times$$

$$f[\underbrace{a_1,\cdots,a_1}_{(k_1+1)\text{ times}};\cdots;\underbrace{a_r,\cdots,a_r}_{(k_r+1)\text{ times}}].$$

2.3 ERROR REPRESENTATIONS

Let $A_{i,j} = x^{(i)}(a_j)$; $0 \le i \le k_j$, $1 \le j \le r$, $\sum_{j=1}^{r} k_j + r = n$, where the function $x(t)$ is assumed to be n times continuously differentiable on $[a,b]$ (however, this restriction is not necessary). In such a case $P_{(2.1.1)}(t)$ is called the Hermite interpolating polynomial of the function $x(t)$. For the associated error $e_{(2.1.1)}(t) = x(t) - P_{(2.1.1)}(t)$ we shall provide four different representations. First three of these representations are classical and find various applications, whereas the fourth representation is new to the literature and its importance to acquire pointwise as well as uniform bounds for $| e^{(k)}_{(2.1.1)}(t) |$, $0 \le k \le n-1$ will be discussed in the next section.

Cauchy's Representation

Theorem 2.3.1. Let $x(t) \in C^{(n-1)}[a,b]$ and suppose that $x^{(n)}(t)$ exists at each point of (a,b). Then, the error function $e_{(2.1.1)}(t)$ can be written as

$$e_{(2.1.1)}(t) = \frac{1}{n!}\omega(t)x^{(n)}(\xi), \tag{2.3.1}$$

where $\xi \in (a, b)$.

Proof. We define an auxiliary function $\phi(\tau)$ as follows

$$\phi(\tau) = e_{(2.1.1)}(\tau) - \psi(t)\omega(\tau); \; t \neq a_i, \; 1 \leq i \leq r \tag{2.3.2}$$

where $\psi(t) = e_{(2.1.1)}(t)/\omega(t)$.

This function $\phi(\tau)$ has $(n + 1)$ zeros at $\tau = a_i$, $1 \leq i \leq r$ and $\tau = t$. Therefore, Rolle's theorem ensures that $\phi'(\tau)$ will have at least n zeros in (a, b). Similarly, $\phi''(\tau)$ has at least $(n - 1)$ zeros in the same interval etc.. Finally, the nth derivative of $\phi(\tau)$, i.e., $\phi^{(n)}(\tau)$ has at least one zero at $\tau = \xi \in (a, b)$. Thus, from (2.3.2), we find

$$\phi^{(n)}(\xi) = x^{(n)}(\xi) - n! \; \psi(t) = 0,$$

which is the same as (2.3.1). ∎

Newton's Representation

From (2.2.30), (2.2.32) and the definition of the divided differences in the case of recurring arguments it is clear that the error in the interpolating polynomial (2.2.36) is

$$e_{(2.1.1)}(t) = \omega(t) \; x[t; \underbrace{a_1, \cdots, a_1}_{(k_1+1) \text{ times}}; \cdots; \underbrace{a_r, \cdots, a_r}_{(k_r+1) \text{ times}}]. \tag{2.3.3}$$

If the function $x(t)$ satisfies the hypotheses of Theorem 2.3.1 then a comparison between (2.3.1) and (2.3.3) leads to the relation

$$x[t; \underbrace{a_1, \cdots, a_1}_{(k_1+1) \text{ times}}; \cdots; \underbrace{a_r, \cdots, a_r}_{(k_r+1) \text{ times}}] = \frac{1}{n!} x^{(n)}(\xi). \tag{2.3.4}$$

Peano's Representation

To obtain yet another representation of $e_{(2.1.1)}(t)$, we need the following :

Lemma 2.3.2. If the function $\phi(t) \in C^{(n)}[a,b]$ is such that $\phi^{(n)}(t) \equiv 0$, and

$$\phi^{(i)}(a_j) = 0;\ 0 \le i \le k_j,\ 1 \le j \le r,\ \sum_{j=1}^{r} k_j + r = n \tag{2.3.5}$$

then $\phi(t) \equiv 0$.

Proof. Obviously, $\phi(t)$ is a polynomial of degree $(n-1)$. The conclusion is now immediate from (2.3.1). ∎

Consider the function

$$y(t) = \sum_{j=1}^{r} \sum_{i=0}^{k_j} H_{ij}(t) c_{ij} + \frac{1}{(n-1)!} \int_a^t (t-s)^{n-1} x^{(n)}(s)\, ds. \tag{2.3.6}$$

It is clear that $y(t) \in C^{(n)}[a,b]$ and $y^{(n)}(t) \equiv x^{(n)}(t)$. Thus, the function $\phi(t) = y(t) - x(t)$ is n times continuously differentiable in $[a,b]$ and that $\phi^{(n)}(t) \equiv 0$. We shall now determine the constants c_{ij} so that $y^{(i)}(a_j) = x^{(i)}(a_j),\ 0 \le i \le k_j,\ 1 \le j \le r$. For this, in view of (2.2.1) we have

$$x^{(i)}(a_j) = c_{ij} + \frac{1}{(n-i-1)!} \int_a^{a_j} (a_j - s)^{n-i-1} x^{(n)}(s)\, ds.$$

Thus, from Lemma 2.3.2 it follows that

$$x(t) = \sum_{j=1}^{r} \sum_{i=0}^{k_j} H_{ij}(t) x^{(i)}(a_j) - \sum_{j=1}^{r} \sum_{i=0}^{k_j} H_{ij}(t) \frac{1}{(n-i-1)!} \int_a^{a_j} (a_j - s)^{n-i-1} x^{(n)}(s)\, ds$$

$$+ \frac{1}{(n-1)!} \int_a^t (t-s)^{n-1} x^{(n)}(s)\, ds$$

and hence

$$e_{(2.1.1)}(t) = -\sum_{j=1}^{r}\int_{a}^{a_j}\sum_{i=0}^{k_j}\frac{(a_j-s)^{n-i-1}}{(n-i-1)!}H_{ij}(t)x^{(n)}(s)\,ds$$

$$+\frac{1}{(n-1)!}\int_a^t (t-s)^{n-1}x^{(n)}(s)\,ds$$

$$= -\sum_{\ell=0}^{r}\int_{a_\ell}^{a_{\ell+1}}\sum_{j=\ell+1}^{r}\sum_{i=0}^{k_j}\frac{(a_j-s)^{n-i-1}}{(n-i-1)!}H_{ij}(t)x^{(n)}(s)\,ds$$

$$+\frac{1}{(n-1)!}\int_a^t (t-s)^{n-1}x^{(n)}(s)\,ds$$

$$(2.3.7)\qquad = \int_{a_0}^{a_{r+1}} g_{(2.1.1)}(t,s)x^{(n)}(s)\,ds,$$

where $a_0 = a$, $a_{r+1} = b$ and $g_{(2.1.1)}(t,s)$ is called *Green's function* (*Peano's kernel*) and for $a_\ell \le s \le a_{\ell+1}$, $\ell = 0, 1, ..., r$ can be written as

$$(2.3.8)\quad g_{(2.1.1)}(t,s) = \begin{cases} -\displaystyle\sum_{j=\ell+1}^{r}\sum_{i=0}^{k_j}\frac{(a_j-s)^{n-i-1}}{(n-i-1)!}H_{ij}(t) + \frac{1}{(n-1)!}(t-s)^{n-1}, \\ \hfill s \le t \\ -\displaystyle\sum_{j=\ell+1}^{r}\sum_{i=0}^{k_j}\frac{(a_j-s)^{n-i-1}}{(n-i-1)!}H_{ij}(t), \hfill s \ge t. \end{cases}$$

However, since in view of (2.3.1), we have

$$\frac{1}{(n-1)!}(t-s)^{n-1} = \sum_{j=1}^{r}\sum_{i=0}^{k_j}\frac{(a_j-s)^{n-i-1}}{(n-i-1)!}H_{ij}(t),$$

and hence Green's function $g_{(2.1.1)}(t,s)$ can be rewritten as

$$(2.3.9)\qquad g_{(2.1.1)}(t,s) = \begin{cases} \displaystyle\sum_{j=1}^{\ell}\sum_{i=0}^{k_j}\frac{(a_j-s)^{n-i-1}}{(n-i-1)!}H_{ij}(t), & s \le t \\ -\displaystyle\sum_{j=\ell+1}^{r}\sum_{i=0}^{k_j}\frac{(a_j-s)^{n-i-1}}{(n-i-1)!}H_{ij}(t), & s \ge t \end{cases}$$

$$a_\ell \le s \le a_{\ell+1},\ \ell = 0, 1, ..., r.$$

Let K be the square $a \le t, s \le b$; the same square with straight lines of the form $s = a_j$ rejected be K_0 and K_0 with rejected diagonal $t = s$ be K_1. The following properties of Green's function $g_{(2.1.1)}(t,s)$ are fundamental :

(i) $\frac{\partial^{(i)} g_{(2.1.1)}(t,s)}{\partial t^i}$, $0 \le i \le n-2$ are continuous in K_0,

(ii) $\frac{\partial^{(n-1)} g_{(2.1.1)}(t,s)}{\partial t^{n-1}}$ is continuous in K_1 and on the diagonal $t = s$ undergoes a discontinuity equal to unity, i.e.,

$$\frac{\partial^{(n-1)} g_{(2.1.1)}(t+0,t)}{\partial t^{n-1}} - \frac{\partial^{(n-1)} g_{(2.1.1)}(t-0,t)}{\partial t^{n-1}} = 1; \; t \neq a_i, \; 1 \le i \le r,$$

(iii) $g_{(2.1.1)}(t,s)$ as a function of t satisfies

$$z^{(n)} = 0$$

$$z^{(i)}(a_j) = 0, \; 0 \le i \le k_j, \; 1 \le j \le r \tag{2.3.10}$$

in K_1, and

(iv) $g_{(2.1.1)}(t,s)$ is uniquely determined in K.

In Section 2.4 we shall be more interested in the case $a_0 = a_1, \; a_{r+1} = a_r$, which corresponds to the interpolation in the exact sense of the word. For this the representation (2.3.7) becomes

$$e_{(2.1.1)}(t) = \int_{a_1}^{a_r} g_{(2.1.1)}(t,s) x^{(n)}(s) \, ds, \tag{2.3.11}$$

where $g_{(2.1.1)}(t,s)$ remains the same as given in (2.3.9) except now ℓ ranges from 1 to $(r-1)$.

Lemma 2.3.3. For Green's function $g_{(2.1.1)}(t,s)$ the following hold:

(i) $g_{(2.1.1)}(t,s) \big/ \omega(t) > 0$ for $a_1 \le t \le a_r, \; a_1 < s < a_r$,

$$\text{(ii)} \; | \, g_{(2.1.1)}(t,s) \, | \le \frac{1}{(n-1)!(a_r - a_1)} \, | \, \omega(t) \, |, \text{ and} \tag{2.3.12}$$

$$\text{(iii)} \int_{a_1}^{a_r} | \, g_{(2.1.1)}(t,s) \, | \, ds = \frac{1}{n!} \, | \, \omega(t) \, | \, . \tag{2.3.13}$$

Proof. Part (i) is a particular case of the known result established by Levin [29] and Pokornyi [36], also see Coppel [18]. Inequality (2.3.12) is due to Beesack [9]; for $r = n$ a short proof of this inequality has also been given by Nehari [34]. To show part (iii), we note that the unique solution of the boundary value problem $z^{(n)} = 1$, (2.3.10) is

$$z(t) = \frac{1}{n!}\omega(t). \tag{2.3.14}$$

But, for this solution the interpolating polynomial $P_{(2.1.1)}(t) \equiv 0$, and hence from (2.3.11), $z(t)$ can also be written as

$$z(t) = \int_{a_1}^{a_r} g_{(2.1.1)}(t,s)\,ds. \tag{2.3.15}$$

From (2.3.14) and (2.3.15), we have the identity

$$\int_{a_1}^{a_r} g_{(2.1.1)}(t,s)\,ds = \frac{1}{n!}\omega(t).$$

The result (2.3.13) now follows from part (i). ∎

Lemma 2.3.4. Corresponding to the Lagrange conditions (2.1.3) Green's function $g_{(2.1.3)}(t,s)$ can be written as

$$g_{(2.1.3)}(t,s) = \frac{1}{(n-1)!}\begin{cases} \displaystyle\sum_{j=1}^{\ell}(a_j - s)^{n-1} \prod_{\substack{k=1\\ k\neq j}}^{n} \left(\frac{t-a_k}{a_j-a_k}\right), & s \le t \\ \displaystyle -\sum_{j=\ell+1}^{n}(a_j - s)^{n-1} \prod_{\substack{k=1\\ k\neq j}}^{n} \left(\frac{t-a_k}{a_j-a_k}\right), & s \ge t \end{cases} \tag{2.3.16}$$

$$a_\ell \le s \le a_{\ell+1},\ \ell = 1,\ 2,\ \ldots,\ n-1. \quad ∎$$

Lemma 2.3.5. Corresponding to the $(m, n-m)$ conditions (2.1.4) Green's function $g_{(2.1.4)}(t,s)$ can be written as

(2.3.17) $g_{(2.1.4)}(t,s)$

$$= \begin{cases} \displaystyle\sum_{j=0}^{m-1}\left[\sum_{p=0}^{m-1-j}\binom{n-m+p-1}{p}\left(\frac{t-a_1}{a_2-a_1}\right)^p\right]\times \\ \displaystyle\frac{(t-a_1)^j(a_1-s)^{n-j-1}}{j!(n-j-1)!}\left(\frac{a_2-t}{a_2-a_1}\right)^{n-m}, \quad a_1\le s\le t\le a_2 \\ \displaystyle-\sum_{i=0}^{n-m-1}\left[\sum_{q=0}^{n-m-i-1}\binom{m+q-1}{q}\left(\frac{a_2-t}{a_2-a_1}\right)^q\right]\times \\ \displaystyle\frac{(t-a_2)^i(a_2-s)^{n-i-1}}{i!(n-i-1)!}\left(\frac{t-a_1}{a_2-a_1}\right)^m, \quad a_1\le t\le s\le a_2. \end{cases} \blacksquare$$

Lemma 2.3.6. Corresponding to the two - point Taylor conditions (2.1.5) Green's function $g_{(2.1.5)}(t,s)$ can be written as

(2.3.18) $g_{(2.1.5)}(t,s)$

$$= \frac{(-1)^m}{(2m-1)!}\begin{cases} \displaystyle p^m(t,s)\sum_{j=0}^{m-1}\binom{m-1+j}{j}(t-s)^{m-1-j}q^j(t,s), \\ \qquad a_1\le s\le t\le a_2 \\ \displaystyle q^m(t,s)\sum_{j=0}^{m-1}\binom{m-1+j}{j}(s-t)^{m-1-j}p^j(t,s), \\ \qquad a_1\le t\le s\le a_2 \end{cases}$$

where

$$p(t,s) = \frac{(s-a_1)(a_2-t)}{(a_2-a_1)} \quad \text{and} \quad q(t,s) = p(s,t). \qquad \blacksquare$$

A New Representation

Theorem 2.3.7. Let the functions $F_k(t)$, $1\le k\le n$ be recursively defined as follows

$$F_n(t) = \int_{a_r}^{t}(a_r-t_n)^{n-1}x^{(n)}(t_n)\,dt_n$$

$$(2.3.19)\quad F_k(t)=\begin{cases}\displaystyle\int_{a_{s+1}}^{t}\frac{(a_{s+1}-t_k)^{k-1}}{(a_{s+2}-t_k)^{k+1}}F_{k+1}(t_k)\,dt_k; \\ \qquad k=p(s+1)+s+1,\ 0\le s\le r-2 \\ \displaystyle\int_{a_{s+1}}^{t}(a_{s+1}-t_k)^{-2}F_{k+1}(t_k)\,dt_k; \\ \qquad p(s)+s+1\le k\le p(s+1)+s,\ 0\le s\le r-1\end{cases}$$

where $p(0)=0$, and for $1\le s\le r$, $p(s)=\sum_{j=1}^{s}k_j$.

Then, the error function $e_{(2.1.1)}(t)$ can be explicitly expressed as

$$(2.3.20)\qquad e_{(2.1.1)}(t)=(a_1-t)^{k_1}\prod_{i=2}^{r}(a_i-t)^{k_i+1}F_1(t).$$

Proof. Since $e^{(k)}_{(2.1.1)}(a_i)=0;\ 0\le k\le k_i,\ 1\le i\le r$ is obvious, it suffices to show that $e^{(n)}_{(2.1.1)}(t)=x^{(n)}(t)$. For this, we shall use induction to deduce that

$$(2.3.21)\ e^{(k)}_{(2.1.1)}(t)$$

$$=\sum_{i=0}^{\min\{k,k_1\}}\binom{k}{i}D^{k-i}\left[(a_1-t)^{k_1-i}\theta_0\right]\frac{F_{i+1}(t)}{(a_1-t)^i}$$

$$+\sum_{\ell=2}^{s-1}\left\{\sum_{i=p(\ell-1)+\ell-1}^{p(\ell)+\ell-1}\binom{k}{i}D^{k-i}\left[(a_\ell-t)^{p(\ell-1)+\ell-2-i}\theta_{\ell-2}\right]\frac{F_{i+1}(t)}{(a_\ell-t)^i}\right\}$$

$$+\sum_{i=p(s-1)+s-1}^{k}\binom{k}{i}D^{k-i}\left[(a_s-t)^{p(s-1)+s-2-i}\theta_{s-2}\right]\frac{F_{i+1}(t)}{(a_s-t)^i};$$

$$p(s-1)+s-1\le k\le p(s)+s-1,\ 1\le s\le r$$

where $D=\frac{d}{dt}$, $D^q=D^{q-1}[D]$, $q>1$, and

$$(2.3.22)\qquad \theta_{-1}=0,\qquad \theta_i=\prod_{j=i+2}^{r}(a_j-t)^{k_j+1},\ 0\le i\le r-2.$$

For $k = 0$, (2.3.21) reduces to (2.3.20). Further, from (2.3.20) it follows that

(2.3.23) $e'_{(2.1.1)}(t)$

$$= D\left[(a_1-t)^{k_1}\theta_0\right] F_1(t) + \theta_0 F_2(t)\begin{Bmatrix} (a_1-t)^{k_1-2}, & \text{if } k_1>0 \\ (a_2-t)^{-2}, & \text{if } k_1=0 \end{Bmatrix}$$

and hence (2.3.21) is true for $k = 1$. Now let (2.3.21) be true for $k = q$, $1 \le q \le k_1 - 1$ so that on differentiating the resulting (2.3.21), we find

$e^{(q+1)}_{(2.1.1)}(t)$

$$= \sum_{i=0}^{q}\binom{q}{i}\left\{D^{q+1-i}\left[(a_1-t)^{k_1-i}\theta_0\right]\frac{F_{i+1}(t)}{(a_1-t)^i}\right.$$

$$\left. + iD^{q-i}\left[(a_1-t)^{k_1-i}\theta_0\right]\frac{F_{i+1}(t)}{(a_1-t)^{i+1}} + D^{q-i}\left[(a_1-t)^{k_1-i}\theta_0\right]\frac{F_{i+2}(t)}{(a_1-t)^{i+2}}\right\}$$

$$= \binom{q+1}{0} D^{q+1}\left[(a_1-t)^{k_1}\theta_0\right] F_1(t) + \sum_{i=1}^{q}\left\{\binom{q}{i} D^{q+1-i}\left[(a_1-t)^{k_1-i}\theta_0\right]\right.$$

$$+ i\binom{q}{i} D^{q-i}\left[(a_1-t)^{k_1-i}\theta_0\right]\frac{1}{a_1-t}$$

$$\left. + \binom{q}{i-1} D^{q+1-i}\left[(a_1-t)^{k_1-i+1}\theta_0\right]\frac{1}{a_1-t}\right\}\frac{F_{i+1}(t)}{(a_1-t)^i}$$

$$+ \binom{q+1}{q+1}(a_1-t)^{k_1-q}\theta_0\frac{F_{q+2}(t)}{(a_1-t)^{q+2}}$$

$$= \binom{q+1}{0} D^{q+1}\left[(a_1-t)^{k_1}\theta_0\right] F_1(t) + \sum_{i=1}^{q}\left\{\binom{q}{i} D^{q+1-i}\left[(a_1-t)^{k_1-i}\theta_0\right]\right.$$

$$\left. + \binom{q}{i-1} D^{q+1-i}\left[(a_1-t)^{k_1-i}\theta_0\right]\right\}\frac{F_{i+1}(t)}{(a_1-t)^i}$$

$$+ \binom{q+1}{q+1}(a_1-t)^{k_1-q}\theta_0\frac{F_{q+2}(t)}{(a_1-t)^{q+2}}$$

$$= \binom{q+1}{0} D^{q+1}\left[(a_1-t)^{k_1}\theta_0\right] F_1(t)$$

$$+\sum_{i=1}^{q}\binom{q+1}{i}D^{q+1-i}\left[(a_1-t)^{k_1-i}\theta_0\right]\frac{F_{i+1}(t)}{(a_1-t)^i}$$

$$+\binom{q+1}{q+1}(a_1-t)^{k_1-q}\theta_0\frac{F_{q+2}(t)}{(a_1-t)^{q+2}}$$

$$=\sum_{i=0}^{q+1}\binom{q+1}{i}D^{q+1-i}\left[(a_1-t)^{k_1-i}\theta_0\right]\frac{F_{i+1}(t)}{(a_1-t)^i}.$$

Thus, (2.3.21) is true for $0 \le k \le k_1$.

Now let $k = k_1$ in (2.3.21) and differentiate the resulting equation to obtain

$$e_{(2.1.1)}^{(k_1+1)}(t)$$

$$=\sum_{i=0}^{k_1}\binom{k_1}{i}\left\{D^{k_1+1-i}\left[(a_1-t)^{k_1-i}\theta_0\right]\frac{F_{i+1}(t)}{(a_1-t)^i}\right.$$

$$+iD^{k_1-i}\left[(a_1-t)^{k_1-i}\theta_0\right]\frac{F_{i+1}(t)}{(a_1-t)^{i+1}}$$

$$\left.+D^{k_1-i}\left[(a_1-t)^{k_1-i}\theta_0\right]F_{i+2}(t)\left\{\begin{array}{ll}\dfrac{1}{(a_1-t)^{i+2}}, & 0\le i\le k_1-1\\ \dfrac{1}{(a_2-t)^{k_1+2}}, & i=k_1\end{array}\right\}\right\}$$

$$=\sum_{i=0}^{k_1}\binom{k_1+1}{i}D^{k_1+1-i}\left[(a_1-t)^{k_1-i}\theta_0\right]\frac{F_{i+1}(t)}{(a_1-t)^i}+\theta_0\frac{F_{k_1+2}(t)}{(a_2-t)^{k_1+2}},$$

which is the same as (2.3.21) for $k = k_1 + 1$.

Next we assume that (2.3.21) is true for $k = q \ge k_1 + 1$, where $p(s-1) + s - 1 \le q \le p(s) + s - 1$ for some fixed s, $2 \le s \le r$. Then, there are two cases to consider, namely

$$p(s-1)+s-1 \le q+1 \le p(s)+s-1 \tag{2.3.24}$$

and

$$q+1 = p(s)+s. \tag{2.3.25}$$

In case of (2.3.24), on differentiating (2.3.21), we obtain

$$e^{(q+1)}_{(2.1.1)}(t)$$

$$= \sum_{i=0}^{k_1} \binom{q}{i} \left\{ D^{q+1-i}\left[(a_1-t)^{k_1-i}\theta_0\right] \frac{F_{i+1}(t)}{(a_1-t)^i} \right.$$

$$\left. + iD^{q-i}\left[(a_1-t)^{k_1-i}\theta_0\right] \frac{F_{i+1}(t)}{(a_1-t)^{i+1}} + D^{q-i}\left[(a_1-t)^{k_1-i}\theta_0\right] \frac{F_{i+2}(t)}{(a_1-t)^{i+2}} \right\}$$

$$-\binom{q}{k_1} D^{q-k_1}[\theta_0] \frac{F_{k_1+2}(t)}{(a_1-t)^{k_1+2}} + \binom{q}{k_1} D^{q-k_1}[\theta_0] \frac{F_{k_1+2}(t)}{(a_2-t)^{k_1+2}}$$

$$+\sum_{\ell=2}^{s-1} \left\{ \binom{q}{p(\ell-1)+\ell-1} D^{q-p(\ell-1)-\ell+2}\left[(a_\ell-t)^{-1}\theta_{\ell-2}\right] \frac{F_{p(\ell-1)+\ell}(t)}{(a_\ell-t)^{p(\ell-1)+\ell-1}} \right.$$

$$+[p(\ell-1)+\ell-1] \binom{q}{p(\ell-1)+\ell-1} D^{q-p(\ell-1)-\ell+1}\left[(a_\ell-t)^{-1}\theta_{\ell-2}\right] \times$$

$$\frac{F_{p(\ell-1)+\ell}(t)}{(a_\ell-t)^{p(\ell-1)+\ell}}$$

$$+\sum_{i=p(\ell-1)+\ell}^{p(\ell)+\ell-1} \left\{ \binom{q}{i} D^{q+1-i}\left[(a_\ell-t)^{p(\ell-1)+\ell-2-i}\theta_{\ell-2}\right] \right.$$

$$+i\binom{q}{i} D^{q-i}\left[(a_\ell-t)^{p(\ell-1)+\ell-2-i}\theta_{\ell-2}\right] \frac{1}{a_\ell-t}$$

$$\left. +\binom{q}{i-1} D^{q+1-i}\left[(a_\ell-t)^{p(\ell-1)+\ell-1-i}\theta_{\ell-2}\right] \frac{1}{a_\ell-t} \right\} \frac{F_{i+1}(t)}{(a_\ell-t)^i}$$

$$\left. +\binom{q}{p(\ell)+\ell-1} D^{q-p(\ell)-\ell+1}[\theta_{\ell-1}] \frac{F_{p(\ell)+\ell+1}(t)}{(a_{\ell+1}-t)^{p(\ell)+\ell+1}} \right\}$$

$$+\binom{q}{p(s-1)+s-1} D^{q-p(s-1)-s+2}\left[(a_s-t)^{-1}\theta_{s-2}\right] \frac{F_{p(s-1)+s}(t)}{(a_s-t)^{p(s-1)+s-1}}$$

$$+[p(s-1)+s-1]\binom{q}{p(s-1)+s-1} D^{q-p(s-1)-s+1}\left[(a_s-t)^{-1}\theta_{s-2}\right] \times$$

$$\frac{F_{p(s-1)+s}(t)}{(a_s-t)^{p(s-1)+s}}$$

$$+\sum_{i=p(s-1)+s}^{q} \left\{ \binom{q}{i} D^{q+1-i}\left[(a_s-t)^{p(s-1)+s-2-i}\theta_{s-2}\right] \right.$$

$$+i\binom{q}{i} D^{q-i}\left[(a_s-t)^{p(s-1)+s-2-i}\theta_{s-2}\right]\frac{1}{a_s-t}$$

$$\left.+\binom{q}{i-1} D^{q+1-i}\left[(a_s-t)^{p(s-1)+s-1-i}\theta_{s-2}\right]\frac{1}{a_s-t}\right\}\frac{F_{i+1}(t)}{(a_s-t)^i}$$

$$+\binom{q+1}{q+1}(a_s-t)^{p(s-1)+s-2-q}\theta_{s-2}\frac{F_{q+2}(t)}{(a_s-t)^{q+2}}$$

$$=\binom{q+1}{0} D^{q+1}\left[(a_1-t)^{k_1}\theta_0\right]F_1(t)+\sum_{i=1}^{k_1}\left\{\binom{q}{i} D^{q+1-i}\left[(a_1-t)^{k_1-i}\theta_0\right]\right.$$

$$+i\binom{q}{i} D^{q-i}\left[(a_1-t)^{k_1-i}\theta_0\right]\frac{1}{a_1-t}$$

$$\left.+\binom{q}{i-1} D^{q+1-i}\left[(a_1-t)^{k_1-i+1}\theta_0\right]\frac{1}{a_1-t}\right\}\frac{F_{i+1}(t)}{(a_1-t)^i}$$

$$+\binom{q}{k_1} D^{q-k_1}\left[\theta_0\right]\frac{F_{k_1+2}(t)}{(a_1-t)^{k_1+2}}-\binom{q}{k_1} D^{q-k_1}\left[\theta_0\right]\frac{F_{k_1+2}(t)}{(a_1-t)^{k_1+2}}$$

$$+\binom{q}{k_1} D^{q-k_1}\left[\theta_0\right]\frac{F_{k_1+2}(t)}{(a_2-t)^{k_1+2}}$$

$$+\sum_{\ell=2}^{s-1}\left\{\binom{q}{p(\ell-1)+\ell-1} D^{q-p(\ell-1)-\ell+2}\left[(a_\ell-t)^{-1}\theta_{\ell-2}\right]\frac{F_{p(\ell-1)+\ell}(t)}{(a_\ell-t)^{p(\ell-1)+\ell-1}}\right.$$

$$+[p(\ell-1)+\ell-1]\binom{q}{p(\ell-1)+\ell-1} D^{q-p(\ell-1)-\ell+1}\left[(a_\ell-t)^{-1}\theta_{\ell-2}\right]\times$$

$$\frac{F_{p(\ell-1)+\ell}(t)}{(a_\ell-t)^{p(\ell-1)+\ell}}$$

$$+\sum_{i=p(\ell-1)+\ell}^{p(\ell)+\ell-1}\binom{q+1}{i} D^{q+1-i}\left[(a_\ell-t)^{p(\ell-1)+\ell-2-i}\theta_{\ell-2}\right]\frac{F_{i+1}(t)}{(a_\ell-t)^i}$$

$$\left.+\binom{q}{p(\ell)+\ell-1} D^{q-p(\ell)-\ell+1}\left[\theta_{\ell-1}\right]\frac{F_{p(\ell)+\ell+1}(t)}{(a_{\ell+1}-t)^{p(\ell)+\ell+1}}\right\}$$

$$+\binom{q}{p(s-1)+s-1} D^{q-p(s-1)-s+2}\left[(a_s-t)^{-1}\theta_{s-2}\right]\frac{F_{p(s-1)+s}(t)}{(a_s-t)^{p(s-1)+s-1}}$$

$$+[p(s-1)+s-1]\binom{q}{p(s-1)+s-1} D^{q-p(s-1)-s+1}\left[(a_s-t)^{-1}\theta_{s-2}\right]\times$$

$$\frac{F_{p(s-1)+s}(t)}{(a_s-t)^{p(s-1)+s}}$$

$$+\sum_{i=p(s-1)+s}^{q}\binom{q+1}{i} D^{q+1-i}\left[(a_s-t)^{p(s-1)+s-2-i}\theta_{s-2}\right]\frac{F_{i+1}(t)}{(a_s-t)^i}$$

$$+\binom{q+1}{q+1}(a_s-t)^{p(s-1)+s-2-q}\theta_{s-2}\frac{F_{q+2}(t)}{(a_s-t)^{q+2}}$$

$$=\binom{q+1}{0} D^{q+1}\left[(a_1-t)^{k_1}\theta_0\right]F_1(t)$$

$$+\sum_{i=1}^{k_1}\binom{q+1}{i} D^{q+1-i}\left[(a_1-t)^{k_1-i}\theta_0\right]\frac{F_{i+1}(t)}{(a_1-t)^i}$$

$$+\binom{q}{k_1} D^{q-k_1}\left[\theta_0\right]\frac{F_{k_1+2}(t)}{(a_2-t)^{k_1+2}}$$

$$+\sum_{\ell=2}^{s-1}\left\{[p(\ell-1)+\ell-1]\binom{q}{p(\ell-1)+\ell-1}\times\right.$$

$$D^{q-p(\ell-1)-\ell+1}\left[(a_\ell-t)^{-1}\theta_{\ell-2}\right]\frac{F_{p(\ell-1)+\ell}(t)}{(a_\ell-t)^{p(\ell-1)+\ell}}$$

$$+\binom{q}{p(\ell-1)+\ell-1} D^{q-p(\ell-1)-\ell+2}\left[(a_\ell-t)^{-1}\theta_{\ell-2}\right]\frac{F_{p(\ell-1)+\ell}(t)}{(a_\ell-t)^{p(\ell-1)+\ell-1}}$$

$$+\sum_{i=p(\ell-1)+\ell}^{p(\ell)+\ell-1}\binom{q+1}{i} D^{q+1-i}\left[(a_\ell-t)^{p(\ell-1)+\ell-2-i}\theta_{\ell-2}\right]\frac{F_{i+1}(t)}{(a_\ell-t)^i}$$

$$\left.+\binom{q}{p(\ell)+\ell-1} D^{q-p(\ell)-\ell+1}\left[\theta_{\ell-1}\right]\frac{F_{p(\ell)+\ell+1}(t)}{(a_{\ell+1}-t)^{p(\ell)+\ell+1}}\right\}$$

$$+[p(s-1)+s-1]\binom{q}{p(s-1)+s-1} D^{q-p(s-1)-s+1}\left[(a_s-t)^{-1}\theta_{s-2}\right]\times$$

$$\frac{F_{p(s-1)+s}(t)}{(a_s-t)^{p(s-1)+s}}$$

$$+\binom{q}{p(s-1)+s-1} D^{q-p(s-1)-s+2}\left[(a_s-t)^{-1}\theta_{s-2}\right]\frac{F_{p(s-1)+s}(t)}{(a_s-t)^{p(s-1)+s-1}}$$

$$+\sum_{i=p(s-1)+s}^{q+1}\binom{q+1}{i} D^{q+1-i}\left[(a_s-t)^{p(s-1)+s-2-i}\theta_{s-2}\right]\frac{F_{i+1}(t)}{(a_s-t)^i}$$

$$=\sum_{i=0}^{k_1}\binom{q+1}{i} D^{q+1-i}\left[(a_1-t)^{k_1-i}\theta_0\right]\frac{F_{i+1}(t)}{(a_1-t)^i}$$

$$+\sum_{\ell=1}^{s-2}\left\{\binom{q}{p(\ell)+\ell-1} D^{q-p(\ell)-\ell+1}\left[(a_{\ell+1}-t)^{-1}\theta_{\ell-1}\right]\frac{F_{p(\ell)+\ell+1}(t)}{(a_{\ell+1}-t)^{p(\ell)+\ell}}\right.$$

$$+\binom{q}{p(\ell)+\ell} D^{q-p(\ell)-\ell+1}\left[(a_{\ell+1}-t)^{-1}\theta_{\ell-1}\right]\frac{F_{p(\ell)+\ell+1}(t)}{(a_{\ell+1}-t)^{p(\ell)+\ell}}$$

$$\left.+\sum_{i=p(\ell)+\ell+1}^{p(\ell+1)+\ell}\binom{q+1}{i} D^{q+1-i}\left[(a_{\ell+1}-t)^{p(\ell)+\ell-1-i}\theta_{\ell-1}\right]\frac{F_{i+1}(t)}{(a_{\ell+1}-t)^i}\right\}$$

$$+\binom{q}{p(s-1)+s-2} D^{q-p(s-1)-s+2}\left[(a_s-t)^{-1}\theta_{s-2}\right]\frac{F_{p(s-1)+s}(t)}{(a_s-t)^{p(s-1)+s-1}}$$

$$+\binom{q}{p(s-1)+s-1} D^{q-p(s-1)-s+2}\left[(a_s-t)^{-1}\theta_{s-2}\right]\frac{F_{p(s-1)+s}(t)}{(a_s-t)^{p(s-1)+s-1}}$$

$$+\sum_{i=p(s-1)+s}^{q+1}\binom{q+1}{i} D^{q+1-i}\left[(a_s-t)^{p(s-1)+s-2-i}\theta_{s-2}\right]\frac{F_{i+1}(t)}{(a_s-t)^i}$$

$$=\sum_{i=0}^{k_1}\binom{q+1}{i} D^{q+1-i}\left[(a_1-t)^{k_1-i}\theta_0\right]\frac{F_{i+1}(t)}{(a_1-t)^i}$$

$$+\sum_{\ell=1}^{s-2}\left\{\binom{q+1}{p(\ell)+\ell} D^{q-p(\ell)-\ell+1}\left[(a_{\ell+1}-t)^{-1}\theta_{\ell-1}\right]\frac{F_{p(\ell)+\ell+1}(t)}{(a_{\ell+1}-t)^{p(\ell)+\ell}}\right.$$

$$\left.+\sum_{i=p(\ell)+\ell+1}^{p(\ell+1)+\ell}\binom{q+1}{i} D^{q+1-i}\left[(a_{\ell+1}-t)^{p(\ell)+\ell-1-i}\theta_{\ell-1}\right]\frac{F_{i+1}(t)}{(a_{\ell+1}-t)^i}\right\}$$

$$+\binom{q+1}{p(s-1)+s-1} D^{q-p(s-1)-s+2}\left[(a_s-t)^{-1}\theta_{s-2}\right]\frac{F_{p(s-1)+s}(t)}{(a_s-t)^{p(s-1)+s-1}}$$

$$+\sum_{i=p(s-1)+s}^{q+1}\binom{q+1}{i}D^{q+1-i}\left[(a_s-t)^{p(s-1)+s-2-i}\theta_{s-2}\right]\frac{F_{i+1}(t)}{(a_s-t)^i}$$

$$=\sum_{i=0}^{k_1}\binom{q+1}{i}D^{q+1-i}\left[(a_1-t)^{k_1-i}\theta_0\right]\frac{F_{i+1}(t)}{(a_1-t)^i}$$

$$+\sum_{\ell=2}^{s-1}\left\{\sum_{i=p(\ell-1)+\ell-1}^{p(\ell)+\ell-1}\binom{q+1}{i}D^{q+1-i}\left[(a_\ell-t)^{p(\ell-1)+\ell-2-i}\theta_{\ell-2}\right]\frac{F_{i+1}(t)}{(a_\ell-t)^i}\right\}$$

$$+\sum_{i=p(s-1)+s-1}^{q+1}\binom{q+1}{i}D^{q+1-i}\left[(a_s-t)^{p(s-1)+s-2-i}\theta_{s-2}\right]\frac{F_{i+1}(t)}{(a_s-t)^i},$$

which is the same as (2.3.21) for $k = q+1$.

In case (2.3.25) holds, then $q = p(s)+s-1$ and as before from (2.3.21) we find that

$$e^{(q+1)}_{(2.1.1)}(t)$$

$$=\sum_{i=0}^{k_1}\binom{q+1}{i}D^{q+1-i}\left[(a_1-t)^{k_1-i}\theta_0\right]\frac{F_{i+1}(t)}{(a_1-t)^i}$$

$$+\binom{q}{k_1}D^{q-k_1}\left[\theta_0\right]\frac{F_{k_1+2}(t)}{(a_2-t)^{k_1+2}}$$

$$+\sum_{\ell=2}^{s-1}\left\{[p(\ell-1)+\ell-1]\binom{q}{p(\ell-1)+\ell-1}\times\right.$$

$$D^{q-p(\ell-1)-\ell+1}\left[(a_\ell-t)^{-1}\theta_{\ell-2}\right]\frac{F_{p(\ell-1)+\ell}(t)}{(a_\ell-t)^{p(\ell-1)+\ell}}$$

$$+\binom{q}{p(\ell-1)+\ell-1}D^{q-p(\ell-1)-\ell+2}\left[(a_\ell-t)^{-1}\theta_{\ell-2}\right]\frac{F_{p(\ell-1)+\ell}(t)}{(a_\ell-t)^{p(\ell-1)+\ell-1}}$$

$$+\sum_{i=p(\ell-1)+\ell}^{p(\ell)+\ell-1}\binom{q+1}{i}D^{q+1-i}\left[(a_\ell-t)^{p(\ell-1)+\ell-2-i}\theta_{\ell-2}\right]\frac{F_{i+1}(t)}{(a_\ell-t)^i}$$

$$\left.+\binom{q}{p(\ell)+\ell-1}D^{q-p(\ell)-\ell+1}\left[\theta_{\ell-1}\right]\frac{F_{p(\ell)+\ell+1}(t)}{(a_{\ell+1}-t)^{p(\ell)+\ell+1}}\right\}$$

$$+[p(s-1)+s-1]\binom{q}{p(s-1)+s-1} D^{q-p(s-1)-s+1}\left[(a_s-t)^{-1}\theta_{s-2}\right]\times$$

$$\frac{F_{p(s-1)+s}(t)}{(a_s-t)^{p(s-1)+s}}$$

$$+\binom{q}{p(s-1)+s-1} D^{q-p(s-1)-s+2}\left[(a_s-t)^{-1}\theta_{s-2}\right]\frac{F_{p(s-1)+s}(t)}{(a_s-t)^{p(s-1)+s-1}}$$

$$+\sum_{i=p(s-1)+s}^{q}\binom{q+1}{i} D^{q+1-i}\left[(a_s-t)^{p(s-1)+s-2-i}\theta_{s-2}\right]\frac{F_{i+1}(t)}{(a_s-t)^{i}}$$

$$+\binom{q+1}{q+1}(a_s-t)^{p(s-1)+s-2-q}\theta_{s-2}\frac{F_{q+2}(t)}{(a_{s+1}-t)^{q+2}}$$

$$=\sum_{i=0}^{k_1}\binom{q+1}{i} D^{q+1-i}\left[(a_1-t)^{k_1-i}\theta_0\right]\frac{F_{i+1}(t)}{(a_1-t)^{i}}$$

$$+\sum_{\ell=1}^{s-2}\left\{\binom{q}{p(\ell)+\ell-1} D^{q-p(\ell)-\ell+1}\left[\theta_{\ell-1}\right]\frac{F_{p(\ell)+\ell+1}(t)}{(a_{\ell+1}-t)^{p(\ell)+\ell+1}}\right.$$

$$+[p(\ell)+\ell]\binom{q}{p(\ell)+\ell} D^{q-p(\ell)-\ell}\left[(a_{\ell+1}-t)^{-1}\theta_{\ell-1}\right]\frac{F_{p(\ell)+\ell+1}(t)}{(a_{\ell+1}-t)^{p(\ell)+\ell+1}}$$

$$+\binom{q}{p(\ell)+\ell} D^{q-p(\ell)-\ell+1}\left[(a_{\ell+1}-t)^{-1}\theta_{\ell-1}\right]\frac{F_{p(\ell)+\ell+1}(t)}{(a_{\ell+1}-t)^{p(\ell)+\ell}}$$

$$\left.+\sum_{i=p(\ell)+\ell+1}^{p(\ell+1)+\ell}\binom{q+1}{i} D^{q+1-i}\left[(a_{\ell+1}-t)^{p(\ell)+\ell-1-i}\theta_{\ell-1}\right]\frac{F_{i+1}(t)}{(a_{\ell+1}-t)^{i}}\right\}$$

$$+\binom{q}{p(s-1)+s-2} D^{q-p(s-1)-s+2}\left[\theta_{s-2}\right]\frac{F_{p(s-1)+s}(t)}{(a_s-t)^{p(s-1)+s}}$$

$$+[p(s-1)+s-1]\binom{q}{p(s-1)+s-1} D^{q-p(s-1)-s+1}\left[(a_s-t)^{-1}\theta_{s-2}\right]\times$$

$$\frac{F_{p(s-1)+s}(t)}{(a_s-t)^{p(s-1)+s}}$$

$$+\binom{q}{p(s-1)+s-1} D^{q-p(s-1)-s+2}\left[(a_s-t)^{-1}\theta_{s-2}\right]\frac{F_{p(s-1)+s}(t)}{(a_s-t)^{p(s-1)+s-1}}$$

$$+\sum_{i=p(s-1)+s}^{q}\binom{q+1}{i}D^{q+1-i}\left[(a_s-t)^{p(s-1)+s-2-i}\theta_{s-2}\right]\frac{F_{i+1}(t)}{(a_s-t)^i}$$

$$+\binom{q+1}{q+1}\theta_{s-1}\frac{F_{q+2}(t)}{(a_{s+1}-t)^{q+2}}$$

$$=\sum_{i=0}^{k_1}\binom{q+1}{i}D^{q+1-i}\left[(a_1-t)^{k_1-i}\theta_0\right]\frac{F_{i+1}(t)}{(a_1-t)^i}$$

$$+\sum_{\ell=2}^{s-1}\left\{\sum_{i=p(\ell-1)+\ell-1}^{p(\ell)+\ell-1}\binom{q+1}{i}D^{q+1-i}\left[(a_\ell-t)^{p(\ell-1)+\ell-2-i}\theta_{\ell-2}\right]\frac{F_{i+1}(t)}{(a_\ell-t)^i}\right\}$$

$$+\sum_{i=p(s-1)+s-1}^{q}\binom{q+1}{i}D^{q+1-i}\left[(a_s-t)^{p(s-1)+s-2-i}\theta_{s-2}\right]\frac{F_{i+1}(t)}{(a_s-t)^i}$$

$$+\sum_{i=q+1}^{q+1}\binom{q+1}{i}D^{q+1-i}\left[(a_{s+1}-t)^{p(s)+s-1-i}\theta_{s-1}\right]\frac{F_{i+1}(t)}{(a_{s+1}-t)^i}$$

$$=\sum_{i=0}^{k_1}\binom{q+1}{i}D^{q+1-i}\left[(a_1-t)^{k_1-i}\theta_0\right]\frac{F_{i+1}(t)}{(a_1-t)^i}$$

$$+\sum_{\ell=2}^{s}\left\{\sum_{i=p(\ell-1)+\ell-1}^{p(\ell)+\ell-1}\binom{q+1}{i}D^{q+1-i}\left[(a_\ell-t)^{p(\ell-1)+\ell-2-i}\theta_{\ell-2}\right]\frac{F_{i+1}(t)}{(a_\ell-t)^i}\right\}$$

$$+\sum_{i=p(s)+s}^{q+1}\binom{q+1}{i}D^{q+1-i}\left[(a_{s+1}-t)^{p(s)+s-1-i}\theta_{s-1}\right]\frac{F_{i+1}(t)}{(a_{s+1}-t)^i},$$

which is the same as (2.3.21) for $k = q+1$. Thus, the proof of (2.3.21) is complete.

Finally, for $k = p(r)+r-1$ $(= n-1)$ the relation (2.3.21) is the same as

$$e^{(n-1)}_{(2.1.1)}(t)$$

$$=\sum_{i=0}^{k_1}\binom{n-1}{i}D^{n-1-i}\left[(a_1-t)^{k_1-i}\theta_0\right]\frac{F_{i+1}(t)}{(a_1-t)^i}$$

$$+\sum_{\ell=2}^{r}\left\{\sum_{i=p(\ell-1)+\ell-1}^{p(\ell)+\ell-1}\binom{n-1}{i}D^{n-1-i}\left[(a_\ell-t)^{p(\ell-1)+\ell-2-i}\theta_{\ell-2}\right]\frac{F_{i+1}(t)}{(a_\ell-t)^i}\right\}$$

$$= \sum_{i=0}^{k_1} \binom{n-1}{i} (-1)^{n-1-i}(n-1-i)! \frac{F_{i+1}(t)}{(a_1-t)^i}$$

$$+\sum_{\ell=2}^{r} \left[\sum_{i=p(\ell-1)+\ell-1}^{p(\ell)+\ell-1} \binom{n-1}{i} (-1)^{n-1-i}(n-1-i)! \frac{F_{i+1}(t)}{(a_\ell-t)^i} \right]$$

$$= (n-1)! \left\{ \sum_{i=0}^{k_1} \frac{(-1)^{n-1-i}}{i!} \frac{F_{i+1}(t)}{(a_1-t)^i} \right. \tag{2.3.26}$$

$$\left. + \sum_{\ell=2}^{r} \left[\sum_{i=p(\ell-1)+\ell-1}^{p(\ell)+\ell-1} \frac{(-1)^{n-1-i}}{i!} \frac{F_{i+1}(t)}{(a_\ell-t)^i} \right] \right\}$$

and hence it follows that

$$\frac{1}{(n-1)!} e^{(n)}_{(2.1.1)}(t)$$

$$= \sum_{i=0}^{k_1-1} \frac{(-1)^{n-1-i}}{i!} \left[i\frac{F_{i+1}(t)}{(a_1-t)^{i+1}} + \frac{F_{i+2}(t)}{(a_1-t)^{i+2}} \right]$$

$$+\frac{(-1)^{n-k_1-1}}{k_1!} \left[k_1 \frac{F_{k_1+1}(t)}{(a_1-t)^{k_1+1}} + \frac{F_{k_1+2}(t)}{(a_2-t)^{k_1+2}} \right]$$

$$+\sum_{\ell=2}^{r-1} \left\{ \sum_{i=p(\ell-1)+\ell-1}^{p(\ell)+\ell-2} \frac{(-1)^{n-1-i}}{i!} \left[i\frac{F_{i+1}(t)}{(a_\ell-t)^{i+1}} + \frac{F_{i+2}(t)}{(a_\ell-t)^{i+2}} \right] \right.$$

$$\left. +\frac{(-1)^{n-p(\ell)-\ell}}{(p(\ell)+\ell-1)!} \left[(p(\ell)+\ell-1)\frac{F_{p(\ell)+\ell}(t)}{(a_\ell-t)^{p(\ell)+\ell}} + \frac{F_{p(\ell)+\ell+1}(t)}{(a_{\ell+1}-t)^{p(\ell)+\ell+1}} \right] \right\}$$

$$+ \sum_{i=n-k_r-1}^{n-2} \frac{(-1)^{n-1-i}}{i!} \left[i\frac{F_{i+1}(t)}{(a_r-t)^{i+1}} + \frac{F_{i+2}(t)}{(a_r-t)^{i+2}} \right]$$

$$+\frac{1}{(n-1)!} \left[(n-1)\frac{F_n(t)}{(a_r-t)^n} + x^{(n)}(t) \right]$$

$$= \sum_{i=1}^{k_1-1} \frac{(-1)^{n-1-i}}{(i-1)!} \frac{F_{i+1}(t)}{(a_1-t)^{i+1}} + \frac{(-1)^{n-k_1}}{(k_1-1)!} \frac{F_{k_1+1}(t)}{(a_1-t)^{k_1+1}}$$

$$+\sum_{i=1}^{k_1-1} \frac{(-1)^{n-i}}{(i-1)!} \frac{F_{i+1}(t)}{(a_1-t)^{i+1}} + \frac{(-1)^{n-k_1-1}}{(k_1-1)!} \frac{F_{k_1+1}(t)}{(a_1-t)^{k_1+1}}$$

$$+\frac{(-1)^{n-k_1-1}}{k_1!}\frac{F_{k_1+2}(t)}{(a_2-t)^{k_1+2}}+\sum_{\ell=2}^{r-1}\left\{\sum_{i=p(\ell-1)+\ell}^{p(\ell)+\ell-2}\frac{(-1)^{n-1-i}}{(i-1)!}\frac{F_{i+1}(t)}{(a_\ell-t)^{i+1}}\right.$$

$$+\frac{(-1)^{n-p(\ell-1)-\ell}}{(p(\ell-1)+\ell-2)!}\frac{F_{p(\ell-1)+\ell}(t)}{(a_\ell-t)^{p(\ell-1)+\ell}}+\frac{(-1)^{n-p(\ell)-\ell+1}}{(p(\ell)+\ell-2)!}\frac{F_{p(\ell)+\ell}(t)}{(a_\ell-t)^{p(\ell)+\ell}}$$

$$+\sum_{i=p(\ell-1)+\ell}^{p(\ell)+\ell-2}\frac{(-1)^{n-i}}{(i-1)!}\frac{F_{i+1}(t)}{(a_\ell-t)^{i+1}}+\frac{(-1)^{n-p(\ell)-\ell}}{(p(\ell)+\ell-2)!}\frac{F_{p(\ell)+\ell}(t)}{(a_\ell-t)^{p(\ell)+\ell}}$$

$$\left.+\frac{(-1)^{n-p(\ell)-\ell}}{(p(\ell)+\ell-1)!}\frac{F_{p(\ell)+\ell+1}(t)}{(a_{\ell+1}-t)^{p(\ell)+\ell+1}}\right\}+\sum_{i=n-k_r}^{n-2}\frac{(-1)^{n-1-i}}{(i-1)!}\frac{F_{i+1}(t)}{(a_r-t)^{i+1}}$$

$$+\frac{(-1)^{k_r}}{(n-k_r-2)!}\frac{F_{n-k_r}(t)}{(a_r-t)^{n-k_r}}+\frac{(-1)^{-1}}{(n-2)!}\frac{F_n(t)}{(a_r-t)^n}$$

$$+\sum_{i=n-k_r}^{n-2}\frac{(-1)^{n-i}}{(i-1)!}\frac{F_{i+1}(t)}{(a_r-t)^{i+1}}+\frac{1}{(n-2)!}\frac{F_n(t)}{(a_r-t)^n}+\frac{1}{(n-1)!}x^{(n)}(t)$$

$$=\frac{(-1)^{n-k_1-1}}{k_1!}\frac{F_{k_1+2}(t)}{(a_2-t)^{k_1+2}}+\sum_{\ell=2}^{r-1}\left\{\frac{(-1)^{n-p(\ell-1)-\ell}}{(p(\ell-1)+\ell-2)!}\frac{F_{p(\ell-1)+\ell}(t)}{(a_\ell-t)^{p(\ell-1)+\ell}}\right.$$

$$\left.+\frac{(-1)^{n-p(\ell)-\ell}}{(p(\ell)+\ell-1)!}\frac{F_{p(\ell)+\ell+1}(t)}{(a_{\ell+1}-t)^{p(\ell)+\ell+1}}\right\}+\frac{(-1)^{k_r}}{(n-k_r-2)!}\frac{F_{n-k_r}(t)}{(a_r-t)^{n-k_r}}$$

$$+\frac{1}{(n-1)!}x^{(n)}(t)$$

$$=\frac{(-1)^{n-k_1-1}}{k_1!}\frac{F_{k_1+2}(t)}{(a_2-t)^{k_1+2}}+\frac{(-1)^{n-k_1-2}}{k_1!}\frac{F_{k_1+2}(t)}{(a_2-t)^{k_1+2}}$$

$$+\frac{(-1)^{n-p(r-1)-r+1}}{(p(r-1)+r-2)!}\frac{F_{p(r-1)+r}(t)}{(a_r-t)^{p(r-1)+r}}+\frac{(-1)^{k_r}}{(n-k_r-2)!}\frac{F_{n-k_r}(t)}{(a_r-t)^{n-k_r}}$$

$$+\frac{1}{(n-1)!}x^{(n)}(t)$$

$$=\frac{1}{(n-1)!}x^{(n)}(t). \qquad \blacksquare$$

Corollary 2.3.8. Let the functions $F_j(t)$, $1\le j\le n$ be recursively defined as follows

(2.3.27)
$$F_n(t) = \int_{a_2}^{t} (a_2 - t_n)^{n-1} x^{(n)}(t_n)\, dt_n$$
$$F_j(t) = \int_{a_2}^{t} (a_2 - t_j)^{-2} F_{j+1}(t_j)\, dt_j,\ m+1 \le j \le n-1$$
$$F_m(t) = \int_{a_1}^{t} \frac{(a_1 - t_m)^{m-1}}{(a_2 - t_m)^{m+1}} F_{m+1}(t_m)\, dt_m$$
$$F_j(t) = \int_{a_1}^{t} (a_1 - t_j)^{-2} F_{j+1}(t_j)\, dt_j,\ 1 \le j \le m-1.$$

Then, the error function $e_{(2.1.4)}(t)$ can be expressed as

(2.3.28)
$$e_{(2.1.4)}(t) = (a_2 - t)^{n-m}(a_1 - t)^{m-1} F_1(t). \quad ∎$$

Corollary 2.3.9. Let the functions $G_j(t),\ 1 \le j \le n$ be recursively defined as follows

(2.3.29)
$$G_n(t) = \int_{a_2}^{t} (a_2 - t_n)^{n-1} x^{(n)}(a_1 + a_2 - t_n)\, dt_n$$
$$G_j(t) = \int_{a_2}^{t} (a_2 - t_j)^{-2} G_{j+1}(t_j)\, dt_j,\ n-m+1 \le j \le n-1$$
$$G_{n-m}(t) = \int_{a_1}^{t} \frac{(a_1 - t_{n-m})^{n-m-1}}{(a_2 - t_{n-m})^{n-m+1}} G_{n-m+1}(t_{n-m})\, dt_{n-m}$$
$$G_j(t) = \int_{a_1}^{t} (a_1 - t_j)^{-2} G_{j+1}(t_j)\, dt_j,\ 1 \le j \le n-m-1.$$

Then, the error function $e_{(2.1.4)}(a_1 + a_2 - t)$ can be expressed as

(2.3.30)
$$e_{(2.1.4)}(a_1 + a_2 - t) = (a_2 - t)^m (a_1 - t)^{n-m-1} G_1(t). \quad ∎$$

Corollary 2.3.10. The error function $e_{(2.1.5)}(t)$ can be written as

(2.3.31)
$$e_{(2.1.5)}(t) = \frac{(t-a_2)^m}{((m-1)!)^2} \int_{a_1}^{a_2} \left[\int_{a_1}^{\min(s,t)} \frac{[(t-r)(s-r)]^{m-1}}{(a_2 - r)^{2m}} dr \right] \times (a_2 - s)^m x^{(2m)}(s)\, ds.$$

Proof. It suffices to note that in this case the functions $F_k(t),\ 1 \le k \le 2m$

defined in (2.3.19) can also be explicitly represented as

$$F_{2m-k+1}(t)$$

$$= \frac{(a_2-t)^{1-k}}{(k-1)!}\int_{a_2}^{t}(t-s)^{k-1}(a_2-s)^{2m-k}x^{(2m)}(s)\,ds,\ 1\le k\le m$$

$$= \frac{(a_1-t)^{1-k}}{(k-1)!}\int_{a_1}^{t}\frac{(t-s)^{k-1}}{(a_2-s)^{m+1}}(a_1-s)^{m-k}F_m(s)\,ds,\ m+1\le k\le 2m. \quad \blacksquare$$

2.4 ERROR ESTIMATES

For $p, \nu \ge 1$ we shall use the relation (2.3.11) to obtain upper bounds of $\| e_{(2.1.1)} \|_p$ in terms of $\| x^{(n)} \|_\nu$. These bounds depend on the distribution of the points a_1, ..., a_r, and also on the multiplicity of zeros of $e_{(2.1.1)}(t)$ at these points. Next, we shall employ the error representation (2.3.20) to find the pointwise as well as uniform upper bounds for $| e^{(k)}_{(2.1.1)}(t) |$, $0 \le k \le n-1$.

Error Estimates in Interpolation

Let μ and ν be conjugate exponents and $1 \le \mu \le \infty$. Using Hölder's inequality in (2.3.11) yields

$$| e_{(2.1.1)}(t) | \le \left(\int_{a_1}^{a_r} | g_{(2.1.1)}(t,s) |^{\mu}\, ds\right)^{\frac{1}{\mu}} \left(\int_{a_1}^{a_r} | x^{(n)}(s) |^{\nu}\, ds\right)^{\frac{1}{\nu}}$$

which, for $p \ge 1$, gives rise to

$$\int_{a_1}^{a_r} | e_{(2.1.1)}(t) |^p\, dt \le \left[\int_{a_1}^{a_r}\left(\int_{a_1}^{a_r} | g_{(2.1.1)}(t,s) |^{\mu}\, ds\right)^{\frac{p}{\mu}} dt\right] \left[\int_{a_1}^{a_r} | x^{(n)}(s) |^{\nu}\, ds\right]^{\frac{p}{\nu}}$$

and therefore

$$\| e_{(2.1.1)} \|_p \le \left[\int_{a_1}^{a_r}\left(\int_{a_1}^{a_r} | g_{(2.1.1)}(t,s) |^{\mu}\, ds\right)^{\frac{p}{\mu}} dt\right] \| x^{(n)} \|_\nu\ . \tag{2.4.1}$$

Since Green's function $g_{(2.1.1)}(t,s)$ is continuous in $[a_1, a_r] \times [a_1, a_r]$, (2.4.1) implies that for all extended real numbers p and ν not less than 1 it is always possible to find finite constants C such that $\| e_{(2.1.1)} \|_p \le C \| x^{(n)} \|_\nu$. In fact, in (2.4.1), Lemma 2.3.3 can be used to obtain

$$\| e_{(2.1.1)} \|_p \le \left[\int_{a_1}^{a_r} \left(\int_{a_1}^{a_r} | g_{(2.1.1)}(t,s) |^{\mu-1} | g_{(2.1.1)}(t,s) | \, ds \right)^{\frac{p}{\mu}} dt \right]^{\frac{1}{p}} \| x^{(n)} \|_\nu$$

$$\le \left[\int_{a_1}^{a_r} \left\{ \left(\frac{| \omega(t) |}{(n-1)!(a_r - a_1)} \right)^{\mu-1} \left(\frac{1}{n!} | \omega(t) | \right) \right\}^{\frac{p}{\mu}} dt \right]^{\frac{1}{p}} \| x^{(n)} \|_\nu$$

$$= \frac{1}{((n-1)!(a_r - a_1))^{1/\nu} (n!)^{1/\mu}} \left(\int_{a_1}^{a_r} | \omega(t) |^p \, dt \right)^{\frac{1}{p}} \| x^{(n)} \|_\nu . \tag{2.4.2}$$

Although, (2.4.2) suffers from the facts that : (i) inequality (2.3.12) is not always the best possible; (ii) (2.4.2) requires the splitting of $\int_{a_1}^{a_r} | g_{(2.1.1)}(t,s) |^\mu ds$ so that both (2.3.12) and (2.3.13) can be used; and (iii) it contains the term $\int_{a_1}^{a_r} | \omega(t) |^p \, dt$ which is unwieldy, it is possible to use it to cover several cases of frequent occurrence. For this, we need the following :

Lemma 2.4.1. For all integers $n \ge 2$ the following inequality holds

$$\frac{(n-1)!}{4} \le \frac{1}{2} \left(\frac{n-1}{2} \right)^{n-1} \le \left(\frac{n-1}{2} \right)^n . \tag{2.4.3}$$

Proof. Since $\frac{1}{2} \left(\frac{n-1}{2} \right)^{n-1} \le \left(\frac{n-1}{2} \right)^n$ is obvious, we need to prove only

$$\frac{(n-1)!}{4} \le \frac{1}{2} \left(\frac{n-1}{2} \right)^{n-1} . \tag{2.4.4}$$

In (2.4.4) equality holds for $n = 2$ and 3, and if it is true for $n = k + 1 \ge 3$, then we find that

$$\begin{aligned} \frac{1}{2} \left(\frac{k+1}{2} \right)^{k+1} &\ge \frac{1}{2^{k+2}} \left[k^{k+1} + (k+1)k^k + \frac{(k+1)k}{2} k^{k-1} \right] \\ &= \frac{1}{2} \left[k \, \frac{1}{2} \left(\frac{k}{2} \right)^k + (k+1) \frac{1}{2} \left(\frac{k}{2} \right)^k + \frac{(k+1)}{2} \frac{1}{2} \left(\frac{k}{2} \right)^k \right] \\ &\ge \frac{1}{2} \left[k + \frac{3}{2}(k+1) \right] \frac{k!}{4} \\ &= \frac{1}{2} \left(\frac{5k+3}{2} \right) \frac{k!}{4} \\ &> \frac{(k+1)!}{4} . \qquad \blacksquare \end{aligned}$$

Lemma 2.4.2. For all integers $n \geq 2$ the following equality holds

$$(2.4.5) \qquad \max_{1\leq i\leq n-1} \prod_{k=1}^{i-1}\left(i-k+\frac{i}{i+1}\right)\prod_{k=i+2}^{n}\left(k-i-\frac{1}{n-i+1}\right)$$

$$=\prod_{k=2}^{n-1}\left(k-\frac{1}{n}\right).$$

Proof. Since, for $1 \leq k \leq i-1$

$$i-k+\frac{i}{i+1}\leq i+1-k-\frac{1}{n}$$

we have

$$(2.4.6) \qquad \prod_{k=1}^{i-1}\left(i-k+\frac{i}{i+1}\right)\leq\prod_{k=1}^{i-1}\left(i+1-k-\frac{1}{n}\right)=\prod_{k=2}^{i}\left(k-\frac{1}{n}\right).$$

Further, since for $i+2 \leq k \leq n$

$$k-i-\frac{1}{n-i+1}\leq k-1-\frac{1}{n}$$

it follows that

$$(2.4.7) \qquad \prod_{k=i+2}^{n}\left(k-i-\frac{1}{n-i+1}\right)\leq\prod_{k=i+2}^{n}\left(k-1-\frac{1}{n}\right)=\prod_{k=i+1}^{n-1}\left(k-\frac{1}{n}\right).$$

From (2.4.6) and (2.4.7), equality (for $i = 1$) (2.4.5) follows. ∎

Case 1 *Equidistant Points* $r = n$, $a_{i+1}-a_i = h = \frac{a_n-a_1}{n-1}$, $1 \leq i \leq n-1$, $p = \infty$

From (2.2.3) we have

$$\omega(t)=\prod_{i=1}^{n}(t-a_i)=\prod_{i=0}^{n-1}(t-a_1-ih).$$

Let $\tau = \frac{t-a_1}{h}$, so that

$$\omega(t)=\omega(a_1+h\tau)=h^n\prod_{i=0}^{n-1}(\tau-i)$$

and hence

$$(2.4.8) \qquad \max_{a_1\leq t\leq a_n}|\,\omega(t)\,|\leq\frac{(a_n-a_1)^n}{(n-1)^n}\max_{0\leq\tau\leq n-1}\prod_{i=0}^{n-1}|\,\tau-i\,|.$$

To compute $\max\limits_{0\leq\tau\leq n-1}\prod\limits_{i=0}^{n-1}|\tau-i|$, we note that the function $\phi_n(\tau)=\prod\limits_{i=0}^{n-1}(\tau-i)$ defined on $[0,n-1]$ is even or odd relative to the point $\tau=\frac{n-1}{2}$ according to $(n-1)$ is odd or even. Further, since $\phi_n(\tau+1)=\frac{\tau+1}{\tau-n+1}\phi_n(\tau)$, the value of $\phi_n(\tau)$ in the interval $[i,i+1]$, $1\leq i\leq n-2$ is obtained from the corresponding value of $\phi_n(\tau)$ in the interval $[i-1,i]$ by multiplying it by $\frac{\tau+1}{\tau-n+1}$. Since $\left|\frac{\tau+1}{\tau-n+1}\right|\leq 1$, $\tau\in\left[0,\frac{n-1}{2}\right]$ the extreme value of $\phi_n(\tau)$ will diminish in absolute value as long as $\tau\in\left[0,\frac{n-1}{2}\right]$, and then will increase by symmetry. Thus, it follows that

$$\max_{0\leq\tau\leq n-1}|\phi_n(\tau)|=\max_{0\leq\tau\leq 1}\tau\prod_{i=1}^{n-1}(i-\tau).$$

In the interval $[0,1]$, we define $\psi_n(\tau)=\tau\prod\limits_{i=1}^{n-1}(i-\tau)$, so that

$$\psi_n'(\tau)=\psi_n(\tau)\left[\frac{1}{\tau}-\sum_{i=1}^{n-1}\frac{1}{i-\tau}\right]. \tag{2.4.9}$$

Let $\tau_n\in(0,1)$ be the unique zero of $\psi_n'(\tau)$, and τ_n^*, τ_n^{**} be any pair of points in (0,1) such that $\frac{1}{\tau_n^*}-\sum\limits_{i=1}^{n-1}\frac{1}{i-\tau_n^*}\geq 0$, and $\frac{1}{\tau_n^{**}}-\sum\limits_{i=1}^{n-1}\frac{1}{i-\tau_n^{**}}\leq 0$. Then, from (2.4.9) it obviously follows that $\tau_n^*\leq\tau_n\leq\tau_n^{**}$.

Since for $i\geq 2$, $-\frac{1}{i-\tau}\geq-\frac{1}{2-\tau}$, we have

$$\frac{1}{\tau}-\sum_{i=1}^{n-1}\frac{1}{i-\tau}\geq\frac{1}{\tau}-\frac{1}{1-\tau}-\frac{n-2}{2-\tau}=\frac{n\tau^2-(n+3)\tau+2}{\tau(1-\tau)(2-\tau)}.$$

Thus, for τ_n^* we can take

$$\tau_n^*=\frac{(n+3)-\sqrt{n^2-2n+9}}{2n}. \tag{2.4.10}$$

Similarly, since

$$\frac{1}{\tau}-\sum_{i=1}^{n-1}\frac{1}{i-\tau}\leq\frac{1}{\tau}-\frac{1}{1-\tau}-\frac{n-2}{n-1-\tau}=\frac{n\tau^2-3(n-1)\tau+n-1}{\tau(\tau-1)(n-1-\tau)}$$

for τ_n^{**} we can take

$$\tau_n^{**}=\frac{3(n-1)-\sqrt{5n^2-14n+9}}{2n}. \tag{2.4.11}$$

Finally, since

$$\max_{0\le\tau\le n-1}\prod_{i=0}^{n-1}|\tau-i| = \max_{0\le\tau\le 1}\tau\prod_{i=1}^{n-1}(i-\tau)$$

$$= \max_{\tau_n^*\le\tau\le\tau_n^{**}}\tau\prod_{i=1}^{n-1}(i-\tau)$$

$$\le \tau_n^{**}(1-\tau_n^{**})\prod_{i=2}^{n-1}(i-\tau_n^*)$$

from (2.4.8), (2.4.10) and (2.4.11) it follows that

$$\max_{a_1\le t\le a_n}|\omega(t)|\le(a_n-a_1)^n\alpha_n, \tag{2.4.12}$$

where

$$\alpha_n=\frac{1}{(n-1)^n}\left(\frac{3(n-1)-\sqrt{5n^2-14n+9}}{2n}\right)\times \tag{2.4.13}$$

$$\left(1-\frac{3(n-1)-\sqrt{5n^2-14n+9}}{2n}\right)\prod_{i=2}^{n-1}\left(i-\frac{(n+3)-\sqrt{n^2-2n+9}}{2n}\right).$$

Therefore, in view of (2.4.2) and (2.4.12), we find

$$\|e_{(2.1.1)}\|_\infty\le\frac{(a_n-a_1)^{n-1/\nu}}{((n-1)!)^{1/\nu}(n!)^{1/\mu}}\alpha_n\|x^{(n)}\|_\nu. \tag{2.4.14}$$

Remark 2.4.1. Since $\tau_n^{**}\le\frac{1}{2}$ and $\tau_n^*\ge\frac{1}{n}$, a simple but crude estimate from (2.4.14) immediately follows

$$\|e_{(2.1.1)}\|_\infty\le\frac{(a_n-a_1)^{n-1/\nu}}{((n-1)!)^{1/\nu}(n!)^{1/\mu}}\frac{1}{4(n-1)^n}\prod_{i=2}^{n-1}\left(i-\frac{1}{n}\right)\|x^{(n)}\|_\nu. \tag{2.4.15}$$

For $\nu=\infty$, (2.4.15) is the same as

$$\|e_{(2.1.1)}\|_\infty\le\frac{(a_n-a_1)^n}{n!}\frac{1}{4(n-1)^n}\prod_{i=2}^{n-1}\left(i-\frac{1}{n}\right)\|x^{(n)}\|_\infty \tag{2.4.16}$$

which is obviously better than the known, e.g., Yakowitz and Szidarovszky [44], error bound

$$\|e_{(2.1.1)}\|_\infty\le\frac{(a_n-a_1)^n}{4(n-1)^n n}\|x^{(n)}\|_\infty. \tag{2.4.17}$$

Further, from Lemma 2.4.1 it is clear that (2.4.17) is better than

$$\| e_{(2.1.1)} \|_\infty \leq \frac{(a_n - a_1)^n}{2^n n!} \| x^{(n)} \|_\infty \tag{2.4.18}$$

which is easily obtained by using the arithmetic-geometric mean inequality, e.g., Phillips and Taylor [35]. ∎

Remark 2.4.2. From the relation $\phi_{n+1}(\tau) = (\tau - n)\phi_n(\tau)$, we have $\phi'_{n+1}(\tau_n) = \phi_n(\tau_n) > 0$, i.e., $\{\tau_n\}$ is a decreasing sequence, and in view of (2.4.9), $\tau_n \to 0$. The lower estimate τ_n^* of τ_n in (2.4.10) also decreases monotonically to 0 as it should, however the upper estimate τ_n^{**} of τ_n in (2.4.11) decreases monotonically to only $\frac{3-\sqrt{5}}{2}$. For large values of n, we can use the following well known inequality

$$\ln n \leq \sum_{i=1}^{n-1} \frac{1}{i}$$

so that

$$\frac{1}{\tau} - \sum_{i=1}^{n-1} \frac{1}{i - \tau} \leq \frac{1}{\tau} - \ln(n-1).$$

Thus, for large values of n (say ≥ 15), we can choose

$$\tau_n^{**} = \frac{1}{\ln(n-1)}. \tag{2.4.19}$$

Finally, we can combine (2.4.11) and (2.4.19), to get

$$\tau_n^{**} = \min\left\{ \frac{3(n-1) - \sqrt{5n^2 - 14n + 9}}{2n}, \frac{1}{\ln(n-1)} \right\}. \tag{2.4.20}$$ ∎

Remark 2.4.3. For $\nu = \infty$ the estimate (2.4.14) is the best possible for $n = 2$ and 3, and for $n \geq 4$ compares favorably with the best possible result obtained by using the computer. ∎

Case 2 *Simple Points* $r = n,\ p = \infty$

We assume no condition on the distribution of the points a_i's, except that $a \leq a_1 < a_2 < \dots < a_n \leq b$. We begin with the observation that

$$\max_{a_1 \leq t \leq a_n} | \omega(t) | = \max_{1 \leq i \leq n-1} \max_{a_i \leq t \leq a_{i+1}} \phi_{i,n}(t), \tag{2.4.21}$$

where

$$\phi_{i,n}(t) = \prod_{k=1}^{i}(t-a_k)\prod_{k=i+1}^{n}(a_k-t).$$

Since in the interval $[a_i, a_{i+1}]$

$$\phi'_{i,n}(t) = \phi_{i,n}(t)\left[\sum_{k=1}^{i}\frac{1}{t-a_k} - \sum_{k=i+1}^{n}\frac{1}{a_k-t}\right] \tag{2.4.22}$$

if $\tau_{i,n} \in (a_i, a_{i+1})$ is the unique zero of $\phi'_{i,n}(t)$, then as in Case 1 it suffices to find any pair of points $\tau^*_{i,n}$, $\tau^{**}_{i,n}$ in (a_i, a_{i+1}) such that $\sum_{k=1}^{i}\frac{1}{\tau^*_{i,n}-a_k} - \sum_{k=i+1}^{n}\frac{1}{a_k-\tau^*_{i,n}} \geq 0$ and $\sum_{k=1}^{i}\frac{1}{\tau^{**}_{i,n}-a_k} - \sum_{k=i+1}^{n}\frac{1}{a_k-\tau^{**}_{i,n}} \leq 0$. Then, it will follow that $\tau^*_{i,n} \leq \tau_{i,n} \leq \tau^{**}_{i,n}$.

Since for $k \geq i+1$, $-\frac{1}{a_k-t} \geq -\frac{1}{a_{i+1}-t}$ we find that

$$\sum_{k=1}^{i}\frac{1}{t-a_k} - \sum_{k=i+1}^{n}\frac{1}{a_k-t} \geq \frac{1}{t-a_i} - \frac{n-i}{a_{i+1}-t} = \frac{a_{i+1}+(n-i)a_i-(n-i+1)t}{(t-a_i)(a_{i+1}-t)}.$$

Thus, for $\tau^*_{i,n}$ we can take

$$\tau^*_{i,n} = \frac{a_{i+1}+(n-i)a_i}{n-i+1}. \tag{2.4.23}$$

Similarly, for $k \leq i$, $\frac{1}{t-a_k} \leq \frac{1}{t-a_i}$ and therefore

$$\sum_{k=1}^{i}\frac{1}{t-a_k} - \sum_{k=i+1}^{n}\frac{1}{a_k-t} \leq \frac{i}{t-a_i} - \frac{1}{a_{i+1}-t} = \frac{i\,a_{i+1}+a_i-(i+1)t}{(t-a_i)(a_{i+1}-t)}.$$

Hence, $\tau^{**}_{i,n}$ can be taken as

$$\tau^{**}_{i,n} = \frac{i\,a_{i+1}+a_i}{i+1}. \tag{2.4.24}$$

Further, since $\max_{a_i\leq t\leq a_{i+1}}(t-a_i)(a_{i+1}-t) \leq \frac{1}{4}(a_{i+1}-a_i)^2$, it follows that

$$\max_{a_i\leq t\leq a_{i+1}}\phi_{i,n}(t) = \max_{\tau^*_{i,n}\leq t\leq \tau^{**}_{i,n}}\phi_{i,n}(t)$$

$$\leq \frac{1}{4}(a_{i+1}-a_i)^2\prod_{k=1}^{i-1}(\tau^{**}_{i,n}-a_k)\prod_{k=i+2}^{n}(a_k-\tau^*_{i,n}). \tag{2.4.25}$$

If $\delta = \max_{1\le i\le n-1}(a_{i+1}-a_i)$, then for $1\le k\le i-1$ we find that

$$\begin{aligned} \tau_{i,n}^{**}-a_k &= \frac{i(a_{i+1}-a_k)+(a_i-a_k)}{i+1} \\ &\le \frac{i(i+1-k)+(i-k)}{i+1}\delta \\ &= \left(i-k+\frac{i}{i+1}\right)\delta. \end{aligned} \tag{2.4.26}$$

Likewise, for $i+2\le k\le n$ we have

$$\begin{aligned} a_k-\tau_{i,n}^{*} &= \frac{(a_k-a_{i+1})+(n-i)(a_k-a_i)}{n-i+1} \\ &\le \frac{(k-i-1)+(n-i)(k-i)}{n-i+1}\delta \\ &= \left(k-i-\frac{1}{n-i+1}\right)\delta. \end{aligned} \tag{2.4.27}$$

Using (2.4.26) and (2.4.27) in (2.4.25), to obtain

$$\max_{1\le i\le n-1}\max_{a_i\le t\le a_{i+1}}\phi_{i,n}(t)\le\frac{1}{4}\delta^n\max_{1\le i\le n-1}\prod_{k=1}^{i-1}\left(i-k+\frac{i}{i+1}\right)\times$$
$$\prod_{k=i+2}^{n}\left(k-i-\frac{1}{n-i+1}\right).$$

Now an application of Lemma 2.4.2 and (2.4.21) gives

$$\max_{a_1\le t\le a_n}|\,\omega(t)\,|\le\frac{1}{4}\delta^n\prod_{k=2}^{n-1}\left(k-\frac{1}{n}\right). \tag{2.4.28}$$

Finally, using (2.4.28) in (2.4.2) it follows that

$$\|\,e_{(2.1.1)}\,\|_\infty\le\frac{1}{((n-1)!(a_r-a_1))^{1/\nu}(n!)^{1/\mu}}\frac{1}{4}\delta^n\prod_{k=2}^{n-1}\left(k-\frac{1}{n}\right)\|\,x^{(n)}\,\|_\nu\,. \tag{2.4.29}$$

Remark 2.4.4. If $\delta=\frac{a_r-a_1}{n-1}$, i.e., the case of equidistant points, then (2.4.29) reduces to (2.4.15). ∎

Remark 2.4.5. Using the arithmetic - geometric mean inequality, Das and Vatsala [20] have obtained

$$\max_{a_1\le t\le a_n}|\,\omega(t)\,|\le(n-1)^{n-1}\left(\frac{\delta}{2}\right)^n. \tag{2.4.30}$$

Since $\frac{1}{4}\prod_{k=2}^{n-1}\left(k-\frac{1}{n}\right) \le \frac{(n-1)!}{4}$ (with equality only for $n = 2$), the estimate (2.4.28) is sharper than (2.4.30) follows from Lemma 2.4.1. ∎

Case 3 *Multiple Points* $p = \infty$

Let $\alpha = \min\{k_1, k_r\}$, $\beta = \min_{2\le i\le r-1}\{k_i\}$, $\gamma = \min\{\alpha, \beta\}$ and $\delta = \max_{1\le i\le r-1}(a_{i+1} - a_i)$. We shall begin with the following

$$\text{(2.4.31)} \qquad \max_{a_1\le t\le a_r} |\,\omega(t)\,| = \max_{1\le i\le r-1}\ \max_{a_i\le t\le a_{i+1}} \phi_{i,r}(t),$$

where

$$\phi_{i,r}(t) = \prod_{\ell=1}^{i}(t-a_\ell)^{k_\ell+1}\prod_{\ell=i+1}^{r}(a_\ell - t)^{k_\ell+1}.$$

Since in the interval $[a_i, a_{i+1}]$

$$\text{(2.4.32)} \quad \phi_{i,r}(t) \le \begin{cases} (t-a_1)^{n-\alpha-(\beta+1)(r-2)-1}\prod_{\ell=2}^{i}(t-a_\ell)^{\beta+1}\prod_{\ell=i+1}^{r-1}(a_\ell-t)^{\beta+1}\times \\ \qquad (a_r-t)^{\alpha+1} = g_{i,r}(t), \qquad t-a_1 \ge a_r - t \\ (t-a_1)^{\alpha+1}\prod_{\ell=2}^{i}(t-a_\ell)^{\beta+1}\prod_{\ell=i+1}^{r-1}(a_\ell-t)^{\beta+1}\times \\ \qquad (a_r-t)^{n-\alpha-(\beta+1)(r-2)-1} = h_{i,r}(t), \quad a_r - t \ge t - a_1 \end{cases}$$

from (2.4.31) it follows that

$$\text{(2.4.33)} \qquad \max_{a_1\le t\le a_r} |\,\omega(t)\,| \le \max_{1\le i\le r-1}\ \max_{a_i\le t\le a_{i+1}} \begin{cases} g_{i,r}(t), & t-a_1 \ge a_r - t \\ h_{i,r}(t), & a_r - t \ge t - a_1. \end{cases}$$

Once again, in the interval $[a_i, a_{i+1}]$, if $t - a_1 \ge a_r - t$ then

$$g'_{i,r}(t) = g_{i,r}(t)\left[\frac{n-\alpha-(\beta+1)(r-2)-1}{t-a_1} + \sum_{\ell=2}^{i}\frac{\beta+1}{t-a_\ell} - \sum_{\ell=i+1}^{r-1}\frac{\beta+1}{a_\ell - t} - \frac{\alpha+1}{a_r - t} = \theta_{i,r}(t)\right]$$

and hence, if $\tau_{i,r} \in (a_i, a_{i+1})$ is the unique zero of $g'_{i,r}(t)$, then as earlier it suffices to find any pair of points $\tau^*_{i,r}$, $\tau^{**}_{i,r}$ in (a_i, a_{i+1}) such that $\theta_{i,r}(\tau^*_{i,r}) \ge 0$ and $\theta_{i,r}(\tau^{**}_{i,r}) \le 0$.

Since $\beta \geq \gamma$ and $n - \alpha - (\beta+1)(r-2) - 1 \geq \gamma + 1$, it follows that

$$\theta_{i,r}(t) \geq \frac{\gamma+1}{t-a_i} - \frac{(r-i-1)(\beta+1)+\alpha+1}{a_{i+1}-t}$$

and hence $\tau^*_{i,r}$ can be taken as

$$\tau^*_{i,r} = a_i + \frac{\gamma+1}{(r-i-1)(\beta+1)+\gamma+1+\alpha+1}(a_{i+1}-a_i). \tag{2.4.34}$$

Similarly, we find that

$$\theta_{i,r}(t) \leq \frac{n-\alpha-(\beta+1)(r-2)-1+(i-1)(\beta+1)}{t-a_i} - \frac{\gamma+1}{a_{i+1}-t}$$

which provides that

$$\tau^{**}_{i,r} = a_{i+1} - \frac{\gamma+1}{n-\alpha-(\beta+1)(r-2)-1+(i-1)(\beta+1)+\gamma+1}\times (a_{i+1}-a_i). \tag{2.4.35}$$

Thus, for $t - a_1 \geq a_r - t$ in view of (2.4.31), (2.4.33), (2.4.34) and (2.4.35) it follows that

$$\begin{aligned} \max_{a_1 \leq t \leq a_r} |\, \omega(t) \,| &\leq \max_{1 \leq i \leq r-1} \max_{\tau^*_{i,r} \leq t \leq \tau^{**}_{i,r}} g_{i,r}(t) \\ &\leq \max_{a_1 \leq t \leq a_r} (t-a_1)^{n-\alpha-(\beta+1)(r-2)-1}(a_r-t)^{\alpha+1} \times \\ &\qquad \max_{1 \leq i \leq r-1} \left[\prod_{\ell=2}^{i} (\tau^{**}_{i,r} - a_\ell) \prod_{\ell=i+1}^{r-1} (a_\ell - \tau^*_{i,r}) \right]^{\beta+1}. \end{aligned} \tag{2.4.36}$$

For $2 \leq \ell \leq i$, from (2.4.35) we get

$$\begin{aligned} &\tau^{**}_{i,r} - a_\ell \\ &= \frac{(a_{i+1}-a_\ell)[n-\alpha-(\beta+1)(r-2)-1+(i-1)(\beta+1)] + (\gamma+1)(a_i-a_\ell)}{n-\alpha-(\beta+1)(r-2)-1+(i-1)(\beta+1)+\gamma+1} \\ &\leq \frac{(i+1-\ell)[n-\alpha-(\beta+1)(r-2)-1+(i-1)(\beta+1)] + (\gamma+1)(i-\ell)}{n-\alpha-(\beta+1)(r-2)-1+(i-1)(\beta+1)+\gamma+1}\delta \\ &= \left[i+1-\ell - \frac{\gamma+1}{n-\alpha-(\beta+1)(r-2)-1+(i-1)(\beta+1)+\gamma+1} \right] \delta \end{aligned}$$

$$\leq \left(i+1-\ell - \frac{\gamma+1}{n-\alpha+\gamma} \right) \delta. \tag{2.4.37}$$

Likewise, for $i+1 \leq \ell \leq r-1$ it follows from (2.4.34) that

$$\begin{aligned}
a_\ell - \tau^*_{i,r} &= \frac{(a_\ell - a_i)[(r-i-1)(\beta+1)+\alpha+1] + (\gamma+1)(a_\ell - a_{i+1})}{(r-i-1)(\beta+1)+\alpha+1+\gamma+1} \\
&\leq \frac{(\ell-i)[(r-i-1)(\beta+1)+\alpha+1] + (\gamma+1)(\ell-i-1)}{(r-i-1)(\beta+1)+\alpha+1+\gamma+1}\delta \\
&= \left[\ell - i - \frac{\gamma+1}{(r-i-1)(\beta+1)+\alpha+1+\gamma+1}\right]\delta
\end{aligned}$$

$$\leq \left(\ell - i - \frac{\gamma+1}{n-\alpha+\gamma}\right)\delta. \tag{2.4.38}$$

A combination of (2.4.37) and (2.4.38) yields

$$\begin{aligned}
&\max_{1\leq i\leq r-1} \prod_{\ell=2}^{i} (\tau^{**}_{i,r} - a_\ell) \prod_{\ell=i+1}^{r-1} (a_\ell - \tau^*_{i,r}) \\
&\leq \max_{1\leq i\leq r-1} \prod_{\ell=1}^{i-1} \left(\ell - \frac{\gamma+1}{n-\alpha+\gamma}\right) \prod_{\ell=i+1}^{r-1} \left(\ell - 1 - \frac{\gamma+1}{n-\alpha+\gamma}\right)\delta^{r-2} \\
&= \max_{1\leq i\leq r-1} \prod_{\ell=1}^{i-1} \left(\ell - \frac{\gamma+1}{n-\alpha+\gamma}\right) \prod_{\ell=i}^{r-2} \left(\ell - \frac{\gamma+1}{n-\alpha+\gamma}\right)\delta^{r-2}
\end{aligned}$$

$$= \prod_{\ell=1}^{r-2} \left(\ell - \frac{\gamma+1}{n-\alpha+\gamma}\right)\delta^{r-2}. \tag{2.4.39}$$

We also need the equality

$$\begin{aligned}
&\max_{a_1\leq t\leq a_r} (t-a_1)^{n-\alpha-(\beta+1)(r-2)-1}(a_r-t)^{\alpha+1} \\
&= \frac{(n-\alpha-(\beta+1)(r-2)-1)^{n-\alpha-(\beta+1)(r-2)-1}(\alpha+1)^{\alpha+1}}{(n-(\beta+1)(r-2))^{n-(\beta+1)(r-2)}} \times \\
&\qquad (a_r - a_1)^{n-(\beta+1)(r-2)}.
\end{aligned} \tag{2.4.40}$$

Using (2.4.39) and (2.4.40) in (2.4.36), we obtain

$$\max_{a_1\leq t\leq a_r} |\,\omega(t)\,| \leq C(n,r,\alpha,\beta,\gamma), \; t - a_1 \geq a_r - t \tag{2.4.41}$$

where

(2.4.42) $C(n,r,\alpha,\beta,\gamma)$

$$= \frac{(n-\alpha-(\beta+1)(r-2)-1)^{n-\alpha-(\beta+1)(r-2)-1}(\alpha+1)^{\alpha+1}}{(n-(\beta+1)(r-2))^{n-(\beta+1)(r-2)}} \times$$

$$\prod_{\ell=1}^{r-2}\left(\ell - \frac{\gamma+1}{n-\alpha+\gamma}\right)^{\beta+1} (a_r - a_1)^{n-(\beta+1)(r-2)}\delta^{(\beta+1)(r-2)}.$$

A similar analysis for the case $t - a_1 \le a_r - t$ also gives

(2.4.43) $$\max_{a_1 \le t \le a_r} |\, \omega(t) \,| \le \max_{1\le i\le r-1} \max_{a_i \le t \le a_{i+1}} h_{i,r}(t) \le C(n,r,\alpha,\beta,\gamma).$$

Hence, on coupling (2.4.41) and (2.4.43), we get

(2.4.44) $$\max_{a_1 \le t \le a_r} |\, \omega(t) \,| \le C(n,r,\alpha,\beta,\gamma).$$

Finally, using (2.4.44) in (2.4.2), we find that

(2.4.45) $$\| \, e_{(2.1.1)} \, \|_\infty \le \frac{1}{((n-1)!(a_r-a_1))^{1/\nu}(n!)^{1/\mu}} C(n,r,\alpha,\beta,\gamma) \, \| \, x^{(n)} \, \|_\nu \, .$$

Remark 2.4.6. If $r = 2$, then (2.4.45) reduces to

(2.4.46) $$\| \, e_{(2.1.1)} \, \|_\infty \le \frac{(n-\alpha-1)^{n-\alpha-1}(\alpha+1)^{\alpha+1}}{((n-1)!(a_r-a_1))^{1/\nu}(n!)^{1/\mu}n^n}(a_r-a_1)^{n-\frac{1}{\nu}} \, \| \, x^{(n)} \, \|_\nu,$$

which is obviously the best possible for $\nu = \infty$ (equality holds for the functions $x_3(t) = (t-a_1)^{n-\alpha-1}(a_r-t)^{\alpha+1}$ and $x_4(t) = (t-a_1)^{\alpha+1}(a_r-t)^{n-\alpha-1}$), and same or sharper than obtained or can be directly deduced from the earlier work in [2,15,32]. ∎

Remark 2.4.7. The estimate (2.4.46) is in fact true for all $2 \le r \le n$. For this, it suffices to note that

(2.4.47) $$g_{i,r}(t) \le (t-a_1)^{n-\alpha-1}(a_r-t)^{\alpha+1}$$

and

(2.4.48) $$h_{i,r}(t) \le (t-a_1)^{\alpha+1}(a_r-t)^{n-\alpha-1}.$$

However, (2.4.45) and (2.4.46) are noncomparable. For example, if $n = 4$, $r = 3$, $\alpha = 0$, $\beta = 1$, $\gamma = 0$ and $\nu = \infty$, then (2.4.45) reduces to

(2.4.49) $$\| \, e_{(2.1.1)} \, \|_\infty \le \frac{3}{512}(a_3-a_1)^2\delta^2 \, \| \, x^{(4)} \, \|_\infty$$

whereas (2.4.46) gives

$$\| e_{(2.1.1)} \|_\infty \leq \frac{9}{2048}(a_3 - a_1)^4 \| x^{(4)} \|_\infty . \tag{2.4.50}$$

Thus, if $\delta < \frac{\sqrt{3}}{2}(a_3 - a_1)$ then (2.4.49) is better than (2.4.50), otherwise (2.4.50) is sharper than (2.4.49). ∎

Remark 2.4.8. If $r = n$, i.e., the case of simple points, then $\alpha = \beta = \gamma = 0$ and (2.4.45) reduces to

$$\| e_{(2.1.1)} \|_\infty \leq \frac{1}{((n-1)!(a_r - a_1))^{1/\nu}(n!)^{1/\mu}} \frac{1}{4} \prod_{\ell=1}^{n-2} \left(\ell - \frac{1}{n}\right) \times (a_n - a_1)^2 \delta^{n-2} \| x^{(n)} \|_\nu . \tag{2.4.51}$$

A comparison between (2.4.29) and (2.4.51) shows that if $\delta < \sqrt{\frac{n-1}{n^2-n-1}}(a_r - a_1)$ then (2.4.29) is sharper than (2.4.51), otherwise (2.4.51) is better than (2.4.29). ∎

Case 4 $p \geq 1$

We observe that $\left(\int_{a_1}^{a_r} | \omega(t) |^p \, dt\right)^{1/p} \leq \max_{a_1 \leq t \leq a_r} | \omega(t) | (a_r - a_1)^{1/p}$, therefore from (2.4.2) it is clear that if the factor $(a_r - a_1)^{1/p}$ is multiplied to the right side of (2.4.12), (2.4.28) and (2.4.44) then the left side of the resulting inequalities are replaced by $\| e_{(2.1.1)} \|_p$, i.e., the error estimates for the Cases 1 - 3 when $p \geq 1$ are readily available. However, we shall now obtain another estimate which is best possible in certain particular cases. For this, (2.4.47), (2.4.48) and a similar analysis of Case 3 provide that

$$\begin{aligned}\left(\int_{a_1}^{a_r} | \omega(t) |^p \, dt\right)^{\frac{1}{p}} &\leq \left(\int_{a_1}^{(a_1+a_r)/2} (t - a_1)^{p(\alpha+1)}(a_r - t)^{p(n-\alpha-1)} \, dt\right. \\ &\quad \left. + \int_{(a_1+a_r)/2}^{a_r} (t - a_1)^{p(n-\alpha-1)}(a_r - t)^{p(\alpha+1)} \, dt\right)^{\frac{1}{p}} \\ &= (a_r - a_1)^{n+\frac{1}{p}} \left(\int_0^{\frac{1}{2}} t^{p(\alpha+1)}(1-t)^{p(n-\alpha-1)} \, dt\right. \\ &\quad \left. + \int_{\frac{1}{2}}^1 t^{p(n-\alpha-1)}(1-t)^{p(\alpha+1)} \, dt\right)^{\frac{1}{p}}\end{aligned}$$

$$(2.4.52) \qquad = (a_r - a_1)^{n+\frac{1}{p}} \left(2B_{\frac{1}{2}}(p(\alpha+1)+1, p(n-\alpha-1)+1)\right)^{\frac{1}{p}},$$

where $B_{\frac{1}{2}}$ is the incomplete Beta function.

Using (2.4.52) in (2.4.2), we get

$$(2.4.53) \qquad \| e_{(2.1.1)} \|_p \leq \frac{1}{((n-1)!)^{1/\nu}(n!)^{1/\mu}} (a_r - a_1)^{n+\frac{1}{p}-\frac{1}{\nu}} \times$$

$$\left(2B_{\frac{1}{2}}(p(\alpha+1)+1, p(n-\alpha-1)+1)\right)^{\frac{1}{p}} \| x^{(n)} \|_\nu .$$

Remark 2.4.9. If $r = 2$, then $| \omega(t) | = (t-a_1)^{k_1+1}(a_2-t)^{k_2+1}$. Since $k_1 + k_2 + 2 = n$, we find that

$$\begin{aligned} \int_{a_1}^{a_2} | \omega(t) |^p \, dt &= \int_{a_1}^{a_2} (t-a_1)^{p(k_1+1)}(a_2-t)^{p(k_2+1)} \, dt \\ &= (a_2 - a_1)^{pn+1} B(p(k_1+1)+1, p(k_2+1)+1) \end{aligned}$$

$$(2.4.54) \qquad = (a_2 - a_1)^{pn+1} B(p(\alpha+1)+1, p(n-\alpha-1)+1).$$

Using (2.4.54) in (2.4.2) yields

$$(2.4.55) \qquad \| e_{(2.1.1)} \|_p \leq \frac{1}{((n-1)!)^{1/\nu}(n!)^{1/\mu}} (a_2 - a_1)^{n+\frac{1}{p}-\frac{1}{\nu}} \times$$

$$(B(p(\alpha+1)+1, p(n-\alpha-1)+1))^{\frac{1}{p}} \| x^{(n)} \|_\nu .$$

Unfortunately, in general for $r = 2$, (2.4.53) does not reduce to (2.4.55). In fact for $\nu = \infty$, (2.4.55) is the best possible. Further, for $p = 1$, $\nu = \infty$ it reduces to

$$(2.4.56) \qquad \| e_{(2.1.1)} \|_1 \leq \frac{(\alpha+1)!(n-\alpha-1)!}{n!(n+1)!} (a_2 - a_1)^{n+1} \| x^{(n)} \|_\infty . \quad \blacksquare$$

Remark 2.4.10. For the two - point Taylor conditions both the inequalities (2.4.53) and (2.4.55) for $p = 1$, $\nu = \infty$ reduce to

$$(2.4.57) \qquad \| e_{(2.1.5)} \|_1 \leq \frac{(m!)^2}{(2m)!(2m+1)!} (a_2 - a_1)^{2m+1} \| x^{(2m)} \|_\infty,$$

which is sharper than Anon's problem [8], and its solution in [41], also it includes Zaidman's problem [45], as well as Iyengar and Mahajani's well known inequalities, e.g., Mitrinović [33]. ∎

Error Estimates For Derivatives

From Rolle's theorem it is clear that for each $0 \le k \le n-1$ the function $e^{(k)}_{(2.1.1)}(t)$ has at least $(n-k)$ zeros in (a,b). Therefore, from Theorem 2.3.1 the following result is immediate.

Theorem 2.4.3. For the error function $e_{(2.1.1)}(t)$ the following inequalities hold

$$(2.4.58)\quad | e^{(k)}_{(2.1.1)}(t) | \le \frac{1}{(n-k)!}(b-a)^{n-k} \max_{a\le t\le b} | x^{(n)}(t) |,\ 0 \le k \le n-1. \quad \blacksquare$$

The constants $\frac{1}{(n-k)!}$ in the above inequalities, obviously, are the best possible in the sense that these constants cannot be replaced by smaller ones. However, if we consider the case $a_1 = a,\ a_r = b$ which corresponds to the interpolation in the exact sense of the word, then the inequalities (2.4.58) can be improved.

Theorem 2.4.4. For the error function $e_{(2.1.1)}(t)$ the following inequalities hold

$$(2.4.59)\quad | e^{(k)}_{(2.1.1)}(t) | \le A_{n,k}(a_r - a_1)^{n-k} \max_{a_1\le t\le a_r} | x^{(n)}(t) |,\ 0 \le k \le n-1$$

where

$$(2.4.60)\quad A_{n,0} = \frac{1}{n!}\frac{(n-1)^{n-1}}{n^n},\ A_{n,k} = \frac{k}{n\,(n-k)!},\ 1 \le k \le n-1. \quad \blacksquare$$

Hukuhara [26] indicates that Tumura [42] proved Theorem 2.4.4. This result has also been mentioned in [1,3,11,12,15,27]. These constants $A_{n,k},\ 0 \le k \le n-1$ are optimal as they are exact for the functions $x_1(t) = (t-a_1)^{n-1}(a_r - t)$ and $x_2(t) = (t-a_1)(a_r-t)^{n-1}$ and only for these functions up to a constant factor.

By introducing the parameter $\alpha = \min\{k_1, k_r\}$ and using the error representation (2.3.11) in [1] Theorem 2.4.4. has been improved. The new constants $A_{n,k}$ in (2.4.59), say, $B_{n,k}$ appear as

$$(2.4.61)\quad B_{n,k} = \begin{cases} \dfrac{(n-\alpha-1)^{n-\alpha-1}(\alpha-k+1)^{\alpha-k+1}}{(n-k)^{n-k}(n-k)!}, & 0 \le k \le \alpha \\ \\ \dfrac{k-\alpha}{(n-\alpha)\,(n-k)!}, & \alpha+1 \le k \le n-1. \end{cases}$$

Obviously, for $\alpha = 0$ the constants $B_{n,k}$ reduce to $A_{n,k}$, $0 \leq k \leq n-1$. Further, the constant $B_{n,0}$ is the best possible as it is exact for the functions $x_3(t) = (t-a_1)^{n-\alpha-1}(a_r - t)^{\alpha+1}$ and $x_4(t) = (t-a_1)^{\alpha+1}(a_r-t)^{n-\alpha-1}$ and only for these functions up to a constant factor.

To acquire pointwise as well as uniform bounds for $| e^{(k)}_{(2.1.1)}(t) |$, $0 \leq k \leq n-1$ the new error representation developed in Theorem 2.3.7 is more helpful than other representations. In conclusion we shall show that the obtained bounds compare sharply with several other known results, more so some of these new bounds are also the best possible. For this, we need the following :

Lemma 2.4.5. (Pascal's triangle)

$$\sum_{i=0}^{k} \binom{n-i}{k-i} = \binom{n+1}{k}. \quad \blacksquare$$

Lemma 2.4.6. (Vandermonde's identity)

$$\sum_{i=0}^{k} \binom{n-\alpha-1}{k-i} \binom{\alpha+1}{i} = \binom{n}{k}. \quad \blacksquare$$

Lemma 2.4.7.

$$\sum_{j=\max\{k-\beta,0\}}^{\min\{\alpha,k\}} \binom{\beta}{k-j} \binom{\alpha}{j} = \binom{\alpha+\beta}{k}.$$

Proof. We have

$$\begin{aligned} & D^k[(1+t)^\alpha(1+t)^\beta] \\ &= \sum_{j=\max\{k-\beta,0\}}^{\min\{\alpha,k\}} \binom{k}{j} [D^j(1+t)^\alpha][D^{k-j}(1+t)^\beta] \\ &= k! \sum_{j=\max\{k-\beta,0\}}^{\min\{\alpha,k\}} \binom{\beta}{k-j} \binom{\alpha}{j} (1+t)^{\alpha-j}(1+t)^{\beta-k+j}. \end{aligned}$$

On the other hand, it is clear that

$$D^k[(1+t)^{\alpha+\beta}] = \frac{(\alpha+\beta)!}{(\alpha+\beta-k)!}(1+t)^{\alpha+\beta-k}.$$

Equating the above relations and substituting $t = 0$ lead to the required result. ∎

Lemma 2.4.8. For the functions $F_k(t)$, $1 \le k \le n$ defined in Theorem 2.3.7 the following inequalities hold

$$| F_k(t) | \le \frac{M_n(k-1)!}{n!} | t - a_{s+1} |^k; \quad p(s) + s + 1 \le k \le p(s+1) + s + 1,$$

$$0 \le s \le r - 1$$

where $M_n = \max_{a_1 \le t \le a_r} | x^{(n)}(t) |$.

Proof. It is clear that

$$| F_n(t) | \le M_n \left| \int_{a_r}^{t} (a_r - t_n)^{n-1} \, dt_n \right| = \frac{M_n(n-1)!}{n!}(a_r - t)^n.$$

Using the above inequality, we find

$$\begin{aligned} | F_{n-1}(t) | &= \left| \int_{a_r}^{t} (a_r - t_{n-1})^{-2} F_n(t_{n-1}) \, dt_{n-1} \right| \\ &\le \frac{M_n}{n} \left| \int_{a_r}^{t} (a_r - t_{n-1})^{n-2} \, dt_{n-1} \right| \\ &= \frac{M_n(n-2)!}{n!}(a_r - t)^{n-1}. \end{aligned}$$

It now follows inductively that

$$| F_k(t) | \le \frac{M_n(k-1)!}{n!}(a_r - t)^k, \; p(r-1) + r \le k \le n$$

i.e., the result follows for $s = r-1$. The proof is similar when $0 \le s \le r-2$. ∎

Theorem 2.4.9. If $(t - a_1) \ge (a_r - t)$, i.e., $\frac{1}{2}(a_1 + a_r) \le t \le a_r$ then the following pointwise error estimates hold

$$\text{(2.4.62)} \qquad \frac{n!}{k! M_n} | e^{(k)}_{(2.1.1)}(t) | \le g_k(t)$$

$$= \begin{cases} (t-a_1)^{n-\alpha-2}(a_r-t)^{\alpha+1-k}\left\{\binom{n-\alpha-1}{k}(t-a_1)\right. \\ \left. +\left[\binom{n}{k}-\binom{n-\alpha-1}{k}\right](a_r-t)\right\}, & 0 \le k \le \alpha \\ (t-a_1)^{n-k-1}\left\{\binom{n-\alpha-1}{k-\alpha-1}(t-a_1)\right. \\ \left. +\left[\binom{n}{k}-\binom{n-\alpha-1}{k-\alpha-1}\right](a_r-t)\right\}, & \alpha+1 \le k \le n-1 \end{cases}$$

where $\alpha = \min\{k_1, k_r\}$.

Proof. First we shall consider the case $0 \le k \le \alpha$. From (2.3.21) and Lemma 2.4.8, we obtain

$$\mid e^{(k)}_{(2.1.1)}(t) \mid \le \sum_{i=0}^{k} \binom{k}{i} \left| D^{k-i}\left[u(t)(a_r-t)^{k_r+1}\right]\right| M_n \frac{i!}{n!}(t-a_1)$$

$$= M_n \sum_{i=0}^{k} \frac{k!}{(k-i)!n!} \left| \sum_{j=0}^{k-i} \binom{k-i}{j} [D^{k-1-j}u(t)] \times \frac{(k_r+1)!}{(k_r+1-j)!}(a_r-t)^{k_r+1-j} \right| (t-a_1), \tag{2.4.63}$$

where $u(t) = (a_1-t)^{k_1-i}\prod_{j=2}^{r-1}(a_j-t)^{k_j+1}$.

Since $u(t)$ is a polynomial of degree $(n-i-k_r-2)$, from the generalized Rolle's theorem, we find that

$$\mid D^{k-i-j}u(t) \mid = \frac{(n-i-k_r-2)!}{(n-k_r-2-k+j)!}(t-a_1)^{k_1-k+j} \prod_{p=1}^{n-k_1-k_r-2} \mid t-t_p \mid,$$

where $t_p \in (a_1, a_{r-1})$, $1 \le p \le n-k_1-k_r-2$, some of which need not be distinct. However, since $(t-a_1) \ge (a_r-t)$ it is clear that

$$\prod_{p=1}^{n-k_1-k_r-2} \mid t-t_p \mid \le (t-a_1)^{n-k_1-k_r-2}.$$

Therefore, we obtain

$$\mid D^{k-i-j}u(t) \mid \le \frac{(n-i-k_r-2)!}{(n-k_r-2-k+j)!}(t-a_1)^{n-k_r-k-2+j}.$$

Using the above inequality in (2.4.63), we get

$$| e^{(k)}_{(2.1.1)}(t) | \le \frac{k!}{n!} M_n \sum_{i=0}^{k} \sum_{j=0}^{k-i} \binom{n-i-k_r-2}{k-i-j} \binom{k_r+1}{j} \times$$

$$(t-a_1)^{n-k_1-k-1+j}(a_r-t)^{k_r+1-j}.$$

Once again using the fact that $(t-a_1) \ge (a_r-t)$ and Lemma 2.4.5, the above inequality successively gives

$$| e^{(k)}_{(2.1.1)}(t) |$$

$$\le \frac{k!}{n!} M_n \sum_{i=0}^{k} \sum_{j=0}^{k-i} \binom{n-i-k_r-2}{k-i-j} \binom{k_r+1}{j} (t-a_1)^{n-\alpha-k-1+j}(a_r-t)^{\alpha+1-j}$$

$$= \frac{k!}{n!} M_n \sum_{j=0}^{k} \left[\sum_{i=0}^{k-j} \binom{n-i-k_r-2}{k-i-j} \right] \binom{k_r+1}{j} (t-a_1)^{n-\alpha-k-1+j} \times$$

$$(a_r-t)^{\alpha+1-j}$$

$$= \frac{k!}{n!} M_n (t-a_1)^{n-\alpha-2}(a_r-t)^{\alpha+1-k} \sum_{j=0}^{k} \binom{n-k_r-1}{k-j} \binom{k_r+1}{j} \times$$

$$(t-a_1)^{j-k+1}(a_r-t)^{k-j}$$

$$\le \frac{k!}{n!} M_n (t-a_1)^{n-\alpha-2}(a_r-t)^{\alpha+1-k} \left[\binom{k_r+1}{k} (t-a_1) \right.$$

$$\left. + \sum_{j=0}^{k-1} \binom{n-k_r-1}{k-j} \binom{k_r+1}{j} (a_r-t) \right]$$

$$\le \frac{k!}{n!} M_n (t-a_1)^{n-\alpha-2}(a_r-t)^{\alpha+1-k} \left\{ \binom{n-\alpha-1}{k} (t-a_1) \right.$$

$$\left. + \left[\binom{n}{k} - \binom{n-\alpha-1}{k} \right] (a_r-t) \right\},$$

where to obtain the last inequality we have also used Lemma 2.4.6.

Now we shall consider the case $\alpha+1 \le k \le n-1$. Once again (2.3.21) and Lemma 2.4.8, and the fact $(t-a_1) \ge (a_r-t)$ determine

$$
\begin{aligned}
&| e^{(k)}_{(2.1.1)}(t) | \\
&\le \frac{k!}{n!} M_n \Bigg\{ \sum_{i=0}^{\min\{k,k_1\}} \sum_{j=\mu}^{\nu(i)} \binom{n-i-k_r-2}{k-i-j} \binom{k_r+1}{j} (t-a_1)^{n-k_r-1-k+j} \times \\
&\qquad (a_r-t)^{k_r+1-j} \\
&+ \sum_{\ell=2}^{s-1} \Bigg[\sum_{i=p(\ell-1)+\ell-1}^{p(\ell)+\ell-1} \sum_{j=\mu}^{\nu(i)} \binom{n-i-k_r-2}{k-i-j} \binom{k_r+1}{j} (t-a_1)^{n-k_r-1-k+j} \times \\
&\qquad (a_r-t)^{k_r+1-j} \Bigg] \\
&+ \sum_{i=p(s-1)+s-1}^{k} \sum_{j=\mu}^{\nu(i)} \binom{n-i-k_r-2}{k-i-j} \binom{k_r+1}{j} (t-a_1)^{n-k_r-1-k+j} \times \\
&\qquad (a_r-t)^{k_r+1-j} \Bigg\},
\end{aligned}
$$

where $\mu = \max\{k-n+k_r+2, 0\}$ and $\nu(i) = \min\{k_r+1, k-i\}$. Since $\mu \le \nu(i)$ implies $i \le k$ and $i \le n-k_r-2$, i.e., $i \le \min\{k, n-k_r-2\} = \xi$, we have

$$
\begin{aligned}
&| e^{(k)}_{(2.1.1)}(t) | \\
&\le \frac{k!}{n!} M_n \Bigg\{ \sum_{i=0}^{\xi} \sum_{j=\mu}^{\nu(i)} \binom{n-i-k_r-2}{k-i-j} \binom{k_r+1}{j} (t-a_1)^{n-k_r-1-k+j} \times \\
&\qquad (a_r-t)^{k_r+1-j} \Bigg\} \\
&\le \frac{k!}{n!} M_n \Bigg\{ \sum_{j=\mu}^{\nu(0)} \Bigg[\sum_{i=0}^{k-j} \binom{n-i-k_r-2}{k-i-j} \Bigg] \binom{k_r+1}{j} (t-a_1)^{n-k_r-1-k+j} \times \\
&\qquad (a_r-t)^{k_r+1-j} \Bigg\}.
\end{aligned}
$$

Thus, on using Lemma 2.4.5, it follows that

$$
\begin{aligned}
| e^{(k)}_{(2.1.1)}(t) | \le \frac{k!}{n!} M_n \Bigg[\sum_{j=\mu}^{\nu(0)} \binom{n-k_r-1}{k-j} \binom{k_r+1}{j} \times \\
(t-a_1)^{n-k_r-1-k+j} (a_r-t)^{k_r+1-j} \Bigg].
\end{aligned}
\tag{2.4.64}
$$

Next, since

$$
\sum_{j=\mu}^{\nu(0)} \binom{n-k_r-1}{k-j} \binom{k_r+1}{j} \le \sum_{j=0}^{k} \binom{n-k_r-1}{k-j} \binom{k_r+1}{j} = \binom{n}{k}
$$

from (2.4.64), we find

$$| e^{(k)}_{(2.1.1)}(t) |$$

$$\leq \frac{k!}{n!}M_n \begin{cases} \binom{n}{k}(t-a_1)^{n-k_r-1}(a_r-t)^{k_r+1-k} = \phi(t), & \text{if } k < k_r+1 \\ \binom{n-k_r-1}{k-k_r-1}(t-a_1)^{n-k} + \left[\binom{n}{k} - \binom{n-k_r-1}{k-k_r-1}\right] \times \\ (t-a_1)^{n-k-1}(a_r-t) = \psi(t), & \text{if } k \geq k_r+1. \end{cases}$$

However, since $\phi(t) \leq \psi(t)$ for all $k < k_r + 1$ is obvious, it follows that

$$| e^{(k)}_{(2.1.1)}(t) | \leq \frac{k!}{n!}M_n\psi(t).$$

The conclusion (2.4.62) for all k, $\alpha + 1 \leq k \leq n-1$ is now immediate. ∎

Theorem 2.4.10. If $(t - a_1) \leq (a_r - t)$, i.e., $a_1 \leq t \leq \frac{1}{2}(a_1 + a_r)$ then the following pointwise error estimates hold

$$\frac{n!}{k!M_n} | e^{(k)}_{(2.1.1)}(t) | \leq g_k(a_1 + a_r - t),\ 0 \leq k \leq n-1. \tag{2.4.65}$$

Proof. For $0 \leq k \leq \alpha$, it follows from (2.3.21) and Lemma 2.4.8 that

$$| e^{(k)}_{(2.1.1)}(t) | \leq \sum_{i=0}^{k} \binom{k}{i} \left| D^{k-i}\left[(a_1 - t)^{k_1 - i}\theta_0\right]\right| M_n \frac{i!}{n!}(t - a_1)$$

$$= M_n \sum_{i=0}^{k} \frac{k!}{(k-i)!n!} \left| \sum_{j=0}^{k-i} \binom{k-i}{j} \frac{(k_1 - i)!}{(k_1 - k + j)!} \times (a_1 - t)^{k_1 - k + j}[D^j\theta_0] \right| (t - a_1), \tag{2.4.66}$$

where $\theta_0 = \prod_{j=2}^{r-1}(a_j - t)^{k_j+1}(a_r - t)^{k_r+1}$.

Using the fact that θ_0 is a polynomial of degree $(n - k_1 - 1)$ and the generalized Rolle's theorem, we get

$$| D^j\theta_0 | = \frac{(n-k_1-1)!}{(n-k_1-1-j)!}(a_r - t)^{k_r+1-j} \prod_{p=1}^{n-k_1-k_r-2} | t - t_p |,$$

where $t_p \in (a_1, a_{r-1})$, $1 \le p \le n - k_1 - k_r - 2$, some of which need not be distinct. From the inequality $(a_r - t) \ge (t - a_1)$ it is obvious that

$$\prod_{p=1}^{n-k_1-k_r-2} | t - t_p | \le (a_r - t)^{n-k_1-k_r-2}$$

and hence

$$| D^j \theta_0 | \le \frac{(n-k_1-1)!}{(n-k_1-1-j)!}(a_r - t)^{n-k_1-1-j}. \tag{2.4.67}$$

Using (2.4.67) in (2.4.66), we obtain

$$| e^{(k)}_{(2.1.1)}(t) | \le \frac{k!}{n!} M_n \sum_{i=0}^{k} \sum_{j=0}^{k-i} \binom{k_1 - i}{k - i - j} \binom{n - k_1 - 1}{j} \times (t - a_1)^{k_1 - k + j + 1}(a_r - t)^{n-k_1-1-j}. \tag{2.4.68}$$

Once again using the inequality $(a_r - t) \ge (t - a_1)$ and Lemmas 2.4.5 and 2.4.6, (2.4.68) provides

$$\begin{aligned}
&| e^{(k)}_{(2.1.1)}(t) | \\
&\le \frac{k!}{n!} M_n \sum_{i=0}^{k} \sum_{j=0}^{k-i} \binom{k_1 - i}{k - i - j} \binom{n - k_1 - 1}{j} (t - a_1)^{\alpha - k + j + 1} \times (a_r - t)^{n-\alpha-1-j} \\
&= \frac{k!}{n!} M_n \sum_{j=0}^{k} \left[\sum_{i=0}^{k-j} \binom{k_1 - i}{k - i - j} \right] \binom{n - k_1 - 1}{j} (t - a_1)^{\alpha - k + j + 1} \times (a_r - t)^{n-\alpha-1-j} \\
&= \frac{k!}{n!} M_n (t - a_1)^{\alpha - k + 1}(a_r - t)^{n-\alpha-2} \sum_{j=0}^{k} \binom{k_1 + 1}{k - j} \binom{n - k_1 - 1}{j} \times (t - a_1)^j (a_r - t)^{1-j} \\
&\le \frac{k!}{n!} M_n (t - a_1)^{\alpha - k + 1}(a_r - t)^{n-\alpha-2} \left[\binom{k_1 + 1}{k} (a_r - t) \right. \\
&\quad \left. + \sum_{j=1}^{k} \binom{k_1 + 1}{k - j} \binom{n - k_1 - 1}{j} (t - a_1) \right]
\end{aligned}$$

$$\leq \frac{k!}{n!} M_n (t-a_1)^{\alpha-k+1} (a_r - t)^{n-\alpha-2} \left\{ \binom{n-\alpha-1}{k} (a_r - t) \right.$$

$$\left. + \left[\binom{n}{k} - \binom{n-\alpha-1}{k} \right] (t-a_1) \right\}.$$

Now we shall consider the case $\alpha + 1 \leq k \leq n-1$. For this, the following term which appears in (2.3.21) satisfies

$$\left| D^{k-i} \left[(a_\ell - t)^{p(\ell-1)+\ell-2-i} \theta_{\ell-2} \right] \right|$$

$$= \left| D^{k-i} \left[(a_\ell - t)^{p(\ell-1)+\ell-2-i} \prod_{j=\ell}^{r-1} (a_j - t)^{k_j+1} (a_r - t)^{k_r+1} \right] \right|$$

$$= \frac{(n-i-1)!}{(n-k-1)!} (a_r - t)^{k_r+1-k+i} \prod_{p=1}^{n-k_r-i-2} | t - t_p |,$$

where $t_p \in (a_1, a_{r-1})$, $1 \leq p \leq n - k_r - i - 2$, some of which need not be distinct. However, since $(a_r - t) \geq (t - a_1)$ it is clear that

$$\prod_{p=1}^{n-k_r-i-2} | t - t_p | \leq (a_r - t)^{n-k_r-i-2}$$

and hence

$$\left| D^{k-i} \left[(a_\ell - t)^{p(\ell-1)+\ell-2-i} \theta_{\ell-2} \right] \right| \leq \frac{(n-i-1)!}{(n-k-1)!} (a_r - t)^{n-k-1}. \tag{2.4.69}$$

Once again (2.3.21), (2.4.67), (2.4.69), Lemma 2.4.8 and the fact $(a_r - t) \geq (t - a_1)$ yield

$$\begin{aligned} (2.4.70) \quad & | e^{(k)}_{(2.1.1)}(t) | \\ & \leq \frac{k!}{n!} M_n \left[\sum_{i=0}^{\min\{k,k_1\}} \sum_{j=\lambda}^{\rho(i)} \binom{k_1 - i}{k-i-j} \binom{n-k_1-1}{j} \times \right. \\ & \qquad (t-a_1)^{k_1-k+j+1} (a_r - t)^{n-k_1-1-j} \\ & \left. + \sum_{i=\min\{k,k_1\}+1}^{k} \binom{n-1-i}{k-i} (a_r - t)^{n-k} \right], \end{aligned}$$

where $\lambda = \max\{k-k_1, 0\}$ and $\rho(i) = \min\{n-k_1-1, k-i\}$. Next, from Lemma 2.4.7 we note that

$$\sum_{j=\lambda}^{\rho(i)} \binom{k_1 - i}{k-i-j}\binom{n-k_1-1}{j} = \binom{n-1-i}{k-i}.$$

Therefore, it follows from (2.4.70) and Lemma 2.4.5 that

$$| e^{(k)}_{(2.1.1)}(t) |$$

$$\leq \frac{k!}{n!}M_n \begin{cases} \binom{n}{k}(t-a_1)^{k_1-k+1}(a_r-t)^{n-k_1-1} = \phi(t), & \text{if } k < k_1 \\ \sum_{i=0}^{k_1}\binom{n-1-i}{k-i}(t-a_1)(a_r-t)^{n-k-1} & \\ \quad + \sum_{i=k_1+1}^{k}\binom{n-1-i}{k-i}(a_r-t)^{n-k} = \psi(t), & \text{if } k \geq k_1. \end{cases}$$

However, since $\phi(t) \leq \psi(t)$ for all $k < k_1$ is obvious, we get

$$| e^{(k)}_{(2.1.1)}(t) | \leq \frac{k!}{n!}M_n\psi(t). \tag{2.4.71}$$

Finally, since

$$\sum_{i=k_1+1}^{k}\binom{n-1-i}{k-i} \leq \sum_{i=\alpha+1}^{k}\binom{n-1-i}{k-i} = \binom{n-\alpha-1}{k-\alpha-1}$$

from (2.4.71), we find

$$\begin{aligned} | e^{(k)}_{(2.1.1)}(t) | \leq \; & \frac{k!}{n!}M_n\left\{\binom{n-\alpha-1}{k-\alpha-1}(a_r-t)^{n-k}\right. \\ & \left. + \left[\binom{n}{k} - \binom{n-\alpha-1}{k-\alpha-1}\right](t-a_1)(a_r-t)^{n-k-1}\right\}. \quad \blacksquare \end{aligned}$$

Remark 2.4.11. For the two - point Taylor interpolating conditions (2.1.5) the error estimates (2.4.62) reduce to

$$\begin{aligned} | e^{(k)}_{(2.1.5)}(t) | \leq \; & \frac{k!M_{2m}}{(2m)!}(t-a_1)^{m-1}(a_2-t)^{m-k}\left\{\binom{m}{k}(t-a_1)\right. \\ & \left. + \left[\binom{2m}{k} - \binom{m}{k}\right](a_2-t)\right\} \end{aligned}$$

$$= \frac{M_{2m}}{(2m)!}\begin{cases} [(t-a_1)(a_2-t)]^m, \text{ if } k=0 \\ k![(t-a_1)(a_2-t)]^{m-k}(t-a_1)^{k-1}\left\{\binom{m}{k}(t-a_1)\right. \\ \quad \left. +\left[\binom{2m}{k}-\binom{m}{k}\right](a_2-t)\right\}, \text{ if } 1\le k\le m-1 \end{cases}$$

$$\le \frac{M_{2m}}{(2m)!}\begin{cases} [(t-a_1)(a_2-t)]^m, \text{ if } k=0 \\ k![(t-a_1)(a_2-t)]^{m-k}(a_2-a_1)^{k-1}\times \\ \quad \max\left\{\binom{m}{k}, \frac{1}{2}\binom{2m}{k}\right\}(a_2-a_1), \text{ if } 1\le k\le m-1 \end{cases}$$

$$(2.4.72) = M_{2m}\begin{cases} \frac{1}{(2m)!}[(t-a_1)(a_2-t)]^m, \text{ if } k=0 \\ \frac{1}{2(2m-k)!}[(t-a_1)(a_2-t)]^{m-k}(a_2-a_1)^k, \\ \qquad \text{if } 1\le k\le m-1. \end{cases}$$

From (2.4.65) the same inequalities hold for $a_1 \le t \le \frac{1}{2}(a_1+a_2)$ also, and hence (2.4.72) indeed holds for all $t \in [a_1, a_2]$.

For the same case Ciarlet, Schultz and Varga [17] obtained the bounds

$$(2.4.73) \qquad |\, e^{(k)}_{(2.1.5)}(t) \,| \le \frac{M_{2m}}{k!(2m-2k)!}[(t-a_1)(a_2-t)]^{m-k}(a_2-a_1)^k, \quad 0\le k\le m-1.$$

A comparison of (2.4.73) with (2.4.72) shows that (2.4.72) is sharper than (2.4.73) by a factor of $1 \Big/ \left[2\binom{2m-k}{k}\right]$, $1\le k\le m-1$. ∎

Motivated by (2.4.72) we shall now consider the problem of computing the best possible constants $C_{2m,k}$ in the inequalities

$$(2.4.74) \quad |\, e^{(k)}_{(2.1.5)}(t) \,| \le C_{2m,k}[(t-a_1)(a_2-t)]^{m-k}(a_2-a_1)^k M_{2m}, \; 0\le k\le m-1.$$

For this, we consider the function $x(t) = x_5(t) = \frac{1}{(2m)!}(t-a_1)^m(t-a_2)^m$ for which $P_{(2.1.5)}(t) \equiv 0$, $M_{2m} = 1$ and hence (2.4.74) reduces to

$$(2.4.75) \quad |\, x_5^{(k)}(t) \,| \le C_{2m,k}[(t-a_1)(a_2-t)]^{m-k}(a_2-a_1)^k, \; 0\le k\le m-1.$$

On the other hand, in the function $x_5(t)$ we use the substitution $(t-a_1) = \frac{1}{2}(a_2-a_1)(u+1)$ so that it can be written as

$$x_5(t(u)) = \frac{1}{(2m)!}\left(\frac{a_2-a_1}{2}\right)^{2m}(u+1)^m(u-1)^m, \ -1 \le u \le 1$$

and hence

$$x_5^{(k)}(t) = \frac{k!}{(2m)!}[(t-a_1)(a_2-t)]^{m-k}(a_2-a_1)^k P_k^{(m-k,m-k)}(u), \quad 0 \le k \le m \tag{2.4.76}$$

where $P_k^{(m-k,m-k)}(u)$ is the Jacobi polynomial. Thus, in view of the well known [40] equality for the Jacobi polynomials

$$\begin{aligned} \max_{-1\le u\le 1} \mid P_n^{(\alpha,\beta)}(u) \mid &= \max\left\{P_n^{(\alpha,\beta)}(\pm 1)\right\} \\ &= \frac{\Gamma(n+q+1)}{\Gamma(q+1)\,n!};\ \alpha > -1,\ \beta > -1,\ q = \max\{\alpha,\beta\} \ge -\frac{1}{2} \end{aligned} \tag{2.4.77}$$

it follows that

$$\mid x_5^{(k)}(t) \mid \le \frac{k!}{(2m)!}[(t-a_1)(a_2-t)]^{m-k}(a_2-a_1)^k \binom{m}{k}, \quad 0 \le k \le m-1. \tag{2.4.78}$$

Hence, a comparison of (2.4.75) with (2.4.78) gives that

$$C_{2m,k} \ge \frac{m!}{(2m)!(m-k)!},\ 0 \le k \le m-1. \tag{2.4.79}$$

We conjecture that the inequalities (2.4.74) hold with

$$C_{2m,k} = \frac{m!}{(2m)!(m-k)!},\ 0 \le k \le m-1. \tag{2.4.80}$$

If our claim is true, then from (2.4.79) it is clear that the inequalities (2.4.74) with these constants $C_{2m,k}$ are the best possible, as in (2.4.74) equality holds for $x(t) = x_5(t)$ and only for this function up to a multiplicative constant.

Related with this conjecture we shall make another conjecture. For this, first we shall prove the following result.

Theorem 2.4.11. In (2.4.74) the constants $C_{2m,k}$, $0 \le k \le m-1$ satisfy the following inequalities

$$C_{2m,k} \le \frac{k!}{(2m)!} \max_{0\le u\le 1} \sum_{i=0}^{k} \left(\frac{1-u}{2}\right)^i \mid P_{k-i}^{(m-k+i,m-k-1)}(u) \mid, \tag{2.4.81}$$

$$0 \le k \le m-1.$$

Proof. From (2.3.21) and Lemma 2.4.8, we successively have

$$\begin{aligned}
&\mid e_{(2.1.5)}^{(k)}(t) \mid \\
&\le \sum_{i=0}^{k-1} \binom{k}{i} \mid D^{k-i}[(a_2-t)^m(a_1-t)^{m-1-i}] \mid \frac{i!}{(2m)!}(t-a_1)M_{2m} \\
&\quad +(a_2-t)^m(t-a_1)^{m-k}\frac{k!}{(2m)!}M_{2m} \\
&= \frac{M_{2m}}{(2m)!}\Bigg[\sum_{i=0}^{k-1}\frac{k!}{(k-i)!}\Bigg|\sum_{\ell=0}^{k-i}\binom{k-i}{\ell}\frac{m!}{(m-\ell)!}\frac{(m-1-i)!}{(m-1-k+\ell)!}(a_2-t)^{m-\ell}\times \\
&\qquad (a_1-t)^{m-k+\ell}(-1)^{k-i}\Bigg| + (a_2-t)^m(t-a_1)^{m-k}k!\Bigg] \\
&= \frac{k!M_{2m}}{(2m)!}(t-a_1)^{m-k}(a_2-t)^{m-k}\Bigg[\sum_{i=0}^{k-1}\Bigg|\sum_{\ell=0}^{k-i}\binom{m}{\ell}\binom{m-1-i}{k-i-\ell}\times \\
&\qquad (a_2-t)^{k-\ell}(a_1-t)^{\ell}\Bigg| + (a_2-t)^k\Bigg]
\end{aligned}$$

$$\begin{aligned}
&= \frac{k!M_{2m}}{(2m)!}[(t-a_1)(a_2-t)]^{m-k}\sum_{i=0}^{k}\Bigg|\sum_{\ell=0}^{k-i}\binom{m}{\ell}\binom{m-1-i}{k-i-\ell}\times \\
&\qquad (a_2-t)^{k-\ell}(a_1-t)^{\ell}\Bigg|
\end{aligned} \tag{2.4.82}$$

$$\begin{aligned}
&= \frac{k!M_{2m}}{(2m)!}[(t-a_1)(a_2-t)]^{m-k}\sum_{i=0}^{k}\frac{(a_2-a_1)^k}{2^i}\times \\
&\qquad \Bigg|\frac{1}{2^{k-i}}\sum_{\ell=0}^{k-i}\binom{m}{\ell}\binom{m-1-i}{k-i-\ell}(u-1)^{k-\ell}(u+1)^{\ell}\Bigg|
\end{aligned}$$

$$= \frac{k!M_{2m}}{(2m)!}[(t-a_1)(a_2-t)]^{m-k}(a_2-a_1)^k \sum_{i=0}^{k}\left(\frac{1-u}{2}\right)^i \times \quad (2.4.83)$$

$$| P_{k-i}^{(m-k+i,m-k-1)}(u) |, \ -1 \le u \le 1$$

where we have used the substitution $(t-a_1) = \frac{1}{2}(a_2-a_1)(u+1)$.

Since $e_{(2.1.5)}(t) = e_{(2.1.5)}(a_1+a_2-t)$, we can assume that $(t-a_1) \ge (a_2-t)$, i.e., $0 \le u \le 1$. Thus, from (2.4.83) it follows that

$$| e_{(2.1.5)}^{(k)}(t) | \le \frac{k!M_{2m}}{(2m)!}[(t-a_1)(a_2-t)]^{m-k}(a_2-a_1)^k \times \quad (2.4.84)$$

$$\max_{0\le u\le 1} \sum_{i=0}^{k}\left(\frac{1-u}{2}\right)^i | P_{k-i}^{(m-k+i,m-k-1)}(u) |,$$

$$0 \le k \le m-1. \quad \blacksquare$$

Remark 2.4.12. Comparing (2.4.84) with (2.4.74), in view of (2.4.80), we conjecture that the constants $C_{2m,k}$ in (2.4.74) are best if

$$\lambda_{2m,k} = \max_{0\le u\le 1} \sum_{i=0}^{k}\left(\frac{1-u}{2}\right)^i | P_{k-i}^{(m-k+i,m-k-1)}(u) | = \binom{m}{k}, \quad (2.4.85)$$

$$0 \le k \le m-1. \quad \blacksquare$$

Remark 2.4.13. For $k=0$, (2.4.85) is obvious. Further, since

$$\begin{aligned} \lambda_{2m,1} &= \frac{1}{2}\max_{0\le u\le 1}[| (m-1)(u-1)+m(u+1) | + | u-1 |] \\ &= \frac{1}{2}\max_{0\le u\le 1}[m(1+u)-(m-1)(1-u)+(1-u)] \\ &= \frac{1}{2}\max\{2m,2\} \\ &= \binom{m}{1} \end{aligned}$$

(2.4.85) is true for $k=1$ also. $\blacksquare$

Remark 2.4.14. Consider

$$
\begin{aligned}
&\alpha(u)\\
&= \frac{1}{4}\Bigg[\bigg|\frac{(m-1)(m-2)}{2}(u-1)^2 + m(m-1)(u-1)(u+1)\\
&\quad + \frac{m(m-1)}{2}(u+1)^2\bigg| + |\,(m-2)(u-1)^2 + m(u-1)(u+1)\,|\\
&\quad + |\,(u-1)^2\,|\Bigg]\\
&= \frac{1}{4}\Bigg[\bigg|\frac{m(m-1)}{2}\{(1+u)^2-(1-u^2)\} - \bigg\{\frac{m(m-1)}{2}(1-u^2)\\
&\quad - \frac{(m-1)(m-2)}{2}(1-u)^2\bigg\}\bigg| + m(1-u^2) - (m-2)(1-u)^2 + (1-u)^2\Bigg]\\
&\le \frac{1}{4}\bigg[\frac{m(m-1)}{2}(1+u)^2 - \frac{(m-1)(m-2)}{2}(1-u)^2 + m(1-u^2)\\
&\quad -(m-2)(1-u)^2 + (1-u)^2\bigg]\\
&\le \frac{1}{4}\bigg[\frac{m(m-1)}{2}(1+u)^2 + m(1-u^2) + \frac{m(m-1)}{2}(1-u)^2\bigg]\\
&\le \frac{1}{4}\frac{m(m-1)}{2}(1+u+1-u)^2\\
&\le \binom{m}{2}.
\end{aligned}
$$

However, since $\alpha(1) = \binom{m}{2}$, (2.4.58) is also true for $k = 2$. ∎

Remark 2.4.15. The analytic proof of (2.4.85) for $k = 3$ is tedious and we omit it here, whereas for the general k we leave it as an open problem. ∎

Remark 2.4.16. Gautschi [23] checked the correctness of (2.4.85) for $k = 1, 2, \ldots, 10$ and $m = k+1,\ k+2,\ \ldots,\ k+10$ on a computer, and he also feels that it is true for the general k. ∎

Remark 2.4.17. From (2.4.82) it is immediate that

$$(2.4.86)\quad |\, e^{(k)}_{(2.1.5)}(t) \,| \le \frac{k! M_{2m}}{(2m)!}[(t-a_1)(a_2-t)]^{m-k} \sum_{i=0}^{k} \binom{m}{i} \binom{m}{k-i} \times (a_2-t)^{k-i}(t-a_1)^i. \qquad \blacksquare$$

In our next result we shall provide uniform error estimates for $|\, e^{(k)}_{(2.1.1)}(t) \,|$, $0 \le k \le n-1$. For this, we need to define the following :

If $0 \le k \le \alpha$ then

$$c_1(k) = \binom{n-\alpha-1}{k}, \quad c_2(k) = \binom{n}{k} - c_1(k),$$

$$\begin{aligned} c_3(k) &= [(n-\alpha-1)(c_2(k)-c_1(k)) - c_2(k)(n-k-1)]^2 \\ &\quad + 4(\alpha+1-k)c_2(k)(c_2(k)-c_1(k)); \end{aligned}$$

and if $\alpha+1 \le k \le n-1$ then

$$d_1(k) = \binom{n-\alpha-1}{k-\alpha-1}, \quad d_2(k) = \binom{n}{k} - d_1(k).$$

Theorem 2.4.12. The following uniform error estimates hold

$$(2.4.87)\qquad \max_{a_1 \le t \le a_r} |\, e^{(k)}_{(2.1.1)}(t) \,| \le \frac{k!}{n!} M_n g_k(t_k^*), \ 0 \le k \le n-1$$

where for $0 \le k \le \alpha$

$$\begin{aligned} t_k^* &= a_1 + (a_r - a_1)\frac{(n-\alpha-1)(c_2(k)-c_1(k)) + c_2(k)(n-k-1) - \sqrt{c_3(k)}}{2(n-k)(c_2(k)-c_1(k))}, \\ &\qquad\qquad\qquad \text{if } c_1(k) \ne c_2(k) \\ &= a_1 + (a_r - a_1)\frac{n-\alpha-2}{n-k-1}, \ \text{if } c_1(k) = c_2(k) \end{aligned}$$

and for $\alpha+1 \le k \le n-1$

$$\begin{aligned} t_k^* &= a_1 + (a_r - a_1)\frac{(n-k-1)d_2(k)}{(n-k)(d_2(k)-d_1(k))}, \\ &\qquad\qquad \text{if it is in the interval } \left(\tfrac{1}{2}(a_1+a_r), a_r\right] \\ &= a_r, \ \text{otherwise.} \end{aligned}$$

Proof. For each $0 \le k \le n-1$, Theorems 2.4.9 and 2.4.10 imply that

$$\max_{a_1 \le t \le a_r} \mid e^{(k)}_{(2.1.1)}(t) \mid \le \frac{k!}{n!} \max_{\frac{1}{2}(a_1+a_r) \le t \le a_r} g_k(t) M_n.$$

Now for $0 \le k \le \alpha$, first we assume that $n-\alpha-2=0$, i.e., $\alpha = n-2$ but since $n \ge 2\alpha+2$ it is possible only when $n=2$, and hence $\alpha=0$, and $k=0$. In this case $c_1(0)=1$, $c_2(0)=0$ and $g_0(t)=(t-a_1)(a_r-t)$, which attains its maximum at $\frac{1}{2}(a_1+a_r)$. Since $c_3(0)=1$, t_0^* is also the same as $\frac{1}{2}(a_1+a_r)$. Now we consider the case $n-\alpha-2>0$. For this, a direct differentiation yields

$$\begin{aligned} g_k'(t) \;&=\; v^{n-\alpha-3}(1-v)^{\alpha-k}(a_r-a_1)^{n-k-1}\left\{(n-k)\left(c_2(k)-c_1(k)\right)v^2\right. \\ &\quad -\left[(n-\alpha-1)\left(c_2(k)-c_1(k)\right)+c_2(k)(n-k-1)\right]v \\ &\quad \left.+c_2(k)(n-\alpha-2)\right\}, \end{aligned}$$

where $(a_r-a_1)v=(t-a_1)$, i.e., $\frac{1}{2} \le v \le 1$. We note that $g_k(a_r)=0$, and

$$\begin{aligned} &g_k'\left(\frac{a_1+a_r}{2}\right) \\ &= \left(\frac{a_r-a_1}{2}\right)^{n-k-1}\frac{1}{k!}\left[(n+k-2\alpha-4)\frac{n!}{(n-k)!}+2\frac{(n-\alpha-1)!}{(n-k-\alpha-1)!}\right] \ge 0. \end{aligned}$$

Here equality holds only when $n=2m,\ \alpha=m-1,\ k=0,1$ (here $c_1(0)=1$, $c_2(0)=0$, and $t_0^*=\frac{1}{2}(a_1+a_r)$ which is the point where $g_0(t)=[(t-a_1)(a_r-t)]^m$ attains its maximum; also $c_1(1)=c_2(1)=m$, and $t_1^*=\frac{1}{2}(a_1+a_r)$ which is the point where $g_1(t)=m(a_r-a_1)[(t-a_1)(a_r-t)]^{m-1}$ attains its maximum). Therefore, except the case $n=2m,\ \alpha=m-1,\ k=0,1$ exactly out of possibly two roots of $g_k'(t)=0$ only one root lies in $\left(\frac{1}{2}(a_1+a_r), a_r\right)$ and at this point $g_k(t)$ attains its maximum. Finding this root is now straightforward.

For $\alpha+1 \le k \le n-1$, we have

$$g_k'(t) = v^{n-k-2}(a_r-a_1)^{n-k-1}\left[(n-k)\left(d_1(k)-d_2(k)\right)v+(n-k-1)d_2(k)\right]$$

where once again $(a_r-a_1)v=(t-a_1)$, i.e., $\frac{1}{2} \le v \le 1$. Since

$$g_k'\left(\frac{a_1+a_r}{2}\right) = \left(\frac{a_r-a_1}{2}\right)^{n-k-1}\left[2\binom{n-\alpha-1}{k-\alpha-1}+(n-k-2)\binom{n}{k}\right] \ge 0$$

with equality only if $n=2m,\ \alpha=m-1,\ k=2m-1$ (in this case $d_1(k)=d_2(k)=m$, and $t^*_{2m-1}=a_r$ which is the point where $g_{2m-1}(t)=m(a_r-a_1)$

attains its maximum), therefore, for each $\alpha+1 \le k \le n-1$, $g_k(t)$ attains its maximum at some point in the interval $\left(\frac{1}{2}(a_1+a_r), a_r\right]$. Finding this point is immediate. ∎

Remark 2.4.18. If $\alpha = 0$, then for $k = 0$, $c_1(0) = 1$, $c_2(0) = 0$, $c_3(0) = (n-1)^2$, $t_0^* = a_1 + \frac{n-1}{n}(a_r - a_1)$; and for $1 \le k \le n-1$, $d_1(k) = \binom{n-1}{k-1}$, $d_2(k) = \binom{n}{k} - \binom{n-1}{k-1}$, $t_k^* = a_r$. Therefore, inequalities (2.4.87) reduce to (2.4.59). ∎

Remark 2.4.19. For $k = 0$, $c_1(0) = 1$, $c_2(0) = 0$, $c_3(0) = (n-\alpha-1)^2$, $t_0^* = a_1 + \frac{n-\alpha-1}{n}(a_r - a_1)$; and for $k = n-1$, $d_1(n-1) = n-\alpha-1$, $d_2(n-1) = \alpha+1$, $t_{n-1}^* = a_r$. Therefore, inequalities (2.4.87) for $k = 0, n-1$ reduce to

$$\max_{a_1 \le t \le a_r} \mid e_{(2.1.1)}(t) \mid \le \frac{1}{n!} \frac{(n-\alpha-1)^{n-\alpha-1}(\alpha+1)^{\alpha+1}}{n^n}(a_r - a_1)^n M_n \tag{2.4.88}$$

and

$$\max_{a_1 \le t \le a_r} \mid e_{(2.1.1)}^{(n-1)}(t) \mid \le \frac{n-\alpha-1}{n}(a_r - a_1)M_n. \tag{2.4.89}$$

Both of these inequalities are the best possible as equalities hold for the functions $x_3(t) = (t-a_1)^{n-\alpha-1}(a_r-t)^{\alpha+1}$ and $x_4(t) = (t-a_1)^{\alpha+1}(a_r-t)^{n-\alpha-1}$, and only for these functions up to a constant factor. ∎

Remark 2.4.20. For all $0 \le k \le \alpha$, and $t \in \left[\frac{1}{2}(a_1+a_r), a_r\right]$, from (2.4.62) we have

$$g_k(t) \le \binom{n}{k}(t-a_1)^{n-\alpha-1}(a_r-t)^{\alpha+1-k}$$

and hence

$$\begin{aligned} g_k(t_k^*) &= \max_{\frac{1}{2}(a_1+a_r) \le t \le a_r} g_k(t) \\ &\le \binom{n}{k} \max_{\frac{1}{2}(a_1+a_r) \le t \le a_r} (t-a_1)^{n-\alpha-1}(a_r-t)^{\alpha+1-k} \\ &= \binom{n}{k} \frac{(n-\alpha-1)^{n-\alpha-1}(\alpha+1-k)^{\alpha+1-k}}{(n-k)^{n-k}}(a_r-a_1)^{n-k}. \end{aligned}$$

Therefore, for all $0 \le k \le \alpha$ from Theorem 2.4.12 it follows that

$$\max_{a_1 \le t \le a_r} | e^{(k)}_{(2.1.1)}(t) | \le \frac{k!}{n!} M_n g_k(t_k^*)$$

$$(2.4.90) \qquad \le \frac{(n-\alpha-1)^{n-\alpha-1}(\alpha+1-k)^{\alpha+1-k}}{(n-k)!(n-k)^{n-k}}(a_r - a_1)^{n-k} M_n.$$

Next, for all $\alpha + 1 \le k \le n-1$ and $t \in \left[\frac{1}{2}(a_1 + a_r), a_r\right]$ it is easy to verify that

$$(2.4.91) \quad \binom{n-\alpha-1}{k-\alpha-1}(t-a_1) + \left[\binom{n}{k} - \binom{n-\alpha-1}{k-\alpha-1}\right](a_r - t)$$
$$\le \binom{n}{k}\left[\frac{n-k}{n-\alpha}(a_r - t) + \frac{k-\alpha}{n-\alpha}(t-a_1)\right].$$

Indeed, (2.4.91) is equivalent to

$$\left[\frac{k-\alpha}{n-\alpha}\binom{n}{k} - \binom{n-\alpha-1}{k-\alpha-1}\right][(t-a_1) - (a_r - t)] \ge 0$$

and since

$$\frac{k-\alpha}{n-\alpha}\binom{n}{k} \ge \binom{n-\alpha-1}{k-\alpha-1}$$

the inequality (2.4.91) holds. Using (2.4.91) in (2.4.62), we get

$$(2.4.92) \quad g_k(t) \le \binom{n}{k}(t-a_1)^{n-k-1}\left[\frac{n-k}{n-\alpha}(a_r - t) + \frac{k-\alpha}{n-\alpha}(t-a_1)\right] = \phi(t).$$

From

$$\phi'(t) = \binom{n}{k}\frac{n-k}{n-\alpha}\, v^{n-k-2}[(n-k-1)(1-v) + (k-\alpha-1)v],$$

where $(a_r - a_1)v = (t - a_1)$, i.e., $\frac{1}{2} \le v \le 1$, we note that $\phi'\left(\frac{a_1+a_r}{2}\right) \ge 0$, $\phi'(a_r) \ge 0$ and so $\phi'(t) \ge 0$ in $\left[\frac{1}{2}(a_1 + a_r), a_r\right]$. The maximum of $\phi(t)$ in $\left[\frac{1}{2}(a_1 + a_r), a_r\right]$ is thus attained when $t = a_r$. On comparing the maximum of both sides of (2.4.92) in the interval $\left[\frac{1}{2}(a_1 + a_r), a_r\right]$, we obtain

$$g_k(t_k^*) \le \binom{n}{k}\frac{k-\alpha}{n-\alpha}(a_r - a_1)^{n-k}.$$

Therefore, for all $\alpha + 1 \le k \le n-1$, Theorem 2.4.12 implies that

$$\max_{a_1 \le t \le a_r} | e^{(k)}_{(2.1.1)}(t) | \le \frac{k!}{n!} M_n g_k(t_k^*)$$

$$\leq \frac{k-\alpha}{(n-k)!(n-\alpha)}(a_r - a_1)^{n-k} M_n. \tag{2.4.93}$$

Thus, inequalities (2.4.87) are better than (2.4.90) and (2.4.93) which are obtained earlier by Agarwal [1] (see (2.4.61)). In inequalities (2.4.87), (2.4.90) and (2.4.93) the constants appearing in the right side decrease as α increases. Thus, in all of these inequalities the right side constants will be maximum when $\alpha = 0$, i.e., the case considered in Remark 2.4.18, and minimum when $n = 2m$, $\alpha = m - 1$. As a consequence, we find that if $\alpha \neq 0$ then the inequalities (2.4.87) are sharper than [11,26,42], and for $\alpha = 0$ reduce to the same known inequalities (2.4.59). ∎

Remark 2.4.21. The first few constants $D^{\alpha}_{n,k} = \frac{k!}{n!} g_k(t^*_k) \big/ (a_r - a_1)^{n-k}$ are given in Table 2.4.1. ∎

Table 2.4.1.

		k					
n	α	0	1	2	3	4	5
2	0	$\frac{1}{8}^*$	$\frac{1}{2}^*$				
3	0	$\frac{2}{81}^*$	$\frac{1}{6}^*$	$\frac{2}{3}^*$			
4	0	$\frac{9}{2048}^*$	$\frac{1}{24}^*$	$\frac{1}{4}^*$	$\frac{3}{4}^*$		
4	1	$\frac{1}{384}^*$	$\frac{1}{48}$	$\frac{25}{192}$	$\frac{1}{2}^*$		
5	0	$\frac{32}{46875}^*$	$\frac{1}{120}^*$	$\frac{1}{15}^*$	$\frac{3}{10}^*$	$\frac{4}{5}^*$	
5	1	$\frac{9}{31250}^*$	$\frac{827-73\sqrt{73}}{61440}$	$\frac{9}{320}$	$\frac{49}{320}$	$\frac{3}{5}^*$	
6	0	$\frac{625}{6718464}^*$	$\frac{1}{720}^*$	$\frac{1}{72}^*$	$\frac{1}{12}^*$	$\frac{1}{3}^*$	$\frac{5}{6}^*$
6	1	$\frac{1}{32805}^*$	$\frac{\sqrt{15}}{7500}$	$\frac{7203}{1406080}$	$\frac{128}{3645}$	$\frac{1}{5}^*$	$\frac{2}{3}^*$
6	2	$\frac{1}{46080}^*$	$\frac{1}{3840}$	$\frac{471+133\sqrt{57}}{552960}$	$\frac{6859}{262440}$	$\frac{2}{15}$	$\frac{1}{2}^*$

(* indicates the best constants)

In our next result once again we shall use the error representation (2.3.20) to obtain corresponding to Table 2.4.1 the best possible table for the $(m, n-m)$, $2 \le n \le 6$ interpolation. For this, the computation does not require any computer usage as compared to that of Birkhoff and Priver [13] for the case $n = 6,\ m = 3$.

Theorem 2.4.13. For $2 \le n \le 6$ the following inequalities hold

$$| e^{(k)}_{(2.1.4)}(t) | \le E^m_{n,k}(a_2 - a_1)^{n-k} M_n,\ 0 \le k \le n-1 \tag{2.4.94}$$

where $M_n = \max_{a_1 \le t \le a_2} | x^{(n)}(t) |$ and the constants $E^m_{n,k}$ are given in Table 2.4.2.

Table 2.4.2.

		k					
n	m	0	1	2	3	4	5
2	1	$\frac{1}{8}$	$\frac{1}{2}$				
3	1,2	$\frac{2}{81}$	$\frac{1}{6}$	$\frac{2}{3}$			
4	1,3	$\frac{9}{2048}$	$\frac{1}{24}$	$\frac{1}{4}$	$\frac{3}{4}$		
4	2	$\frac{1}{384}$	$\frac{1}{72\sqrt{3}}$	$\frac{1}{12}$	$\frac{1}{2}$		
5	1,4	$\frac{32}{46875}$	$\frac{1}{120}$	$\frac{1}{15}$	$\frac{3}{10}$	$\frac{4}{5}$	
5	2,3	$\frac{9}{31250}$	$\frac{3+8\sqrt{6}}{20000}$	$\frac{1}{60}$	$\frac{3}{20}$	$\frac{3}{5}$	
6	1,5	$\frac{625}{6718464}$	$\frac{1}{720}$	$\frac{1}{72}$	$\frac{1}{12}$	$\frac{1}{3}$	$\frac{5}{6}$
6	2,4	$\frac{1}{32805}$	$\frac{25+34\sqrt{10}}{911250}$	$\frac{1}{360}$	$\frac{1}{30}$	$\frac{1}{5}$	$\frac{2}{3}$
6	3	$\frac{1}{46080}$	$\frac{\sqrt{5}}{30000}$	$\frac{1}{1920}$	$\frac{1}{120}$	$\frac{1}{10}$	$\frac{1}{2}$

In inequalities (2.4.94) the constants $E^m_{n,k}$ are optimal, as equality holds for the function $x_6(t) = (t-a_1)^m(a_2-t)^{n-m}$ whose $(m, n-m)$ interpolating polynomial $P_{(2.1.4)}(t) \equiv 0$, and only for this function up to a constant factor.

Proof. From the transformation $u = a_1 + a_2 - t$ it is clear that the function $e_{(2.1.4)}(a_1 + a_2 - t)$ satisfies the following conditions

$$e^{(i)}_{(2.1.4)}(a_1) = 0, \ 0 \le i \le n-m-1$$

$$e^{(i)}_{(2.1.4)}(a_2) = 0, \ 0 \le i \le m-1$$

and hence in inequalities (2.4.94) the constants

$$E^m_{n,k} = E^{n-m}_{n,k}, \ 0 \le k \le n-1. \tag{2.4.95}$$

Thus, in inequalities (2.4.94) it suffices to consider $1 \le m \le \left[\frac{n}{2}\right]$, or $\left[\frac{n+1}{2}\right] \le m \le n-1$. Further, since for the two - point Taylor conditions $e_{(2.1.5)}(t) = e_{(2.1.5)}(a_1 + a_2 - t)$, it follows that

$$\begin{aligned}\max_{a_1 \le t \le a_2} \mid e^{(k)}_{(2.1.5)}(t) \mid &= \max_{a_1 \le t \le \frac{1}{2}(a_1+a_2)} \mid e^{(k)}_{(2.1.5)}(t) \mid \\ &= \max_{\frac{1}{2}(a_1+a_2) \le t \le a_2} \mid e^{(k)}_{(2.1.5)}(t) \mid, \ 0 \le k \le 2m-1\end{aligned}$$

i.e., it suffices to obtain bounds for $\mid e^{(k)}_{(2.1.5)}(t) \mid$ over $\left[a_1, \frac{1}{2}(a_1+a_2)\right]$ or $\left[\frac{1}{2}(a_1 + a_2), a_2\right]$.

In view of Lemma 2.4.8 for the functions $F_j(t)$, $1 \le j \le n$ defined in Corollary 2.3.8 the following inequalities hold

$$\mid F_j(t) \mid \le \frac{M_n(j-1)!}{n!} \begin{cases} (a_2-t)^j, \ m+1 \le j \le n \\ (t-a_1)^j, \ 1 \le j \le m. \end{cases} \tag{2.4.96}$$

Similarly, for the functions $G_j(t)$, $1 \le j \le n$ defined in Corollary 2.3.9 it follows that

$$\mid G_j(t) \mid \le \frac{M_n(j-1)!}{n!} \begin{cases} (a_2-t)^j, \ n-m+1 \le j \le n \\ (t-a_1)^j, \ 1 \le j \le n-m. \end{cases} \tag{2.4.97}$$

The proof of Theorem 2.4.13 is contained in the following cases :

Case 1 $k = 0$

From (2.3.28) and (2.4.96), we find

$$| e_{(2.1.4)}(t) | \le \frac{1}{n!}(a_2 - t)^{n-m}(t - a_1)^m M_n.$$

Since the function $(a_2 - t)^{n-m}(t - a_1)^m$ attains its maximum at $t = \frac{1}{n}[ma_2 + (n-m)a_1]$, and its maximum value is $\frac{(n-m)^{n-m}m^m}{n^n}(a_2 - a_1)^n$, it follows that

$$| e_{(2.1.4)}(t) | \le \frac{(n-m)^{n-m}m^m}{n!\ n^n}(a_2 - a_1)^n M_n. \tag{2.4.98}$$

Case 2 $k = n - 1$

For this problem (2.3.26) reduces to

$$e^{(n-1)}_{(2.1.4)}(t) = (n-1)!\left[\sum_{i=0}^{m-1} \frac{(-1)^{n-1-i}}{i!}\frac{F_{i+1}(t)}{(a_1 - t)^i} + \sum_{i=m}^{n-1} \frac{(-1)^{n-1-i}}{i!}\frac{F_{i+1}(t)}{(a_2 - t)^i}\right]$$

and therefore from (2.4.96) it follows that

$$\begin{aligned} | e^{(n-1)}_{(2.1.4)}(t) | &\le \frac{1}{n}\left[\sum_{i=0}^{m-1}(t - a_1) + \sum_{i=m}^{n-1}(a_2 - t)\right] M_n \\ &= \frac{1}{n}\left[m(t - a_1) + (n-m)(a_2 - t)\right] M_n \\ &\le \frac{1}{n}\max\{m, n-m\}(a_2 - a_1)M_n. \end{aligned} \tag{2.4.99}$$

Case 3 $k = 1,\ m = n - 1$

For this problem (2.3.21) reduces to

$$e'_{(2.1.4)}(t) = D\left[(a_1 - t)^{n-2}(a_2 - t)\right] F_1(t) + (a_2 - t)(a_1 - t)^{n-4}F_2(t)$$

and thus from (2.4.96), we find that

$$\begin{aligned} &| e'_{(2.1.4)}(t) | \\ &\le | -(a_1 - t)^{n-2} - (n-2)(a_1 - t)^{n-3}(a_2 - t) | \frac{1}{n!}(t - a_1)M_n \\ &\quad +(a_2 - t)(a_1 - t)^{n-4}\frac{1}{n!}(t - a_1)^2 M_n \end{aligned}$$

$$= \frac{M_n}{n!}\begin{cases} (n-1)(t-a_1)^{n-2}(a_2-t)-(t-a_1)^{n-1}, & (t-a_1)\le(n-2)(a_2-t) \\ (t-a_1)^{n-1}-(n-3)(t-a_1)^{n-2}(a_2-t), & (t-a_1)\ge(n-2)(a_2-t) \end{cases}$$

$$= \frac{M_n}{n!}(a_2-a_1)^{n-1}\begin{cases} (n-1)v^{n-2}-nv^{n-1}=\phi_1(v), & 0\le v\le \dfrac{n-2}{n-1} \\ v^{n-2}[(n-2)v-(n-3)], & \dfrac{n-2}{n-1}\le v\le 1 \end{cases}$$

$$\le \frac{M_n}{n!}(a_2-a_1)^{n-1}\begin{cases} \phi_1(v), & 0\le v\le \dfrac{n-2}{n-1} \\ 1, & \dfrac{n-2}{n-1}\le v\le 1 \end{cases}$$

where $(a_2-a_1)v=t-a_1$. Since $\phi_1'(v_1)=0$, where $v_1=\frac{n-2}{n}$, we get

$$\begin{aligned} \phi_1(v) &\le \max\left\{\phi_1(0),\ \phi_1(v_1),\ \phi_1\left(\frac{n-2}{n-1}\right)\right\} \\ &= \max\left\{0,\ \left(\frac{n-2}{n}\right)^{n-2},\ \left(\frac{n-2}{n-1}\right)^{n-2}\frac{1}{n-1}\right\} \\ &< 1. \end{aligned}$$

Therefore, it follows that

$$\text{(2.4.100)} \qquad |\ e'_{(2.1.4)}(t)\ | \le \frac{1}{n!}(a_2-a_1)^{n-1}M_n.$$

Case 4 $k=1,\ m=1$

From (2.4.95) and Case 3, it is clear that in this case also the inequality (2.4.100) holds.

Case 5 $k=2,\ m=n-1\ (n\ge 4)$

For this problem (2.3.21) and (2.4.96) give

$|\ e''_{(2.1.4)}(t)\ |$

$$\begin{aligned} &\le (n-2)\ |\ 2(a_1-t)^{n-3}+(n-3)(a_2-t)(a_1-t)^{n-4}\ |\ \frac{1}{n!}(t-a_1)M_n \\ &+2\ |\ -(a_1-t)^{n-3}-(n-3)(a_2-t)(a_1-t)^{n-4}\ |\ \frac{1}{n!}(t-a_1)M_n \\ &+(a_2-t)(t-a_1)^{n-3}\frac{2}{n!}M_n \end{aligned}$$

$$= \frac{M_n}{n!}(t-a_1)^{n-3}\,[2(a_2-t)+2\mid(t-a_1)-(n-3)(a_2-t)\mid$$

$$+(n-2)\mid(n-3)(a_2-t)-2(t-a_1)\mid]$$

$$= \frac{M_n}{n!}(a_2-a_1)^{n-2}\begin{cases} v^{n-3}\{n(n-3)+2-[(n-2)(n+1)+2]v\} \\ \quad = \phi_2(v), \qquad 0 \le v \le (n-3)/(n-1) \\ v^{n-3}\{-(n-3)(n-4)+2[(n-2)(n-3)-2]v\} \\ \quad = \phi_3(v), \quad (n-3)/(n-1) \le v \le (n-3)/(n-2) \\ v^{n-3}\{-n(n-3)+2+[(n+1)(n-2)-2]v\} \\ \quad = \phi_4(v), \qquad (n-3)/(n-2) \le v \le 1. \end{cases}$$

Obviously, $\phi_4(v) \le \phi_4(1) = 2n-2$, and

$$\begin{aligned} \phi_3(v) &< -(n-3)(n-4)+2+[(n-2)(n-3)-2]\,\frac{n-3}{n-2} \\ &= 2+\frac{(n-3)(n-4)}{n-2} \\ &< 2n-2. \end{aligned}$$

Further, since $\phi_2'(v_1) = 0$, where $v_1 = \frac{(n-3)[n(n-3)+2]}{(n-2)[(n-2)(n+1)+2]}$, we have

$$\begin{aligned} \phi_2(v) &\le \max\left\{\phi_2(0),\ \phi_2(v_1),\ \phi_2\left(\frac{n-3}{n-1}\right)\right\} \\ &= \max\left\{0,\ \left(\frac{n-3}{n}\right)^{n-3}(n-1),\ 2\left(\frac{n-3}{n-1}\right)^{n-3}\right\} \\ &< 2n-2. \end{aligned}$$

Therefore, we obtain

$$\mid e''_{(2.1.4)}(t)\mid \le \frac{2(n-1)}{n!}(a_2-a_1)^{n-2}M_n. \tag{2.4.101}$$

Case 6 $k = 2,\ m = 1\ (n \ge 4)$

From (2.4.95) and Case 5, it is clear that in this case also the inequality (2.4.101) holds.

Case 7 $n = 2m,\ m = m,\ k = 1$

For this case we shall show that

$$(2.4.102)\qquad |\ e'_{(2.1.5)}(t)\ | \le \frac{1}{2^m\sqrt{2m-1}(2m-1)!}\left(\frac{m-1}{2m-1}\right)^{m-1}\times$$

$$(a_2-a_1)^{2m-1}M_{2m}.$$

For this problem from (2.3.28) and (2.3.27) we successively have

$$e'_{(2.1.5)}(t)$$

$$= [-m(a_2-t)^{m-1}(a_1-t)^{m-1} - (m-1)(a_2-t)^m(a_1-t)^{m-2}]\times$$

$$\left[(a_1-t_1)^{-1}\ F_2(t_1)|_{a_1}^{t} - \int_{a_1}^{t}(a_1-t_1)^{-3}F_3(t_1)\ dt_1\right]$$

$$+(a_2-t)^m(a_1-t)^{m-1}F_2(t)$$

$$= (a_2-t)^{m-1}(a_1-t)^{m-3}\{[-m(a_1-t)-(m-1)(a_2-t)+(a_2-t)]F_2(t)$$

$$-[-m(a_1-t)^2-(m-1)(a_2-t)(a_1-t)]\int_{a_1}^{t}(a_1-t_1)^{-3}F_3(t_1)\ dt_1\}$$

$$= (a_2-t)^{m-1}(a_1-t)^{m-3}\int_{a_1}^{t}\{[-m(a_1-t)-(m-2)(a_2-t)](a_1-t_1)^{-2}$$

$$+[m(a_1-t)^2+(m-1)(a_2-t)(a_1-t)](a_1-t_1)^{-3}\}F_3(t_1)\ dt_1$$

and hence from (2.4.96) it follows that

$$|\ e'_{(2.1.5)}(t)\ |\ \le\ \frac{2M_{2m}(a_2-t)^{m-1}(t-a_1)^{m-3}}{(2m)!}\times$$

$$\int_{a_1}^{t}|\ [m(t-a_1)-(m-2)(a_2-t)](a_1-t_1)$$

$$+[m(t-a_1)-(m-1)(a_2-t)](t-a_1)\ |\ dt_1$$

$$(2.4.103)\qquad = \frac{2M_{2m}(a_2-a_1)^{2m-1}}{(2m)!}v^{m-3}(1-v)^{m-1}\times$$

$$\int_0^v|\ [(m-2)(1-v)-mv]y+[mv-(m-1)(1-v)]v\ |\ dy,$$

where $(t-a_1) = v(a_2-a_1)$ so that $0 \le v \le 1$, and $(t_1-a_1) = y(a_2-a_1)$, and hence $0 \le y \le v$.

Since $e_{(2.1.5)}(t) = e_{(2.1.5)}(a_1+a_2-t)$, we can consider only $(a_2-t) \geq (t-a_1)$, i.e., $0 \leq v \leq \frac{1}{2}$. Thus, from (2.4.103) we find

$$| e'_{(2.1.5)}(t) |$$

$$\leq \frac{2M_{2m}(a_2-a_1)^{2m-1}}{(2m)!}\times$$

$$\begin{cases} \frac{m}{2}[v(1-v)]^{m-1}(1-2v) = \phi_5(v), & 0 \leq v \leq (m-1)/(2m-1) \\ \frac{1}{2}[v(1-v)]^{m-1}\left\{m(1-2v) + \frac{2[mv-(m-1)(1-v)]^2}{mv-(m-2)(1-v)}\right\} = \phi_6(v), & \\ & (m-1)/(2m-1) \leq v \leq 1/2. \end{cases}$$

By direct computation, we find

$$\max_{0\leq v\leq \frac{m-1}{2m-1}} \phi_5(v) = \phi_5\left(\frac{1}{2}\left(1-\frac{1}{\sqrt{2m-1}}\right)\right) = \frac{m}{2^m\sqrt{2m-1}}\left(\frac{m-1}{2m-1}\right)^{m-1}.$$

Next, since

$$m(1-2v) + \frac{2[mv-(m-1)(1-v)]^2}{mv-(m-2)(1-v)} \leq \frac{m}{2m-1}, \quad \frac{m-1}{2m-1} \leq v \leq \frac{1}{2}$$

it follows that

$$\phi_6(v) \leq \frac{1}{2^{2m-1}}\frac{m}{2m-1}, \quad \frac{m-1}{2m-1} \leq v \leq \frac{1}{2}.$$

We shall show that

$$\frac{1}{2^{2m-1}}\frac{m}{2m-1} \leq \frac{m}{2^m\sqrt{2m-1}}\left(\frac{m-1}{2m-1}\right)^{m-1}$$

which is the same as

$$1 \leq \sqrt{2m-1}\left(\frac{2m-2}{2m-1}\right)^{m-1}.$$

For this, we note that in the above inequality the minimum of the right side is $\frac{2}{\sqrt{3}}$. Therefore

$$\phi_6(v) \leq \frac{m}{2^m\sqrt{2m-1}}\left(\frac{m-1}{2m-1}\right)^{m-1}, \quad \frac{m-1}{2m-1} \leq v \leq \frac{1}{2}.$$

Case 8 $n = 2m,\ m = m,\ k = 2m-2\ (m \geq 2)$

From (2.3.21) and (2.4.96), we find

$$\mid e^{(2m-2)}_{(2.1.5)}(t) \mid$$

$$\leq \sum_{i=0}^{m-1} \binom{2m-2}{i} \mid D^{2m-2-i}[(a_2-t)^m(a_1-t)^{m-1-i}] \mid \frac{i!}{(2m)!}(t-a_1)M_{2m}$$

$$+\sum_{i=m}^{2m-2} \binom{2m-2}{i} \mid D^{2m-2-i}[(a_2-t)^{2m-1-i}] \mid \frac{i!}{(2m)!}(a_2-t)M_{2m}$$

$$= \frac{M_{2m}}{2m(2m-1)} \left\{ \sum_{i=0}^{m-1} \mid m(a_2-a_1)-(2m-1-i)(t-a_1) \mid (t-a_1) \right.$$

$$\left. + \sum_{i=m}^{2m-2} (2m-1-i)(a_2-t)^2 \right\}$$

$$\leq \frac{M_{2m}}{2m(2m-1)}(a_2-a_1)^2 \max_{\frac{1}{2}\leq v\leq 1} \left\{ \sum_{i=0}^{m-1} \mid m-(2m-1-i)v \mid v \right.$$

$$\left. + \frac{1}{2}m(m-1)(1-v)^2 \right\}$$

$$= \frac{M_{2m}}{2m(2m-1)}(a_2-a_1)^2 \max_{0\leq j\leq m-1} \max_{v_j\leq v\leq v_{j+1}} \left\{ \sum_{i=0}^{j-1}[(2m-1-i)v-m]v \right.$$

$$\left. + \sum_{i=j}^{m-1}[m-(2m-1-i)v]v + \frac{1}{2}m(m-1)(1-v)^2 \right\}$$

$$= \frac{M_{2m}}{2m(2m-1)}(a_2-a_1)^2 \max_{0\leq j\leq m-1} \max_{v_j\leq v\leq v_{j+1}} \left\{ \frac{1}{2}m(m-1)+(m-2mj)v \right.$$

$$\left. + (4mj-j^2-j-m^2)v^2 = \phi_7^j(v) \right\},$$

where $v_j = \frac{m}{2m-j}$. We shall show that $\max_{0\leq j\leq m-1} \max_{v_j\leq v\leq v_{j+1}} \phi_7^j(v) = \frac{1}{2}m(m-1)$.

For this, if $j = 0$ then we have $\phi_7^0(v) = \frac{1}{2}m(m-1)+mv(1-mv)$, however, since $mv \geq 1$, for all $v \in [v_0, v_1]$ it is obvious that $\phi_7^0(v) \leq \frac{1}{2}m(m-1)$. If $1 \leq j \leq m-1$ and $4mj-j^2-j-m^2 \leq 0$, then for all $v \in [v_j, v_{j+1}]$, $\phi_7^j(v) \leq \frac{1}{2}m(m-1)$ is immediate. Also, if $1 \leq j \leq m-1$ and $4mj-j^2-j-m^2 > 0$, then $\phi_7^j(v^*) = 0$, where $v^* = -\frac{m-2mj}{2(4mj-j^2-j-m^2)} \geq \frac{1}{2}$, and since

$$\phi_7^j(v^*) = \frac{1}{2}m(m-1) + v^*\frac{1}{2}(m-2mj) \leq \frac{1}{2}m(m-1)$$

it follows that for all $v \in [v_j, v_{j+1}]$, $\phi_7^j(v) \leq \max\left\{\phi_7^j(v_j), \phi_7^j(v_{j+1})\right\}$. However,

$$\phi_7^j(v_j) = \frac{1}{2}m(m-1) + v_j^2(m+j-2)(j-m) \leq \frac{1}{2}m(m-1)$$

and

$$\phi_7^j(v_{j+1}) = \frac{1}{2}m(m-1) + v_{j+1}^2\left[j^2-(m-1)^2\right] \leq \frac{1}{2}m(m-1)$$

(with equality only for $j = m-1$)

imply that for all $v \in [v_j, v_{j+1}]$, $\phi_7^j(v) \leq \frac{1}{2}m(m-1)$. Finally, it remains to note that $\phi_7^{m-1}(1) = \frac{1}{2}m(m-1)$.

Therefore, we obtain

$$| e_{(2.1.5)}^{(2m-2)}(t) | \leq \frac{1}{4}\left(\frac{m-1}{2m-1}\right)(a_2-a_1)^2 M_{2m}. \tag{2.4.104}$$

Case 9 $k = n-2,\ m = n-1\ (n \geq 5)$

From (2.3.21) and (2.4.96), we get

$$\begin{aligned} &| e_{(2.1.4)}^{(n-2)}(t) | \\ &\leq \sum_{i=0}^{n-3}\binom{n-2}{i} | D^{n-2-i}[(a_2-t)(a_1-t)^{n-2-i}] | \frac{i!}{n!}(t-a_1)M_n \\ &\quad +(a_2-t)\frac{(n-2)!}{n!}(t-a_1)M_n \\ &= \frac{M_n}{n(n-1)}\left\{\sum_{i=0}^{n-3} | (a_2-a_1)+(n-1-i)(a_1-t) | (t-a_1)\right. \\ &\qquad\qquad \left. +(t-a_1)(a_2-t)\right\} \\ &\leq \frac{M_n(a_2-a_1)^2}{n(n-1)} \max_{0\leq v\leq 1} v\left\{\sum_{i=0}^{n-3} | 1-(n-1-i)v | +(1-v)\right\} \end{aligned}$$

$$= \frac{M_n(a_2-a_1)^2}{n(n-1)} \max \begin{cases} \max\limits_{0\le v\le \frac{1}{n-1}} v\left[(n-1)-\frac{1}{2}n(n-1)v\right] \\ \max\limits_{\frac{1}{n-j-1}\le v\le \frac{1}{n-j-2}} v\Big\{(n-2j-3)+\frac{1}{2}[(j+1)(2n-j-2) \\ \qquad -(n-j)(n-j-3)-2]v = \phi_8^j(v)\Big\}, \\ \qquad\qquad 0\le j\le n-4 \\ \max\limits_{\frac{1}{2}\le v\le 1} v\left[\frac{1}{2}(n^2-n-4)v-(n-3)\right] \end{cases}$$

$$\le \frac{M_n(a_2-a_1)^2}{n(n-1)} \max \begin{cases} 1 \\ \max\limits_{0\le j\le n-4} \max\left\{\phi_8^j\left(\frac{1}{n-j-1}\right), \phi_8^j\left(\frac{1}{n-j-2}\right)\right\}\times \\ \qquad \frac{1}{n-j-2} \\ \frac{1}{2}(n-1)(n-2) \qquad \text{(equality at } v=1) \end{cases}$$

$$= \frac{M_n(a_2-a_1)^2}{n(n-1)} \times \frac{1}{2}(n-1)(n-2)$$

$$(2.4.105) \qquad = \left(\frac{n-2}{2n}\right)(a_2-a_1)^2 M_n.$$

Case 10 $k=n-2,\ m=1\ (n\ge 5)$

From (2.4.95) and Case 9, it follows that in this case also the inequality (2.4.105) holds.

Case 11 $k=1,\ m=n-2\ (n\ge 5)$

From (2.3.30) and (2.3.29), we have

$$-e'_{(2.1.5)}(a_1+a_2-t)$$

$$= [-(a_2-t)^{n-2}-(n-2)(a_2-t)^{n-3}(a_1-t)]G_1(t)$$

$$+(a_2-t)^{n-2}(a_1-t)\frac{1}{(a_2-t)^2}G_2(t)$$

$$= [-(a_2-t)^{n-2}-(n-2)(a_2-t)^{n-3}(a_1-t)]\left\{\frac{G_2(t)}{a_1-t}-\int_{a_1}^{t}\frac{G_2'(t_1)}{a_1-t_1}\,dt_1\right\}$$

$$+\frac{(a_2-t)^{n-2}}{a_1-t}G_2(t)$$

$$= [(a_2-t)^{n-2}+(n-2)(a_2-t)^{n-3}(a_1-t)]\int_{a_1}^{t}\frac{1}{(a_2-t_1)^3}G_3(t_1)\,dt_1$$

$$-(n-2)(a_2-t)^{n-3}\int_{a_1}^{t}\frac{a_1-t_1}{(a_2-t_1)^3}G_3(t_1)\,dt_1.$$

Thus, from (2.4.97) it follows that

$$|\, e'_{(2.1.5)}(a_1+a_2-t)\,|$$

$$\leq M_n(a_2-t)^{n-3}\int_{a_1}^{t}\frac{1}{(a_2-t_1)^3}\times$$
$$|\,(a_2-t)+(n-2)(a_1-t)-(n-2)(a_1-t_1)\,|\,\frac{2}{n!}(a_2-t_1)^3\,dt_1$$

$$= \frac{2M_n}{n!}(a_2-a_1)^{n-1}(1-v)^{n-3}\int_0^v |\,(n-2)v_1+1-(n-1)v\,|\,dv_1$$

$$= \frac{M_n}{n!}(a_2-a_1)^{n-1}\begin{cases}(1-v)^{n-3}v(2-nv)=\phi_9(v), & 0\leq v\leq \dfrac{1}{n-1}\\ (1-v)^{n-3}\left[v(nv-2)+\dfrac{2}{n-2}(1-v)^2\right]=\phi_{10}(v), & \\ & \dfrac{1}{n-1}\leq v\leq 1\end{cases}$$

$$\leq \frac{M_n}{n!}(a_2-a_1)^{n-1}\phi_9\left(\frac{2(n-1)-\sqrt{2(n-1)(n-2)}}{n(n-1)}\right).$$

Therefore, we have

$$(2.4.106)\quad |\, e'_{(2.1.5)}(t)\,|$$

$$\leq \frac{1}{n^{n-2}(n-1)^{n-2}n!}\left[(n-1)(n-2)+\sqrt{2(n-1)(n-2)}\right]^{n-3}\times$$
$$\left[2(n-1)-\sqrt{2(n-1)(n-2)}\right]\sqrt{\frac{2(n-2)}{n-1}}(a_2-a_1)^{n-1}M_n.$$

Case 12 $k = 1,\ m = 2\ (n \geq 5)$

From (2.4.95) and Case 11, the inequality (2.4.106) holds in this case also.

Case 13 $k = n-3,\ m = \left\{ \begin{array}{cc} n-2 & \text{if } n=5 \\ 2 & \text{if } n>5 \end{array} \right\}$

For this case we shall show that

$$(2.4.107) \qquad |\, e_{(2.1.5)}^{(n-3)}(t)\,| \leq \frac{1}{6}\frac{(n-3)(n-4)}{n(n-1)}(a_2-a_1)^3 M_n.$$

If $n = 5$, then from (2.3.21) and (2.4.96), we have

$$\begin{aligned}
|\, e''_{(2.1.5)}(t)\,| \;\leq\; & \sum_{i=0}^{1}\binom{2}{i}\,|\, D^{2-i}[(a_2-t)^2(a_1-t)^{2-i}]\,|\,\frac{i!}{5!}(t-a_1)M_5 \\
& +(a_2-t)^2(t-a_1)\frac{1}{60}M_5 \\
=\; & \frac{M_5}{60}(a_2-a_1)^3 v\left[|\,1-6v+6v^2\,|+|\,1-4v+3v^2\,|+(1-v)^2\right] \\
=\; & \frac{M_5}{60}(a_2-a_1)^3 v\begin{cases} 10v^2-12v+3, & 0\leq v\leq (3-\sqrt{3})/6 \\ -2v^2+1, & (3-\sqrt{3})/6\leq v\leq 1/3 \\ -8v^2+8v-1, & 1/3\leq v\leq (3+\sqrt{3})/6 \\ 4v^2-4v+1, & (3+\sqrt{3})/6\leq v\leq 1 \end{cases} \\
\leq\; & \frac{1}{60}(a_2-a_1)^3 M_5.
\end{aligned}$$

If $n > 5$, then from (2.3.21) and (2.4.96), we find

$$\begin{aligned}
& |\, e_{(2.1.5)}^{(n-3)}(t)\,| \\
& \leq \frac{M_n}{n(n-1)}\left[\left|\frac{n-1}{2}(a_2-t)^2-(a_2-t)(a_2-a_1)\right|(t-a_1)\right. \\
& \quad +\frac{n-3}{2}(a_2-t)^2(t-a_1) \\
& \quad \left. +\frac{1}{2(n-2)}\sum_{i=2}^{n-3}(n-1-i)(n-2-i)(a_2-t)^3\right]
\end{aligned}$$

$$= \frac{M_n}{n(n-1)}(a_2 - a_1)^3 \begin{cases} \frac{1}{6}(n-3)(n-4)(1-v)^3 + (n-3)v(1-v)^2 \\ \quad -v^2(1-v) = \phi_{11}(v), \ 0 \le v \le (n-3)/(n-1) \\ \frac{1}{6}(n-3)(n-4)(1-v)^3 + v^2(1-v) \\ \quad = \phi_{12}(v), \qquad (n-3)/(n-1) \le v \le 1. \end{cases}$$

We note that for $n > 5$

$$\begin{aligned} \phi_{12}(v) &\le \frac{1}{6}(n-3)(n-4)\left(1 - \frac{n-3}{n-1}\right) + \left(1 - \frac{n-3}{n-1}\right) \\ &= \frac{1}{6}(n-3)(n-4)\left(\frac{2}{n-1}\right) + \left(\frac{2}{n-1}\right) \\ &\le \frac{1}{6}(n-3)(n-4). \end{aligned}$$

Further, from

$$\phi'_{11}(v) = -\frac{1}{2}(n-3)(n-6)(1-v)^2 - v[2n-4-v(2n-3)]$$

it follows that $\phi'_{11}(0) \le 0$ only if $n > 5$, and $\phi'_{11}(v) < 0$ for $v \in \left[0, \frac{n-3}{n-1}\right]$. Thus,

$$\max_{0 \le v \le \frac{n-3}{n-1}} \phi_{11}(v) = \phi_{11}(0) = \frac{1}{6}(n-3)(n-4).$$

Case 14 $k = n-3, \ m = \left\{ \begin{array}{cl} 2 & \text{if } n = 5 \\ n-2 & \text{if } n > 5 \end{array} \right\}$

From (2.4.95) and Case 13, the inequality (2.4.107) holds for this case also.

Case 15 $k = n-2, \ m = n-2 \ (n \ge 5)$

From (2.3.21) and (2.4.96), we find

$\mid e^{(n-2)}_{(2.1.5)}(t) \mid$

$$\le \frac{M_n}{n(n-1)} \left[\sum_{i=0}^{n-3} (t-a_1) \mid 2(a_2-a_1) - (n-1-i)(t-a_1) \mid + (a_2-t)^2 \right]$$

$$\leq \frac{M_n}{n(n-1)}(a_2-a_1)^2 \begin{cases} 1+2v(n-3)-\frac{1}{2}v^2(n^2-n-4), \ 0 \leq v \leq \frac{2}{n-1} \\ 1+2v(n-3-2j)+\frac{1}{2}v^2[j(2n-j-1) \\ \qquad -(n-2-j)(n-j+1)+2], \\ \qquad \frac{2}{n-j} \leq v \leq \frac{2}{n-j-1}, \ 1 \leq j \leq n-3. \end{cases}$$

Therefore, it follows that

$$(2.4.108) \qquad | e^{(n-2)}_{(2.1.5)}(t) | \leq \frac{(n-2)(n-3)}{2n(n-1)}(a_2-a_1)^2 M_n.$$

Case 16 $k = n-2, \ m = 2 \ (n \geq 5)$

From (2.4.95) and Case 15, the inequality (2.4.108) holds in this case also.

Case 17 $k = n-3, \ m = n-1 \ (n \geq 6)$

From (2.3.21) and (2.4.96), it follows that

$$| e^{(n-3)}_{(2.1.5)}(t) |$$

$$\leq \frac{M_n}{n(n-1)(n-2)}\left[\sum_{i=0}^{n-4}(n-2-i) \mid (a_2-a_1)-\frac{1}{2}(n-1-i)(t-a_1) \mid \right.$$
$$\left. +(a_2-t)\right](t-a_1)^2$$

$$= \frac{M_n}{n(n-1)(n-2)}(a_2-a_1)^3 \times$$

$$\begin{cases} v^2\left[1-\frac{1}{6}n(n-1)(n-2)v+\frac{1}{2}n(n-3)\right] = \phi_{13}(v), \ 0 \leq v \leq \frac{2}{n-1} \\ v^2\left\{\frac{1}{6}[n(n-1)(n-2)-2(n-j)(n-j-1)(n-j-2)]v \right. \\ \qquad \left. +\frac{1}{2}(n^2-3n+2+2j^2-4nj+6j)\right\} = \phi_{14}(v), \\ \qquad \frac{2}{n-j} \leq v \leq \frac{2}{n-j-1}, \ 1 \leq j \leq n-4 \\ v^2\left[1-2v+\frac{1}{6}n(n-1)(n-2)v-\frac{1}{2}n(n-3)\right] = \phi_{15}(v), \ \frac{2}{3} \leq v \leq 1. \end{cases}$$

Clearly,

$$\phi_{15}(v) \le \phi_{15}(1) = \frac{1}{6}(n-1)(n-2)(n-3).$$

Further, since

$$\phi_{13}(v) \le \left(\frac{2}{n-1}\right)^2 \left[1 + \frac{1}{2}n(n-3)\right] = \frac{2(n-2)}{n-1} < 2$$

and $\phi_{15}(1) \ge 10$ for $n \ge 6$, we find that $\phi_{13}(v) < \phi_{15}(1)$. Next, since

$$n^2 - 3n + 2 + 2j^2 - 4nj + 6j \le n^2 - 7n + 10$$

it follows that

$$\phi_{14}(v) \le \left(\frac{2}{3}\right)^2 \left[\frac{1}{6}n(n-1)(n-2) + \frac{1}{2}(n^2 - 7n + 10)\right] < \phi_{15}(1).$$

Therefore, we get

$$| e^{(n-3)}_{(2.1.5)}(t) | \le \frac{n-3}{6n}(a_2 - a_1)^3 M_n. \tag{2.4.109}$$

Case 18 $k = n-3, \; m = 1 \; (n \ge 6)$

From (2.4.95) and Case 17, the inequality (2.4.109) holds in this case as well.

Case 19 $n = 2m, \; m = 2m-2, \; k = 2m-4 \; (m \ge 3)$

From (2.3.21) and (2.4.96), we obtain

$$| e^{(2m-4)}_{(2.1.5)}(t) |$$

$$\le \frac{M_{2m}}{2m(2m-1)(2m-2)(2m-3)} \times$$

$$\left[\sum_{i=0}^{2m-5} | (a_2 - a_1)^2(2m-3-i)(a_1 - t)\right.$$

$$+ (a_2 - a_1)(2m-2-i)(2m-3-i)(a_1 - t)^2$$

$$+ \frac{1}{6}(2m-1-i)(2m-2-i)(2m-3-i)(a_1 - t)^3 \, | \, (t - a_1)$$

$$\left. + (a_2 - t)^2(t - a_1)^2\right]$$

$$(2.4.110)\ = \frac{M_{2m}}{2m(2m-1)(2m-2)(2m-3)}(a_2-a_1)^4v^2\left[\sum_{i=0}^{2m-5}(2m-3-i)\times\right.$$
$$\mid -1+(2m-2-i)v-\frac{1}{6}(2m-1-i)(2m-2-i)v^2\mid$$
$$\left.+(1-v)^2\right],\qquad 0\le v\le 1.$$

In particular, for $m=3$, (2.4.110) reduces to

$$\mid e''_{(2.1.5)}(t)\mid$$
$$\le \frac{M_6}{360}(a_2-a_1)^4v^2\left[\mid 10v^2-12v+3\mid+\mid 4v^2-6v+2\mid+(1-v)^2\right]$$
$$=\frac{M_6}{360}(a_2-a_1)^4v^2\begin{cases}6-20v+15v^2, & 0\le v\le \frac{6-\sqrt{6}}{10}\\ 4v-5v^2, & \frac{6-\sqrt{6}}{10}\le v\le \frac{1}{2}\\ -4+16v-13v^2, & \frac{1}{2}\le v\le\frac{6+\sqrt{6}}{10}\\ 2-8v+7v^2, & \frac{6+\sqrt{6}}{10}\le v\le 1\end{cases}$$

$$(2.4.111)\ \le \frac{1}{360}(a_2-a_1)^4M_6.$$

Case 20 $n=2m,\ m=m,\ k=2\ (m\ge 3)$

From (2.3.21), we have

$$e''_{(2.1.5)}(t)$$
$$=\left[m(m-1)(a_2-t)^{m-2}(a_1-t)^{m-1}\right.$$
$$+2m(m-1)(a_2-t)^{m-1}(a_1-t)^{m-2}$$
$$\left.+(m-1)(m-2)(a_2-t)^m(a_1-t)^{m-3}\right]\times$$
$$\left[\left.\frac{F_2(t_1)}{(a_1-t_1)}\right|_{a_1}^{t}-\int_{a_1}^{t}\frac{F_2'(t_1)}{(a_1-t_1)}\,dt_1\right]$$
$$+2[-m(a_2-t)^{m-1}(a_1-t)^{m-2}-(m-2)(a_2-t)^m(a_1-t)^{m-3}]\frac{F_2(t)}{a_1-t}$$

$$+(a_2-t)^m(a_1-t)^{m-5}F_3(t)$$

$$\begin{aligned}(2.4.112)\quad &= [m(m-1)(a_2-t)^{m-2}(a_1-t)^{m-2}\\ &+2m(m-2)(a_2-t)^{m-1}(a_1-t)^{m-3}\\ &+(m-2)(m-3)(a_2-t)^m(a_1-t)^{m-4}]F_2(t)\\ &-[m(m-1)(a_2-t)^{m-2}(a_1-t)^{m-1}\\ &+2m(m-1)(a_2-t)^{m-1}(a_1-t)^{m-2}\\ &+(m-1)(m-2)(a_2-t)^m(a_1-t)^{m-3}]\int_{a_1}^{t}\frac{F_3(t_1)}{(a_1-t_1)^3}\,dt_1\\ &+(a_2-t)^m(a_1-t)^{m-5}F_3(t).\end{aligned}$$

For $m = 3$, (2.4.112) reduces to

$$e''_{(2.1.5)}(t)$$

$$= (a_2-a_1)^4\Big\{[-6(1-v)v+6(1-v)^2]F_2(v)+[6(1-v)v^2-12(1-v)^2v$$

$$+2(1-v)^3]\int_0^v\frac{F_3(v_1)}{v_1^3}\,dv_1+\frac{(1-v)^3}{v^2}F_3(v)\Big\}$$

$$= (a_2-a_1)^4\left\{(1-v)\int_0^v\left(\frac{6-12v}{v_1^2}+\frac{20v^2-16v+2}{v_1^3}\right)F_3(v_1)\,dv_1\right.$$

$$\left.+\frac{(1-v)^3}{v^2}F_3(v)\right\}$$

$$= (a_2-a_1)^4\left\{(1-v)\int_0^v\left(\frac{6-12v}{v_1^2}+\frac{20v^2-16v+2}{v_1^3}\right)\times\right.$$

$$\left.\left(\int_0^{v_1}v_2^2(1-v_2)^{-4}F_4(v_2)dv_2\right)dv_1+\frac{(1-v)^3}{v^2}F_3(v)\right\}$$

$$= (a_2-a_1)^4\left\{(1-v)\int_0^v\left[\int_{v_2}^v\left(\frac{6-12v}{v_1^2}+\frac{20v^2-16v+2}{v_1^3}\right)dv_1\right]\times\right.$$

$$\left.v_2^2(1-v_2)^{-4}F_4(v_2)\,dv_2+\frac{(1-v)^3}{v^2}F_3(v)\right\}$$

$$= (a_2 - a_1)^4 \Big\{ (1-v) \int_0^v \Big(\frac{12v-6}{v} - \frac{10v^2-8v+1}{v^2}$$

$$+ \frac{6-12v}{v_2} + \frac{10v^2-8v+1}{v_2^2} \Big) v_2^2 (1-v_2)^{-4} F_4(v_2)\, dv_2$$

$$+ \frac{(1-v)^3}{v^2} \int_0^v v_2^2 (1-v_2)^{-4} F_4(v_2)\, dv_2 \Big\}$$

$$= (a_2 - a_1)^4 (1-v) \int_0^v \Big(3 + \frac{6-12v}{v_2} + \frac{10v^2-8v+1}{v_2^2} \Big) v_2^2 (1-v_2)^{-4} F_4(v_2)\, dv_2$$

$$= -(a_2 - a_1)^4 (1-v) \int_0^v \Big(3 + \frac{6-12v}{v_2} + \frac{10v^2-8v+1}{v_2^2} \Big) v_2^2 (1-v_2)^{-4} \times$$

$$\Big(\int_{v_2}^v (1-v_3)^{-2} F_5(v_3)\, dv_3 + \int_v^1 (1-v_3)^{-2} F_5(v_3)\, dv_3 \Big) dv_2$$

$$= -(a_2 - a_1)^4 (1-v) \Big\{ \int_0^v \Big[\int_0^{v_3} \Big(3 + \frac{6-12v}{v_2} + \frac{10v^2-8v+1}{v_2^2} \Big) \times$$

$$v_2^2 (1-v_2)^{-4}\, dv_2 \Big] (1-v_3)^{-2} F_5(v_3)\, dv_3$$

$$- F_4(v) \int_0^v \Big(3 + \frac{6-12v}{v_2} + \frac{10v^2-8v+1}{v_2^2} \Big) v_2^2 (1-v_2)^{-4}\, dv_2 \Big\}$$

$$= -(a_2 - a_1)^4 \Big\{ (1-v) \int_0^v \Big[\frac{10}{3} \frac{(1-v)^2}{(1-v_3)^3} - \frac{6(1-v)}{(1-v_3)^2} + \frac{3}{1-v_3}$$

$$- \frac{1}{3}(10v^2 - 2v + 1) \Big] (1-v_3)^{-2} F_5(v_3)\, dv_3 + \frac{1}{3}(3v - 12v^2 + 10v^3) F_4(v) \Big\}$$

which on using (2.4.96) gives

(2.4.113) $\mid e''_{(2.1.5)}(t) \mid$

$$\leq (a_2 - a_1)^4 \Big\{ \frac{1}{90} (1-v) \int_0^v \mid v_3^3 (10v^2 - 2v + 1)$$

$$+ v_3^2(-30v^2 + 6v + 6) + v_3(30v^2 - 24v + 3) \mid dv_3$$

$$+ \frac{1}{360} \mid 3v - 12v^2 + 10v^3 \mid (1-v)^4 \Big\} M_6, \; 0 \leq v \leq \frac{1}{2}.$$

For $v \in \left[0, \frac{4-\sqrt{6}}{10}\right]$, (2.4.113) becomes

$$| e''_{(2.1.5)}(t) | \le \frac{1}{120}(a_2 - a_1)^4 v(1-v)(5v^2 - 5v + 1)M_6$$

$$\le \frac{1}{2400}(a_2 - a_1)^4 M_6. \tag{2.4.114}$$

For $v \in \left[\frac{4-\sqrt{6}}{10}, \frac{6-\sqrt{6}}{10}\right]$, (2.4.113) is the same as

$$| e''_{(2.1.5)}(t) |$$
$$\le (a_2 - a_1)^4 \Big\{ -\frac{1}{180}(1-v)[\alpha^2(60v^2 - 48v + 6) + \alpha^3(-40v^2 + 8v + 8)$$
$$+\alpha^4(10v^2 - 2v + 1)] + \frac{1}{120}v(1-v)(5v^2 - 5v + 1) = \phi_{16}(v) \Big\} M_6$$

where $\alpha = \frac{15v^2 - 3v - 3 + (1-v)\sqrt{-75v^2 + 60v + 6}}{10v^2 - 2v + 1}$. The maximum of $\phi_{16}(v)$ is attained when $v = \frac{4-\sqrt{6}}{10}$, and the maximum value is $\frac{1}{2655.3468}$. Thus, in this interval

$$| e''_{(2.1.5)}(t) | \le \frac{1}{2655.3468}(a_2 - a_1)^4 M_6. \tag{2.4.115}$$

For $v \in \left[\frac{6-\sqrt{6}}{10}, \frac{1}{2}\right]$, (2.4.113) reduces to

$$| e''_{(2.1.5)}(t) | \le \frac{1}{120}(a_2 - a_1)^4 v(1-v)(-5v^2 + 5v - 1)M_6$$

$$\le \frac{1}{1920}(a_2 - a_1)^4 M_6. \tag{2.4.116}$$

Combining (2.4.113) - (2.4.116), we find that

$$| e''_{(2.1.5)}(t) | \le \frac{1}{1920}(a_2 - a_1)^4 M_6. \tag{2.4.117}$$

Case 21 $n = 2m,\ m = m,\ k = 2m - 3\ (m \ge 3)$

From (2.3.21) and (2.4.96), we find

$$| e^{(2m-3)}_{(2.1.5)}(t) |$$
$$\le \frac{M_{2m}}{2m(2m-1)(2m-2)} \Bigg[\sum_{i=0}^{m-1} \Big| \frac{1}{2}(2m-1-i)(2m-2-i)(a_1 - t)^2$$
$$+m(2m-2-i)(a_1 - t)(a_2 - a_1) + \frac{1}{2}m(m-1)(a_2 - a_1)^2 \Big| (t - a_1)$$
$$+ \sum_{i=m}^{2m-3} \frac{(2m-1-i)(2m-2-i)}{2}(a_2 - t)^3 \Bigg]$$

$$(2.4.118) = \frac{M_{2m}}{2m(2m-1)(2m-2)}(a_2-a_1)^3\left[\sum_{i=0}^{m-1}\left|\frac{1}{2}(2m-1-i)(2m-2-i)v^2 - m(2m-2-i)v + \frac{1}{2}m(m-1)\right|v + \frac{1}{6}m(m-1)(m-2)(1-v)^3\right],$$
$$\frac{1}{2} \le v \le 1.$$

For $m = 3$, (2.4.118) is the same as

$$\begin{aligned} \mid e'''_{(2.1.5)}(t) \mid &\le \frac{M_6}{120}(a_2-a_1)^3\left[\mid 10v^2 - 12v + 3 \mid v + \mid 6v^2 - 9v + 3 \mid v \right. \\ &\qquad \left. + \mid 3v^2 - 6v + 3 \mid v + (1-v)^3\right] \\ &= \frac{M_6}{120}(a_2-a_1)^3 \begin{cases} 1 - 6v + 18v^2 - 14v^3, & \frac{1}{2} \le v \le \frac{6+\sqrt{6}}{10} \\ 1 - 6v^2 + 6v^3, & \frac{6+\sqrt{6}}{10} \le v \le 1 \end{cases} \end{aligned}$$

$$(2.4.119) \qquad \le \frac{1}{120}(a_2-a_1)^3 M_6. \qquad ■$$

Now we shall consider the problem of determining the best possible constants $F_{2m,k}$ in the inequalities

$$(2.4.120) \qquad \mid e^{(k)}_{(2.1.5)}(t) \mid \le F_{2m,k}(a_2-a_1)^{2m-k}M_{2m}, \; 0 \le k \le 2m-1.$$

For this, as earlier in (2.4.74) we expect that in (2.4.120) also equalities hold for the function $x(t) = x_5(t) = \frac{1}{(2m)!}(t-a_1)^m(t-a_2)^m$. Thus, in view of (2.4.76) it is clear that for $0 \le k \le m-1$ the maximum of $\mid x_5^{(k)}(t) \mid$ on $[a_1, a_2]$ is attended at one of the zeros $u(i, k+1)$, $1 \le i \le k+1$ of $P_{k+1}^{(m-k-1,m-k-1)}(u)$, i.e., the best constants $F_{2m,k}$, $0 \le k \le m-1$ we claim are

$$(2.4.121) \qquad F_{2m,k} = \frac{k!}{2^{2m-2k}(2m)!}\max_{0 \le i \le k+1}\left[\left(1-u^2(i,k+1)\right)^{m-k} \times \mid P_k^{(m-k,m-k)}(u(i,k+1)) \mid\right], \; 0 \le k \le m-1.$$

An elementary computation gives that

$$F_{2m,0} = \frac{1}{2^{2m}(2m)!},$$

(2.4.122)
$$F_{2m,1} = \frac{1}{2^m\sqrt{2m-1}(2m-1)!}\left(\frac{m-1}{2m-1}\right)^{m-1},$$

$$F_{2m,2} = \frac{1}{2^{2m-2}(2m-1)!}.$$

However, for $3 \le k \le m-1$ it does not seem to be possible to get meaningful expressions for (2.4.121), and hence one should expect only estimates.

For $m \le k \le 2m-1$ we use Rodrigues' formula [31]

$$\frac{d^{m-r}(u^2-1)^m}{du^{m-r}} = \frac{(m-r)!}{(m+r)!}(u^2-1)^r\ \frac{d^{m+r}(u^2-1)^m}{du^{m+r}}$$

to obtain

$$x_5^{(m+r)}(t(u)) = \frac{(m+r)!}{(2m)!}(a_2-a_1)^{m-r}P_{m-r}^{(r,r)}(u)$$

and hence in view of (2.4.77) it follows that

$$\mid x_5^{(m+r)}(t) \mid \le \frac{(m+r)!m!}{(2m)!(m-r)!r!}(a_2-a_1)^{m-r},\ 0 \le r \le m-1.$$

Thus, the best constants $F_{2m,k}$, $m \le k \le 2m-1$ we conjecture are

(2.4.123)
$$F_{2m,k} = \frac{k!m!}{(2m)!(2m-k)!(k-m)!},\ m \le k \le 2m-1.$$

Inequalities (2.4.88), (2.4.102), (2.4.104) and (2.4.89) justify our claim that the constants $F_{2m,0}$, $F_{2m,1}$, $F_{2m,2m-2}$ and $F_{2m,2m-1}$ given in (2.4.122) and (2.4.123) are best in (2.4.120). While such justification for $2 \le k \le 2m-3$ remains unresolved, we shall prove the following :

Theorem 2.4.14. In (2.4.120) the constants $F_{2m,k}$, $2 \le k \le 2m-3$ satisfy the following inequalities

(2.4.124)
$$F_{2m,k} \le \frac{k!}{(2m-k)^{2m-k}(2m)!} \times \begin{cases} m^m(m-k)^{m-k}F(m,k),\ 2 \le k \le m-1 \\[2ex] \dfrac{(2m-k-1)^{2m-k-1}[F(m,k)]^{2m-k}}{\left[F(m,k) - \begin{pmatrix} m \\ k-m \end{pmatrix}\right]^{2m-k-1}},\ m \le k \le 2m-3 \end{cases}$$

where

$$F(m,k) = \sum_{i=0}^{\min\{k,m-1\}} \max\left\{ \sum_{\substack{\ell=0\\ \ell:\,\text{even}}}^{k-i} \binom{m}{\ell}\binom{m-1-i}{k-i-\ell}, \right.$$

$$\left. \sum_{\substack{\ell=1\\ \ell:\,\text{odd}}}^{k-i} \binom{m}{\ell}\binom{m-1-i}{k-i-\ell} \right\}, \ 2 \le k \le 2m-3.$$

Proof. Since $e_{(2.1.5)}(t) = e_{(2.1.5)}(a_1 + a_2 - t)$, it suffices to prove (2.4.124), $2 \le k \le m-1$ for $(t-a_1) \ge (a_2 - t)$. For this, from (2.4.82) successively we have

$$\begin{aligned}
&| e^{(k)}_{(2.1.5)}(t) | \\
\le\ & \frac{k!M_{2m}}{(2m)!}[(t-a_1)(a_2-t)]^{m-k} \sum_{i=0}^{k} \left| \sum_{\substack{\ell=0\\ \ell:\,\text{even}}}^{k-i} \binom{m}{\ell}\binom{m-1-i}{k-i-\ell} \times \right. \\
& \left. (t-a_1)^{\ell}(a_2-t)^{k-\ell} - \sum_{\substack{\ell=1\\ \ell:\,\text{odd}}}^{k-i} \binom{m}{\ell}\binom{m-1-i}{k-i-\ell} (t-a_1)^{\ell}(a_2-t)^{k-\ell} \right| \\
\le\ & \frac{k!M_{2m}}{(2m)!}[(t-a_1)(a_2-t)]^{m-k} \sum_{i=0}^{k} \max\left\{ \sum_{\substack{\ell=0\\ \ell:\,\text{even}}}^{k-i} \binom{m}{\ell}\binom{m-1-i}{k-i-\ell} \times \right. \\
& \left. (t-a_1)^{k}, \ \sum_{\substack{\ell=1\\ \ell:\,\text{odd}}}^{k-i} \binom{m}{\ell}\binom{m-1-i}{k-i-\ell} (t-a_1)^{k} \right\} \\
=\ & \frac{k!M_{2m}}{(2m)!}(t-a_1)^{m}(a_2-t)^{m-k}F(m,k) \\
\le\ & \frac{k!M_{2m}}{(2m)!} \frac{m^m(m-k)^{m-k}}{(2m-k)^{2m-k}} F(m,k)(a_2-a_1)^{2m-k}, \ 2 \le k \le m-1.
\end{aligned}$$

Once again, since $e_{(2.1.5)}(t) = e_{(2.1.5)}(a_1+a_2-t)$, it suffices to prove (2.4.124), $m \le k \le 2m-3$ for $(a_2-t) \ge (t-a_1)$. For this, we use Lemma 2.4.8 in (2.3.21) to obtain

$$| e^{(k)}_{(2.1.5)}(t) |$$

$$\leq \sum_{i=0}^{m-1} \binom{k}{i} \left| D^{k-i}\left[(a_2-t)^m(a_1-t)^{m-1-i}\right] \right| \frac{i!}{(2m)!}(t-a_1)M_{2m}$$

$$+\sum_{i=m}^{k} \binom{k}{i} \mid D^{k-i}(a_2-t)^{2m-1-i} \mid \frac{i!}{(2m)!}(a_2-t)M_{2m}$$

$$= \frac{k!M_{2m}}{(2m)!}\left\{ \sum_{i=0}^{m-1} \left| \sum_{\ell=0}^{k-i} \binom{m}{\ell} \binom{m-1-i}{k-i-\ell} (a_2-t)^{m-\ell}(a_1-t)^{m+\ell-k} \right| \right.$$

$$\left. +\sum_{i=m}^{k} \binom{2m-1-i}{k-i} (a_2-t)^{2m-k} \right\}$$

$$\leq \frac{k!M_{2m}}{(2m)!}\left\{ F(m,k)(t-a_1)(a_2-t)^{2m-k-1} + \binom{m}{k-m}(a_2-t)^{2m-k} \right\}$$

$$\leq \frac{k!M_{2m}}{(2m)!}\frac{(2m-k-1)^{2m-k-1}}{(2m-k)^{2m-k}} \frac{[F(m,k)]^{2m-k}}{\left[F(m,k)-\binom{m}{k-m}\right]^{2m-k-1}}(a_2-a_1)^{2m-k},$$

$$m \leq k \leq 2m-3. \quad \blacksquare$$

Remark 2.4.22. The error representation (2.3.11) can also be used to prove (2.4.120) for $k = 2m-1$ with the best constant $F_{2m,2m-1} = \frac{1}{2}$. Indeed, from (2.3.18) we have

$$g^{(2m-1)}_{(2.1.5)}(t,s)$$

$$= \begin{cases} -\left(\frac{s-a_1}{a_2-a_1}\right)^m \sum_{j=0}^{m-1} \binom{m-1+j}{j} \left(\frac{a_2-s}{a_2-a_1}\right)^j, & a_1 \leq s \leq t \leq a_2 \\ \left(\frac{a_2-s}{a_2-a_1}\right)^m \sum_{j=0}^{m-1} \binom{m-1+j}{j} \left(\frac{s-a_1}{a_2-a_1}\right)^j, & a_1 \leq t \leq s \leq a_2 \end{cases}$$

and hence $g^{(2m-1)}_{(2.1.5)}(t,s)$ is nonnegative (nonpositive) as long as $t \leq s$ ($s \leq t$). Thus, from (2.3.11) it follows that

$$| e^{(2m-1)}_{(2.1.5)}(t) | \leq M_{2m} \max_{a_1 \leq t \leq a_2} \psi(t),$$

where

$$\psi(t) = \sum_{j=0}^{m-1} \binom{m-1+j}{j} \frac{1}{(a_2-a_1)^{m+j}} \left[\int_{a_1}^{t} (s-a_1)^m (a_2-s)^j ds + \int_{t}^{a_2} (a_2-s)^m (s-a_1)^j ds\right].$$

For the functions

$$\psi_j(t) = \int_{a_1}^{t} (s-a_1)^m (a_2-s)^j ds + \int_{t}^{a_2} (a_2-s)^m (s-a_1)^j ds, \ 0 \le j \le m-1$$

we have

$$\psi_j'(t) = (t-a_1)^j (a_2-t)^j \left[(t-a_1)^{m-j} - (a_2-t)^{m-j}\right]$$

and hence $\psi_j'(t) \le 0$ (≥ 0) if $(t-a_1) \le (a_2-t)$ ($(t-a_1) \ge (a_2-t)$). Thus, $\psi_j(t)$ has a maximum at a_1 or a_2. However, since $\psi_j(a_1) = \psi_j(a_2)$, we find that

$$\begin{aligned}
\psi(t) &\le \sum_{j=0}^{m-1} \binom{m-1+j}{j} \frac{1}{(a_2-a_1)^{m+j}} \int_{a_1}^{a_2} (s-a_1)^m (a_2-s)^j ds \\
&= (a_2-a_1) \sum_{j=0}^{m-1} \binom{m-1+j}{j} \int_0^1 u^m (1-u)^j du \\
&= (a_2-a_1) \sum_{j=0}^{m-1} \frac{(m-1+j)!}{j!(m-1)!} \times \frac{m!j!}{(m+j+1)!} \\
&= m(a_2-a_1) \sum_{j=0}^{m-1} \frac{1}{(m+j)(m+j+1)} \\
&= m(a_2-a_1) \sum_{j=0}^{m-1} \left(\frac{1}{m+j} - \frac{1}{m+j+1}\right) \\
&= m(a_2-a_1) \left(\frac{1}{m} - \frac{1}{2m}\right) \\
&= \frac{1}{2}(a_2-a_1). \quad \blacksquare
\end{aligned}$$

2.5 SOME APPLICATIONS

While the results of Sections 2.2 - 2.4 are of immense value in every aspect of numerical mathematics, here we shall demonstrate their importance in the theory of ordinary differential equations.

Generalized Maximum Principle

Here we shall establish the following generalized maximum principle.

Theorem 2.5.1. Let $1 \leq m \leq n-1 (n \geq 2)$ and $x \in C^{(n)}[a_1, a_2]$ be such that

$$x^{(n)}(t) \geq 0, \; t \in (a_1, a_2), \tag{2.5.1}$$

$$(-1)^{n-m} x^{(i)}(a_1) \geq 0, \; i=1,\, 2,\, ...,\, m-1 \text{ (if such } i \text{ exist)}, \tag{2.5.2}$$

$$(-1)^{n-m+i} x^{(i)}(a_2) \geq 0, \; i=1,\, 2,\, ...,\, n-m-1 \text{ (if such } i \text{ exist)}. \tag{2.5.3}$$

Then, in the case $(n-m)$ even x attains its minimum and in the case $(n-m)$ odd x attains its maximum at a_1 or a_2.

If $n = 2$, then obviously $m = 1$. In this case our theorem gives the classical maximum principle "If $x \in C^{(2)}[a_1, a_2]$, $x''(t) \geq 0$, $t \in (a_1, a_2)$ and attains its maximum at an interior point of $[a_1, a_2]$, then x is identically a constant on $[a_1, a_2]$." The proof of this maximum principle and several of its applications are available in Protter and Weinberger [37].

For $n = 4$, $m = 2$ our theorem reduces to the maximum principle of Chow, Dunninger and Lasota [16]. It reads as "If $x \in C^{(4)}[a_1, a_2]$ and satisfies the inequalities

$$x^{(4)}(t) \geq 0, \; t \in (a_1, a_2),$$

$$x'(a_1) \geq 0, \; x'(a_2) \leq 0,$$

then x attains its minimum at a_1 or a_2." An alternative proof of this maximum principle and its applications to fourth order boundary value problems has been given by Kuttler [28].

For $n = 2m,\ m = m$ our theorem becomes the maximum principle of Šeda [38], which is stated as "If $x \in C^{(2m)}[a_1, a_2]$ and satisfies the inequalities

$$x^{(2m)}(t) \geq 0,\ t \in (a_1, a_2),$$

$$(-1)^m x^{(i)}(a_1) \geq 0,\ i=1,\ 2,\ ...,\ m-1 \text{ (if such } i \text{ exist)},$$

$$(-1)^{m+i} x^{(i)}(a_2) \geq 0,\ i=1,\ 2,\ ...,\ m-1 \text{ (if such } i \text{ exist)},$$

then in the case m even (m odd) x attains its minimum (maximum) at either a_1 or a_2." In the same paper, Šeda also separately proves two more particular cases of our theorem namely when $m = n - 1$ and $m = 1$.

Thus, Theorem 2.5.1 accommodates several known maximum principles. Further, the novelty is in its proof which is rather constructive and does not require any sign property of Green's functions of the related boundary value problems, cf. see [38].

Proof of the Theorem. In view of Theorem 2.3.1 and Corollary 2.2.5 any function $x \in C^{(n)}[a_1, a_2]$ can be written as

$$x(t) = \sum_{i=0}^{m-1} \tau_i(t) x^{(i)}(a_1) + \sum_{i=0}^{n-m-1} \eta_i(t) x^{(i)}(a_2) \tag{2.5.4}$$

$$+ \frac{1}{n!}(t - a_1)^m (t - a_2)^{n-m} x^{(n)}(\xi),$$

where $\xi \in (a_1, a_2)$ and the polynomials $\tau_i(t),\ \eta_i(t)$ are defined in (2.2.19) and (2.2.20) respectively.

Thus, if $(n - m)$ is even, then in view of (2.5.1) the representation (2.5.4) gives

$$x(t) \geq \sum_{i=0}^{m-1} \tau_i(t) x^{(i)}(a_1) + \sum_{i=0}^{n-m-1} \eta_i(t) x^{(i)}(a_2). \tag{2.5.5}$$

Next, from (2.2.19) and (2.2.20) it is clear that for all $t \in [a_1, a_2]$, $\tau_i(t) \geq 0$, $0 \leq i \leq m - 1$ and $(-1)^i \eta_i(t) \geq 0$, $0 \leq i \leq n - m - 1$. Therefore, conditions (2.5.2) and (2.5.3) in (2.5.5) lead to

$$x(t) \geq \tau_0(t) x(a_1) + \eta_0(t) x(a_2). \tag{2.5.6}$$

Let $x(t) \equiv 1$ in (2.5.4), to obtain

$$1 = \tau_0(t) + \eta_0(t). \tag{2.5.7}$$

Using (2.5.7) in (2.5.6), we obtain

$$x(t) \geq (\tau_0(t) + \eta_0(t)) \min\{x(a_1), x(a_2)\} = \min\{x(a_1), x(a_2)\}.$$

The proof for the case $(n-m)$ odd is identical. ∎

Hermite Boundary Value Problems

For the Hermite boundary value problem

$$x^{(n)}(t) = f(t, \mathbf{x}(t)) \tag{2.5.8}$$

$$x^{(i)}(a_j) = A_{i,j};\ 0 \leq i \leq k_j,\ 1 \leq j \leq r,\ \sum_{j=1}^{r} k_j + r = n, \tag{2.5.9}$$

$$a \leq a_1 < a_2 < \cdots < a_r \leq b$$

where $\mathbf{x}(t) = (x(t), x'(t), ..., x^{(q)}(t))$, $0 \leq q \leq n-1$ but fixed, and f is continuous at least in the interior of the domain of interest, we shall state several results which are analogous to those of Lidstone boundary value problem (1.6.1), (1.6.2). For this, we shall assume that $G_{n,k}$ are the best 'available' constants in the inequalities

$$|\, e^{(k)}_{(2.1.1)}(t) \,| \leq G_{n,k}(a_r - a_1)^{n-k} M_n,\ 0 \leq k \leq n-1. \tag{2.5.10}$$

Theorem 2.5.2. Suppose that

(i) $K_i > 0$, $0 \leq i \leq q$ are given real numbers and let Q be the maximum of $|\, f(t, x_0, x_1, ..., x_q) \,|$ on the compact set $[a_1, a_r] \times D_0$, where D_0 is the same as in the condition (i) of Theorem 1.6.1;

(ii) $(a_r - a_1) \leq \left(\dfrac{K_i}{QG_{n,i}}\right)^{1/(n-i)}$, $0 \leq i \leq q$;

(iii) $\max_{a_1 \le t \le a_r} | P^{(i)}_{(2.1.1)}(t) | \le K_i,\ 0 \le i \le q.$

Then, the boundary value problem (2.5.8), (2.5.9) has a solution in D_0. ∎

Corollary 2.5.3. Suppose that the function $f(t, x_0, x_1, ..., x_q)$ on $[a_1, a_r] \times \Re^{q+1}$ satisfies the condition (1.6.7). Then, the boundary value problem (2.5.8), (2.5.9) has a solution. ∎

Theorem 2.5.4. Suppose that the function $f(t, x_0, x_1, ..., x_q)$ on $[a_1, a_r] \times D_1$ satisfies the condition (1.6.8), where

$$D_1 = \Big\{(x_0, x_1, ..., x_q)\ :\ | x_i | \le \max_{a_1 \le t \le a_r} | P^{(i)}_{(2.1.1)}(t) | + G_{n,i}(a_r - a_1)^{n-i} \frac{L + C}{1 - \theta},\quad 0 \le i \le q\Big\}$$

and

$$C = \max_{a_1 \le t \le a_r} \sum_{i=0}^{q} L_i \, | P^{(i)}_{(2.1.1)}(t) |,$$

$$\theta = \sum_{i=0}^{q} G_{n,i} L_i (a_r - a_1)^{n-i} < 1. \tag{2.5.11}$$

Then, the boundary value problem (2.5.8), (2.5.9) has a solution in D_1. ∎

Theorem 2.5.5. Suppose that the differential equation (2.5.8) together with the conditions

$$x^{(i)}(a_j) = 0;\ 0 \le i \le k_j,\ 1 \le j \le r \tag{2.5.12}$$

has a nontrivial solution $x(t)$ and the condition (1.6.8) with $L = 0$ on $[a_1, a_r] \times D_2$ is satisfied, where

$$D_2 = \Big\{(x_0, x_1, ..., x_q)\ :\ | x_i | \le G_{n,i}(a_r - a_1)^{n-i} \max_{a_1 \le t \le a_r} | x^{(n)}(t) |,\ 0 \le i \le q\Big\}.$$

Then, it is necessary that $\theta \ge 1$. ∎

Theorem 2.5.6. Suppose that the boundary value problem (2.5.8), (2.5.12) has a nontrivial solution $x(t)$ and on $[a_1, a_r] \times \Re^{q+1}$ the condition

$$| f(t, x_0, x_1, ..., x_q) | \le b(t) \, | x_0 | \tag{2.5.13}$$

is satisfied, where $b(t) \not\equiv 0$ is a nonnegative continuous function on $[a_1, a_r]$. Then, it is necessary that

$$(2.5.14) \qquad 1 < \frac{1}{(n-1)!(a_r - a_1)} \max_{a_1 \le t \le a_r} |\, \omega(t) \,| \int_{a_1}^{a_r} b(s)\, ds.$$

Proof. From (2.3.11) it is clear that

$$x(t) = \int_{a_1}^{a_r} g_{(2.1.1)}(t,s) f(s, \mathbf{x}(s))\, ds.$$

Thus, in view of (2.5.13) it follows that

$$(2.5.15) \qquad |\, x(t) \,| \le \int_{a_1}^{a_r} |\, g_{(2.1.1)}(t,s) \,|\, b(s) \,|\, x(s) \,|\, ds.$$

Let t be the point - or one of the points - at which $|\, x(t) \,|$ assumes its maximum value on $[a_1, a_r]$, we obtain

$$(2.5.16) \qquad 1 < \int_{a_1}^{a_r} |\, g_{(2.1.1)}(t,s) \,|\, b(s)\, ds.$$

The inequality (2.5.14) now follows from (2.3.12). ∎

Remark 2.5.1. Depending on the nature of the multiplicities k_j and the distribution of the points a_j, $1 \le j \le r$ in (2.5.12) we can estimate the right side of (2.5.14) by using the inequalities (2.4.12), (2.4.28), (2.4.44), or as in Remarks 2.4.6 and 2.4.7 deduce that

$$(2.5.17) \quad 1 < \frac{1}{(n-1)!} \frac{(n-\alpha-1)^{n-\alpha-1}(\alpha+1)^{\alpha+1}}{n^n} (a_r - a_1)^{n-1} \int_{a_1}^{a_r} b(s)\, ds. \ ∎$$

Remark 2.5.2. If $x(t)$ is a solution of the boundary value problem (2.5.8) with $n = 2m$,

$$(2.5.18) \qquad x^{(i)}(a_1) = x^{(i)}(a_2) = 0, \ 0 \le i \le m-1$$

then the inequality (2.5.17) can be improved. Indeed, in this case Green's function $g_{(2.1.5)}(t,s)$ is symmetric, i.e., $g_{(2.1.5)}(t,s) = g_{(2.1.5)}(s,t)$ and hence (2.5.16) can be written as

$$1 < \int_{a_1}^{a_2} |\, g_{(2.1.5)}(s,t) \,|\, b(s)\, ds$$

and now (2.3.12) gives that

$$1 < \frac{1}{(2m-1)!(a_2-a_1)} \int_{a_1}^{a_2} (s-a_1)^m (a_2-s)^m b(s)\, ds. \quad \blacksquare \tag{2.5.19}$$

Theorem 2.5.7. Suppose that the function $f(t, x_0, x_1, ..., x_q)$ on $[a_1, a_r] \times D_1'$ satisfies the Lipschitz condition (1.6.18) where D_1' is the same as D_1 with $L = \max_{a_1 \le t \le a_r} | f(t, 0, 0, ..., 0) |$. Then, the boundary value problem (2.5.8), (2.5.9) has a unique solution in D_1'. ∎

Definition 2.5.1. A function $\bar{x}(t) \in C^{(n)}[a_1, a_r]$ is called an *approximate solution* of (2.5.8), (2.5.9) if there exist nonnegative constants δ and ϵ such that

$$\max_{a_1 \le t \le a_r} | \bar{x}^{(n)}(t) - f(t, \bar{\mathbf{x}}(t)) | \le \delta \tag{2.5.20}$$

and

$$\max_{a_1 \le t \le a_r} | P^{(i)}_{(2.1.1)}(t) - \bar{P}^{(i)}_{(2.5.22)}(t) | \le \epsilon\, G_{n,i}(a_r - a_1)^{n-i}, \ 0 \le i \le q \tag{2.5.21}$$

where $P_{(2.1.1)}(t)$ and $\bar{P}_{(2.5.22)}(t)$ are polynomials of degree $(n-1)$ satisfying (2.1.1) and

$$\bar{P}^{(i)}_{(2.5.22)}(a_j) = \bar{x}^{(i)}(a_j); \ 0 \le i \le k_j, \ 1 \le j \le r \tag{2.5.22}$$

respectively.

The inequality (2.5.20) means that there exists a continuous function $\eta(t)$ such that

$$\bar{x}^{(n)}(t) = f(t, \bar{\mathbf{x}}(t)) + \eta(t) \tag{2.5.23}$$

and

$$\max_{a_1 \le t \le a_r} | \eta(t) | \le \delta. \tag{2.5.24}$$

Thus, from (2.3.11) the approximate solution $\bar{x}(t)$ can be expressed as

$$\bar{x}(t) = \bar{P}_{(2.5.22)}(t) + \int_{a_1}^{a_r} g_{(2.1.1)}(t, s)[f(s, \bar{\mathbf{x}}(s)) + \eta(s)]\, ds. \tag{2.5.25}$$

In the following results we shall consider the Banach space $C^{(q)}[a_1, a_r]$ and for $x(t) \in C^{(q)}[a_1, a_r]$

$$\| x \| = \max_{0 \le j \le q} \left\{ \frac{G_{n,0}(a_r - a_1)^j}{G_{n,j}} \max_{a_1 \le t \le a_r} | x^{(j)}(t) | \right\}.$$

Theorem 2.5.8. With respect to the boundary value problem (2.5.8), (2.5.9) we assume that there exists an approximate solution $\bar{x}(t)$ and

(i) the function $f(t, x_0, x_1, ..., x_q)$ satisfies the Lipschitz condition (1.6.18) on $[a_1, a_r] \times D_3$, where

$$D_3 = \left\{ (x_0, x_1, ..., x_q) \ : | x_j - \bar{x}^{(j)}(t) | \le N \frac{G_{n,j}}{G_{n,0}(a_r - a_1)^j}, \ 0 \le j \le q \right\}$$

(ii) $N_0 = (1 - \theta)^{-1}(\epsilon + \delta) G_{n,0}(a_r - a_1)^n \le N$.

Then, the following hold

(1) there exists a solution $x^*(t)$ of (2.5.8), (2.5.9) in $\bar{S}(\bar{x}, N_0)$;

(2) $x^*(t)$ is the unique solution of (2.5.8), (2.5.9) in $\bar{S}(\bar{x}, N)$;

(3) the Picard iterative sequence $\{x_m(t)\}$ defined by

$$x_{m+1}(t) = P_{(2.1.1)}(t) + \int_{a_1}^{a_r} g_{(2.1.1)}(t,s) f(s, \mathbf{x}_m(s)) \, ds$$

$$x_0(t) = \bar{x}(t); \ m = 0, \ 1, \ ...$$

converges to $x^*(t)$ with $\| x^* - x_m \| \le \theta^m N_0$, and

$$\| x^* - x_m \| \le \theta(1 - \theta)^{-1} \| x_m - x_{m-1} \|;$$

(4) for any $x_0(t) = x(t) \in \bar{S}(\bar{x}, N_0)$, $x^*(t) = \lim_{m \to \infty} x_m(t)$. ∎

Now we shall consider the convergence of the approximate Picard's iterative scheme

$$y_{m+1}(t) = P_{(2.1.1)}(t) + \int_{a_1}^{a_r} g_{(2.1.1)}(t,s) f_m(s, \mathbf{y}_m(s)) \, ds; \ m = 0, \ 1, \ ... \tag{2.5.26}$$

where $y_0(t) = x_0(t) = \bar{x}(t)$.

Theorem 2.5.9. With respect to the boundary value problem (2.5.8), (2.5.9) we assume that there exists an approximate solution $\bar{x}(t)$, and for each f_m appearing in (2.5.26) the Condition C_1 (cf. Theorem 1.6.9) is satisfied. Further, we assume that

(i) condition (i) of Theorem 2.5.8;

(ii) condition (ii) of Theorem 1.6.9;

(iii) $N_1 = (1-\theta_1)^{-1}(\epsilon+\delta+\Delta F)G_{n,0}(a_r-a_1)^n \leq N$, where $F = \max_{a_1\leq t\leq a_r} \mid f(t,\bar{\mathbf{x}}(t)) \mid$.

Then,

(1) all the conclusions (1) - (4) of Theorem 2.5.8 hold;

(2) the sequence $\{y_m(t)\}$ obtained from (2.5.26) remains in $\bar{S}(\bar{x}, N_1)$;

(3) the sequence $\{y_m(t)\}$ converges to $x^*(t)$, the solution of (2.5.8), (2.5.9) if and only if $\lim_{m\to\infty} a_m = 0$, where

$$a_m = \| y_{m+1}(t) - P_{(2.1.1)}(t) - \int_{a_1}^{a_r} g_{(2.1.1)}(t,s)f(s,\mathbf{y}_m(s))\, ds \| \tag{2.5.27}$$

and the following error estimate holds

$$\| x^* - y_{m+1} \| \leq (1-\theta)^{-1}\Big[\theta \| y_{m+1} - y_m \| + \Delta G_{n,0}(a_r-a_1)^n \times \max_{a_1\leq t\leq a_r} \mid f(t,\mathbf{y}_m(t)) \mid \Big]. \tag{2.5.28}$$

■

Theorem 2.5.10. With respect to the boundary value problem (2.5.8), (2.5.9) we assume that there exists an approximate solution $\bar{x}(t)$, and for each f_m appearing in (2.5.26) the Condition C_2 (cf. Theorem 1.6.10) is satisfied. Further, we assume that

(i) condition (i) of Theorem 2.5.8;

(ii) $N_2 = (1-\theta)^{-1}(\epsilon+\delta+\nabla)G_{n,0}(a_r-a_1)^n \leq N$.

Then,

(1) all the conclusions (1) - (4) of Theorem 2.5.8 hold;

(2) the sequence $\{y_m(t)\}$ obtained from (2.5.26) remains in $\bar{S}(\bar{x}, N_2)$;

(3) the sequence $\{y_m(t)\}$ converges to $x^*(t)$, the solution of (2.5.8), (2.5.9) if and only if $\lim_{m\to\infty} a_m = 0$, and the following error estimate holds

$$\| x^* - y_{m+1} \| \le (1-\theta)^{-1} \left[\theta \| y_{m+1} - y_m \| + \nabla G_{n,0}(a_r - a_1)^n\right]. \quad (2.5.29)$$ ∎

Definition 2.5.2. A function $\mu(t) \in C^{(n)}[a_1, a_r]$ is called a *lower solution* of (2.5.8) with $q = 0$ provided

$$\mu^{(n)}(t) \ge f(t, \mu(t)), \; t \in [a_1, a_r].$$

Similarly, a function $\nu(t) \in C^{(n)}[a_1, a_r]$ is said to be an *upper solution* of (2.5.8) with $q = 0$ if

$$\nu^{(n)}(t) \le f(t, \nu(t)), \; t \in [a_1, a_r].$$

In the space $C[a_1, a_r]$ we shall consider the maximum norm and introduce a partial ordering as follows : For $x(t), y(t) \in C[a_1, a_r]$ we say that $x \le y$ if and only if $(-1)^{k_{i+1}+\cdots+k_r+(r-i)}[y(t) - x(t)] \le 0$ for all $t \in [a_i, a_{i+1}]$, $1 \le i \le r-1$. Thus, in view of Lemma 2.3.3 the set of indices 1, 2, ..., $r-1$ can be divided into two groups $H_1 = \{i \in \{1, ..., r-1\} : k_{i+1} + \cdots + k_r + (r-i) \text{ is odd}\}$, $H_2 = \{i \in \{1, ..., r-1\} : k_{i+1} + \cdots + k_r + (r-i) \text{ is even}\}$ such that if $i \in H_1$ (H_2), then $g_{(2.1.1)}(t, s) \le 0$, $a_i \le t \le a_{i+1}$ and $y(t) \ge x(t)$ on the same interval ($g_{(2.1.1)}(t, s) \ge 0$, $y(t) \le x(t)$ for $a_i \le t \le a_{i+1}$).

Theorem 2.5.11. With respect to the boundary value problem (2.5.8), (2.5.9) with $q = 0$ we assume that

(i) the function $f(t, x)$ is nonincreasing in x for each $t \in [a_i, a_{i+1}]$, $i \in H_1$ and nondecreasing in x for each $t \in [a_i, a_{i+1}]$, $i \in H_2$;

(ii) there exist lower and upper solutions $x_0(t)$, $y_0(t)$ of (2.5.8) such that $x_0 \le y_0$ and

$$P_{(2.5.30)} \le P_{(2.1.1)} \le P_{(2.5.31)}$$

where $P_{(2.5.30)}(t)$ and $P_{(2.5.31)}(t)$ are the polynomials of degree $(n-1)$, satisfying

$$P^{(i)}_{(2.5.30)}(a_j) = x_0^{(i)}(a_j) \tag{2.5.30}$$

and

$$P^{(i)}_{(2.5.31)}(a_j) = y_0^{(i)}(a_j);\ 0 \le i \le k_j,\ 1 \le j \le r \tag{2.5.31}$$

respectively.

Then, the sequences $\{x_m\}$, $\{y_m\}$ where $x_m(t)$ and $y_m(t)$ are defined by the iterative schemes

$$x_{m+1}(t) = P_{(2.1.1)}(t) + \int_{a_1}^{a_r} g_{(2.1.1)}(t,s) f(s, x_m(s))\, ds$$

$$y_{m+1}(t) = P_{(2.1.1)}(t) + \int_{a_1}^{a_r} g_{(2.1.1)}(t,s) f(s, y_m(s))\, ds;\ m = 0,\ 1,\ \ldots$$

are well defined and $\{x_m\}$ converges to an element $x = x(t) \in C[a_1, a_r]$, $\{y_m\}$ converges to an element $y = y(t) \in C[a_1, a_r]$ (the convergence being in the norm of $C[a_1, a_r]$). Further,

$$x_0 \le x_1 \le \cdots \le x_m \le \cdots \le x \le y \le \cdots \le y_m \le \cdots \le y_1 \le y_0,$$

$x(t)$ and $y(t)$ are solutions of the boundary value problem (2.5.8), (2.5.9) with $q = 0$, and each solution $z(t)$ of this problem which is such that $z \in [x_0, y_0]$ satisfies $x \le z \le y$. ∎

Theorem 2.5.12. With respect to the boundary value problem (2.5.8), (2.5.9) we assume that there exists an approximate solution $\bar{x}(t)$, and

(i) the function $f(t, x_0, x_1, \ldots, x_q)$ is continuously differentiable with respect to all x_i, $0 \le i \le q$ on $[a_1, a_r] \times D_3$;

(ii) there exist nonnegative constants L_i, $0 \le i \le q$ such that for all $(t, x_0, x_1, ..., x_q) \in [a_1, a_r] \times D_3$

$$\left|\frac{\partial}{\partial x_i} f(t, x_0, x_1, ..., x_q)\right| \le L_i;$$

(iii) condition (iii) of Theorem 1.6.14;

(iv) $N_3 = (1 - 3\theta)^{-1}(\epsilon + \delta)G_{n,0}(a_r - a_1)^n \le N$.

Then, the following hold

(1) the sequence $\{x_m(t)\}$ generated by the quasilinear iterative scheme

$$x_{m+1}^{(n)}(t) = f(t, \mathrm{x}_m(t)) + \sum_{i=0}^{q} \left(x_{m+1}^{(i)}(t) - x_m^{(i)}(t)\right) \frac{\partial}{\partial x_m^{(i)}(t)} f(t, \mathrm{x}_m(t)) \tag{2.5.32}$$

$$x_{m+1}^{(i)}(a_j) = A_{i,j};\ 0 \le i \le k_j,\ 1 \le j \le r;\ m = 0,\ 1,\ ... \tag{2.5.33}$$

where $x_0(t) = \bar{x}(t)$ remains in $\bar{S}(\bar{x}, N_3)$;

(2) the sequence $\{x_m(t)\}$ converges to the unique solution $x^*(t)$ of the boundary value problem (2.5.8), (2.5.9);

(3) a bound on the error is given by

(2.5.34) $\| x_m - x^* \|$

$$\le \left(\frac{2\theta}{1-\theta}\right)^m \left(1 - \frac{2\theta}{1-\theta}\right)^{-1} \| x_1 - \bar{x} \|$$

$$\le \left(\frac{2\theta}{1-\theta}\right)^m \left(1 - \frac{2\theta}{1-\theta}\right)^{-1} (1-\theta)^{-1}(\epsilon + \delta)G_{n,0}(a_r - a_1)^n. \quad ■$$

Theorem 2.5.13. Let the conditions of Theorem 2.5.12 be satisfied. Further, let $f(t, x_0, x_1, ..., x_q)$ be twice continuously differentiable with respect to all x_i, $0 \le i \le q$ on $[a_1, a_r] \times D_3$, and

$$\left|\frac{\partial^2}{\partial x_i \partial x_j} f(t, x_0, x_1, ..., x_q)\right| \le L_i L_j K,\ 0 \le i, j \le q.$$

Then,

$$\text{(2.5.35)} \qquad \| x_{m+1} - x_m \| \leq \alpha \| x_m - x_{m-1} \|^2 \leq \frac{1}{\alpha} (\alpha \| x_1 - x_0 \|)^{2^m}$$

$$\leq \frac{1}{\alpha} \left[\frac{1}{2} K(\epsilon + \delta) \left(\frac{\theta}{1-\theta} \right)^2 \right]^{2^m},$$

where $\alpha = [K\theta^2/2(1-\theta)G_{n,0}(a_r - a_1)^n]$. Thus, the convergence is quadratic if

$$\frac{1}{2} K(\epsilon + \delta) \left(\frac{\theta}{1-\theta} \right)^2 < 1. \quad \blacksquare$$

Next we shall consider the convergence of the approximate quasilinear iterative scheme

$$\text{(2.5.36)} \quad y_{m+1}^{(n)}(t) = f_m(t, \mathbf{y}_m(t)) + \sum_{i=0}^{q} \left(y_{m+1}^{(i)}(t) - y_m^{(i)}(t) \right) \frac{\partial}{\partial y_m^{(i)}(t)} f_m(t, \mathbf{y}_m(t))$$

$$\text{(2.5.37)} \qquad y_{m+1}^{(i)}(a_j) = A_{i,j};\ 0 \leq i \leq k_j,\ 1 \leq j \leq r;\ m = 0,\ 1,\ \ldots$$

where $y_0(t) = x_0(t) = \bar{x}(t)$.

Theorem 2.5.14. With respect to the boundary value problem (2.5.8), (2.5.9) we assume that there exists an approximate solution $\bar{x}(t)$, and for each f_m appearing in (2.5.36) the Condition C_3 (cf. Theorem 1.6.16) on $[a_1, a_r] \times D_3$ is satisfied. Further, we assume that

(i) conditions (i) and (ii) of Theorem 2.5.12;

(ii) condition (ii) of Theorem 1.6.16;

(iii) $N_4 = (1-\theta_2)^{-1}(\epsilon + \delta + \Delta F)G_{n,0}(a_r - a_1)^n \leq N$, where $F = \max_{a_1 \leq t \leq a_r} | f(t, \bar{\mathbf{x}}(t)) |$.

Then,

(1) all the conclusions (1) - (3) of Theorem 2.5.12 hold;

(2) the sequence $\{y_m(t)\}$ obtained from (2.5.36), (2.5.37) remains in $\bar{S}(\bar{x}, N_4)$;

(3) the sequence $\{y_m(t)\}$ converges to $x^*(t)$, the unique solution of (2.5.8), (2.5.9) if and only if $\lim_{m\to\infty} a_m = 0$, and the following error estimate holds

$$\| x^* - y_{m+1} \| \leq (1-\theta)^{-1}\Big[2\theta \| y_{m+1} - y_m \| + \Delta G_{n,0}(a_r - a_1)^n \times \max_{a_1 \leq t \leq a_r} | f(t, y_m(t)) | \Big]. \tag{2.5.38}$$

■

Theorem 2.5.15. Let the conditions of Theorem 2.5.14 be satisfied. Further, let $f_m = f_0$ for $m = 1, 2, \ldots$ and $f_0(t, x_0, x_1, \ldots, x_q)$ be twice continuously differentiable with respect to all x_i, $0 \leq i \leq q$ on $[a_1, a_r] \times D_3$, and

$$\left| \frac{\partial^2}{\partial x_i \partial x_j} f_0(t, x_0, x_1, \ldots, x_q) \right| \leq L_i L_j K, \; 0 \leq i, j \leq q.$$

Then,

$$\| y_{m+1} - y_m \| \leq \alpha \| y_m - y_{m-1} \|^2 \leq \frac{1}{\alpha} (\alpha \| y_1 - y_0 \|)^{2^m} \leq \frac{1}{\alpha} \left[\frac{1}{2} K(\epsilon + \delta + \Delta F) \left(\frac{\theta}{1-\theta} \right)^2 \right]^{2^m}, \tag{2.5.39}$$

where α is the same as in Theorem 2.5.13. ■

Generalized Liapunoff's Inequality

Let the function $b(t)$ be continuous on $[a_1, a_2]$. A necessary condition for the differential equation

$$x''(t) + b(t)\, x = 0 \tag{2.5.40}$$

to have a solution $x(t) \not\equiv 0$ possessing zeros at a_1 and a_2 is that

$$\int_{a_1}^{a_2} | b(s) | \, ds > \frac{4}{a_2 - a_1}. \tag{2.5.41}$$

This oscillation criterion is originally due to Liapunoff (cf. [14]). We can restate this result as follows : If $x''(t)$ and $x''(t)x^{-1}(t)$ are continuous on $[a_1, a_2]$ with $x(a_1) = x(a_2) = 0$, then

$$\int_{a_1}^{a_2} | x''(s)x^{-1}(s) | \, ds > \frac{4}{a_2 - a_1}. \tag{2.5.42}$$

By taking $f = x^{(n)}(t)x^{-1}(t)x$ in (2.5.8), $b(t) = | x^{(n)}(t)x^{-1}(t) |$ in (2.5.13), (2.5.17) leads to the following extension of (2.5.42) : If $x^{(n)}(t)$ and $x^{(n)}(t)x^{-1}(t)$ are continuous on $[a_1, a_r]$ and $x(t)$ satisfies (2.5.12), then

$$(2.5.43)\quad \int_{a_1}^{a_r} | x^{(n)}(s)x^{-1}(s) | \, ds > \frac{(n-1)!\, n^n}{(n-\alpha-1)^{n-\alpha-1}(\alpha+1)^{\alpha+1}(a_r - a_1)^{n-1}}.$$

This reduces to (2.5.42) when $n = 2$. Similarly, in view of (2.5.19) we have : If $x^{(2m)}(t)$ and $x^{(2m)}(t)x^{-1}(t)$ are continuous on $[a_1, a_2]$ and $x(t)$ satisfies (2.5.18), then

$$(2.5.44)\quad \int_{a_1}^{a_2} (s - a_1)^m (a_2 - s)^m | x^{(2m)}(s)x^{-1}(s) | \, ds > (2m-1)!(a_2 - a_1).$$

Generalized Hartman's Inequality

Once again let the function $b(t)$ be continuous on $[a_1, a_2]$. Hartman [24] has shown that a necessary condition for the differential equation (2.5.40) to have a solution $x(t) \not\equiv 0$ possessing two zeros in $[a_1, a_2]$ is that

$$(2.5.45)\qquad \int_{a_1}^{a_2} (s - a_1)(a_2 - s) b_+(s) \, ds > (a_2 - a_1)$$

where $b_+(t) = \max \{b(t), 0\}$. It is clear that Liapunoff's inequality (2.5.41) follows from (2.5.45).

An extension of Hartman's inequality for the differential equation

$$(2.5.46)\qquad x^{(2m)} + (-1)^{m-1} b(t)\, x = 0$$

is embodied in the following :

Theorem 2.5.16. Let the function $b(t)$ be continuous on $[a_1, a_2]$ and positive somewhere in $[a_1, a_2]$. Then, a necessary condition for the differential equation (2.5.46) to have a solution $x(t) \not\equiv 0$ satisfying

$$(2.5.47)\qquad x^{(i)}(t_1) = x^{(i)}(t_2) = 0; \ 0 \le i \le m-1, \ a_1 \le t_1 < t_2 \le a_2$$

is that

$$\int_{a_1}^{a_2} (s-a_1)^{2m-1}(a_2-s)^{2m-1} b_+(s)\, ds \geq (2m-1)[(m-1)!]^2 (a_2-a_1)^{2m-1}. \tag{2.5.48}$$

Proof. Following Das and Vatsala [21] consider the auxiliary equation

$$y^{(2m)} + (-1)^{m-1} b_+(t)\, y = 0. \tag{2.5.49}$$

From the given hypotheses it follows that (cf. Hinton [25]) the differential equation (2.5.49) has a nontrivial solution $y(t)$ satisfying

$$y^{(i)}(t_1') = y^{(i)}(t_2') = 0;\ 0 \leq i \leq m-1,\ a_1 \leq t_1' < t_2' \leq a_2. \tag{2.5.50}$$

Now, consider the differential equation

$$y^{(2m)} + (-1)^{m-1} \lambda\, b_+(t)\, y = 0 \tag{2.5.51}$$

together with the boundary conditions (2.5.50). The eigenvalue problem (2.5.51), (2.5.50) is equivalent to the integral equation

$$z(t) = \lambda \int_{t_1'}^{t_2'} K(t,s) z(s)\, ds \tag{2.5.52}$$

where

$$K(t,s) = (-1)^{m-1} \bar{g}_{(2.1.5)}(t,s)\, [b_+(t) b_+(s)]^{\frac{1}{2}}$$

$$z(t) = (b_+(t))^{\frac{1}{2}}\, y(t)$$

and $\bar{g}_{(2.1.5)}(t,s)$ is the same as $g_{(2.1.5)}(t,s)$ with $a_1 = t_1',\ a_2 = t_2'$.

Since the k th eigenvalue λ_k of (2.5.51), (2.5.50) is the minimum of

$$\int_{t_1'}^{t_2'} \left(\omega^{(m)}(t)\right)^2 dt \Big/ \int_{t_1'}^{t_2'} b_+(t) \omega^2(t)\, dt$$

where $\omega(t)$ satisfies the boundary conditions $\omega^{(i)}(t_1') = \omega^{(i)}(t_2') = 0,\ 0 \leq i \leq m-1$ as well as the orthogonality conditions related to Courant's mini - max principle [19], $K(t,s)$ is nonnegative definite. Thus, Mercer's theorem [19] leads to (note that $\lambda_1 = 1$)

$$(2.5.53) \qquad 1 \leq \sum_{k=1}^{\infty} \frac{1}{\lambda_k} = \int_{t_1'}^{t_2'} \mid \bar{g}_{(2.1.5)}(s,s) \mid b_+(s)\; ds$$

$$= \frac{1}{(2m-1)[(m-1)!]^2} \int_{t_1'}^{t_2'} \left[\frac{(s-t_1')(t_2'-s)}{t_2'-t_1'}\right]^{2m-1} b_+(s)\; ds.$$

However, since $(s-a_1)(a_2-s)/\,(a_2-a_1) \geq (s-t_1')(t_2'-s)/\,(t_2'-t_1')$ for $a_1 \leq t_1' \leq s \leq t_2' \leq a_2$ the inequality (2.5.48) is immediate from (2.5.53). ∎

Remark 2.5.3. If in (2.5.46) the function $b(t)$ is positive, then the eigenvalues of (2.5.51), (2.5.50) are all positive and thus the kernel $K(t,s)$ is such that 0 is not an eigenvalue of

$$\mu\; z(t) = \int_{t_1'}^{t_2'} K(t,s) z(s)\; ds.$$

Further, since $K(t,s)$ defines a completely continuous symmetric operator, each eigenvalue μ_k is of finite multiplicity. Thus, in place of (2.5.53) we have

$$1 < \int_{t_1'}^{t_2'} \mid \bar{g}_{(2.1.5)}(s,s) \mid b_+(s)\; ds$$

and hence (2.5.48) holds with the strict inequality. This inequality reduces to (2.5.45) when $m = 1$. ∎

Remark 2.5.4. The inequality

$$(2.5.54) \qquad \int_{a_1}^{a_2} b_+(s)\; ds > \frac{4^{2m-1}(2m-1)[(m-1)!]^2}{(a_2-a_1)^{2m-1}}$$

due to Levin [30] follows from (2.5.48) in view of the inequality $(s-a_1)(a_2-s) < (a_2-a_1)^2/4,\;\; s \in [a_1, a_2]$ and $s \neq (a_1+a_2)/2$. ∎

A Lower Bound for the Zeros of the Solutions

The following result provides a lower bound for the m th zero of solutions of the linear differential equation

$$(2.5.55) \qquad x^{(n)} + b(t)\; x = 0.$$

Theorem 2.5.17. Let the function $b(t)$ be continuous, $b(t) \not\equiv 0$ on $[t_0, \infty)$, and

$$\int_{t_0}^{\infty} | b(s) | \, ds = K. \tag{2.5.56}$$

Further, let $a_1 \leq a_2 \leq \cdots \leq a_m$ be m consecutive zeros of any solution of (2.5.55) on the interval $[t_0, \infty)$, then for $m \geq n$

$$a_m > a_1 + \frac{n}{n-1}\left(\frac{(m-n+1)(n-1)!}{K}\right)^{\frac{1}{n-1}}. \tag{2.5.57}$$

Proof. It is clear that no solution of (2.5.55) can have a zero of multiplicity greater than $(n-1)$ at any point of $[t_0, \infty)$. Hence, if $a_i \leq a_{i+1} \leq \cdots \leq a_{i+n-1}$ are n consecutive zeros of a solution of (2.5.55) on $[t_0, \infty)$ then $a_i < a_{i+n-1}$, and (2.5.17) with $\alpha = 0$ and $a_1 = a_i$, $a_r = a_{i+n-1}$ applies to give

$$\left(\frac{n}{n-1}\right)^{n-1} n! < (a_{i+n-1} - a_i)^{n-1} \int_{a_i}^{a_{i+n-1}} | b(s) | \, ds. \tag{2.5.58}$$

Suppose $m = qn + p$, where $q \geq 1$, $0 \leq p \leq n-1$, so $a_m \geq a_{qn}$. Taking $i = 1,\ n+1,\ \ldots,\ (q-1)n+1$ in (2.5.58) and adding these inequalities gives

$$\left(\frac{n}{n-1}\right)^{n-1} q \ n! < \sum_{i=1}^{q} (a_{in} - a_{(i-1)n+1})^{n-1} \int_{a_{(i-1)n+1}}^{a_{in}} | b(s) | \, ds,$$

which in view of $qn = m - p \geq m - n + 1$ leads to

$$\left(\frac{n}{n-1}\right)^{n-1} (m-n+1)(n-1)! < (a_m - a_1)^{n-1} \int_{a_1}^{a_m} | b(s) | \, ds. \tag{2.5.59}$$

The inequality (2.5.57) follows at once from (2.5.59). ∎

Remark 2.5.5. If $m > 2n-1$ then the inequality (2.5.57) can be improved. Indeed, if $m = qn + p$, with $q \geq 2$, $0 \leq p \leq n-1$ then there exists precisely one integer $\tau \geq 2$ such that $\tau n - (\tau - 1) < m \leq (\tau + 1)n - \tau$. Now taking $i = 1,\ n,\ 2n-1,\ \ldots,\ (\tau-1)n - (\tau-2)$ in (2.5.58), and proceeding as above, we obtain

$$\left(\frac{n}{n-1}\right)^{n-1} \tau \ n! < (a_{\tau n - (\tau-1)} - a_1)^{n-1} \int_{a_1}^{a_{\tau n-(\tau-1)}} | b(s) | \, ds.$$

Since $\tau(n-1) + n \geq m$ and $a_m > a_{\tau n - (\tau - 1)}$, we have

$$\left(\frac{n}{n-1}\right)^{n-1} \frac{m-n}{n-1} \, n! < (a_m - a_1)^{n-1} \int_{a_1}^{a_m} | b(s) | \, ds;$$

this yields the estimate

$$a_m > a_1 + \frac{n}{n-1}\left(\frac{(m-n)\, n!}{(n-1)K}\right)^{\frac{1}{n-1}}, \tag{2.5.60}$$

which is an improvement on (2.5.57) for $m > 2n-1$. ∎

Remark 2.5.6. If in Theorem 2.5.17 instead of (2.5.56) we assume that $|\, b(t) \,| \le K$ for all $t \in [t_0, \infty)$, then the inequality (2.5.16) with $a_1 = a_i$, $a_r = a_{i+(n-1)}$ can be used to obtain

$$1 < K \int_{a_i}^{a_{i+(n-1)}} |\, g_{(2.1.1)}(t,s) \,|\, ds,$$

which in view of (2.3.13) is the same as

$$1 < \frac{K}{n!} \prod_{j=i}^{i+(n-1)} |\, t - a_j \,| \,.$$

Therefore, as earlier we find that

$$1 < \frac{K}{n!} \frac{(n-1)^{n-1}}{n^n} (a_{i+(n-1)} - a_i)^n.$$

Now following as in Theorem 2.5.17, we easily obtain

$$a_m > a_1 + \left[\frac{(m-n+1)\, n!}{K}\left(\frac{n}{n-1}\right)^{n-1}\right]^{\frac{1}{n}}. \tag{2.5.61}$$

Further, if $m > 2n-1$ then as in Remark 2.5.5 the inequality (2.5.60) can be replaced by

$$a_m > a_1 + \frac{n}{n-1}\left(\frac{(m-n)\, n!}{K}\right)^{\frac{1}{n}}. \tag{2.5.62}$$ ∎

A Test for Disconjugacy

The linear differential equation

$$x^{(n)} + a_{n-1}(t)\, x^{(n-1)} + \cdots + a_0(t)\, x = 0 \tag{2.5.63}$$

is said to be $(k_1+1,\ ...,\ k_r+1)$ *disconjugate* on an interval $[a,b]$ if the only solution of (2.5.63) satisfying (2.5.12) where $a \le a_1 < a_2 < \cdots < a_r \le b$ is the trivial solution. For example, the differential equation $x^{(n)} = 0$ is $(k_1+1,\ ...,\ k_r+1)$ disconjugate on any interval.

Theorem 2.5.18. Suppose that the functions $a_i(t),\ 0 \le i \le n-1$ are continuous on $[a,b]$ and $|\ a_i(t)\ | \le L_i,\ 0 \le i \le n-1$ for all $t \in [a,b]$. Then, the differential equation (2.5.63) is $(k_1+1,\ ...,\ k_r+1)$ disconjugate on $[a,b]$ provided

$$\bar{\theta} = \sum_{i=0}^{n-1} G_{n,i} L_i (b-a)^{n-i} < 1. \tag{2.5.64}$$

Proof. Suppose on the contrary that (2.5.63) has a nontrivial solution $x(t)$ satisfying (2.5.12) where $a \le a_1 < a_2 < \cdots < a_r \le b$. Then, there exists a $\tau \in [a_1, a_r]$ such that $M_n = \max_{a_1 \le t \le a_r} |\ x^{(n)}(t)\ | = |\ x^{(n)}(\tau)\ |$. Therefore, in view of (2.5.10) it follows that

$$\begin{aligned} M_n &= |\ x^{(n)}(\tau)\ | = |\ a_0(\tau)x(\tau) + \cdots + a_{n-1}(\tau)x^{(n-1)}(\tau)\ | \\ &\le \sum_{i=0}^{n-1} L_i\ |\ x^{(i)}(\tau)\ | \\ &\le \theta\ M_n \ \le\ \bar{\theta}\ M_n. \end{aligned}$$

Evidently $M_n > 0$, since otherwise $x(t)$ would coincide in $[a_1, a_r]$ with a polynomial of degree $m < n$ and $x^{(m)}(t)$ would not vanish in $[a_1, a_r]$. Hence $\bar{\theta} \ge 1$. This contradiction completes the proof. ∎

Remark 2.5.7. In view of Remark 2.4.20 the condition (2.5.64) can be written as

$$\begin{aligned} &\sum_{i=0}^{\alpha} \frac{(n-\alpha-1)^{n-\alpha-1}(\alpha+1-i)^{\alpha+1-i}}{(n-i)!(n-i)^{n-i}} L_i (b-a)^{n-i} \\ &\quad + \sum_{i=\alpha+1}^{n-1} \frac{(i-\alpha)}{(n-i)!(n-\alpha)} L_i (b-a)^{n-i} < 1. \end{aligned} \tag{2.5.65}$$

This inequality for $\alpha = 0$ (which corresponds to the ordinary disconjugacy, i.e., no nontrivial solution of (2.5.63) has n zeros in [a,b]) reduces to the same criterion established earlier by Bessmertnyh and Levin [11]. ∎

REFERENCES

1. **R.P.Agarwal,** Some inequalities for a function having n zeros, in General Inequalities 3, Ed. E.F.Beckenbach and W.Walter, International Series of Numerical Mathematics, Birkhäuser Verlag, 64(1983), 371-378.

2. **R.P.Agarwal,** Boundary value problems for higher order integro - differential equations, Nonlinear Analysis 7(1983), 259-270.

3. **R.P.Agarwal,** Boundary Value Problems for Higher Order Differential Equations, World Scientific, Singapore • Philadelphia, 1986.

4. **R.P.Agarwal,** Better error estimates in polynomial interpolation, J. Math. Anal. Appl. 161(1991), 241-257.

5. **R.P.Agarwal** and **P.J.Y.Wong,** Optimal error bounds for the derivatives of two point Hermite interpolation, Computers Math. Applic. 21(1991), 21-35.

6. **R.P.Agarwal,** Sharp Hermite interpolation error bounds for derivatives, Nonlinear Analysis, 17(1991), 773-786.

7. **R.P.Agarwal,** Sharp inequalities in polynomial interpolation, in General Inequalities 6, Ed. W.Walter, International Series of Numerical Mathematics, Birkhäuser Verlag, 103(1992), 73-92.

8. **S.Anon,** Problem E 2155, Amer. Math. Monthly 76(1969), 188.

9. **P.R.Beesack,** On the Green's function of an N - point boundary value problem, Pacific J. Math. 12(1962), 801-812.

10. **I.S.Berezin** and **N.P.Zhidkov,** Computing Methods, Pergamon Press, Oxford, 1965.

11. **G.A.Bessmertnyh** and **A.Ju.Levin,** Some inequalities satisfied by differentiable functions of one variable, Soviet Math. Dokl. 3(1962), 737-740.

12. **G.A.Bessmertnyh,** The existence and uniqueness of the solutions of a multi-point de la Vallée Poussin problem for nonlinear differential equations (Russian), Differential Equations 6(1970), 298-310.

13. **G.Birkhoff** and **A.Priver,** Hermite interpolation errors for derivatives, J. Math. Phys. 46(1967), 440-447.

14. **G.Borg,** On a Liapunoff criterion of stability, Amer. J. Math. 71(1949), 67-70.

15. **J.Brink,** Inequalities involving $\| f \|_p$ and $\| f^{(n)} \|_q$ for f with n zeros, Pacific J. Math. 42(1972), 289-311.

16. **S.N.Chow, D.R.Dunninger** and **A.Lasota,** A maximum principle for fourth order ordinary differential equations, J. Differential Equations 14(1973), 101-105.

17. **P.G.Ciarlet, M.H.Schultz** and **R.S.Varga,** Numerical methods of high order accuracy for nonlinear boundary value problems, I. one dimensional problem, Numerische Mathematik 9(1967), 394-430.

18. **W.A.Coppel,** Disconjugacy, Lecture Notes in Mathematics 220, Springer-Verlag, New York, 1971.

19. **R.Courant** and **D.Hilbert,** Methods of Mathematical Physics, Vol. 1, Interscience Pub., New York, 1953.

20. **K.M.Das** and **A.S.Vatsala,** On Green's function of an n - point boundary value problem, Trans. Amer. Math. Soc. 182(1973), 469-480.

21. **K.M.Das** and **A.S.Vatsala,** Green's function for n - n boundary value problem and an analogue of Hartman's result, J. Math. Anal. Appl. 51(1975), 670-677.

22. **P.J.Davis,** Interpolation and Approximation, Blaisdell Publishing Co., Boston, 1961.

23. **W.Gautschi,** Private communication (September 1987).

24. **P.Hartman,** Ordinary Differential Equations, John Wiley, New York, 1964.

25. **D.B.Hinton,** A criterion for n - n oscillation in differential equations of order $2n$, Proc. Amer. Math. Soc. 19(1968), 511-518.

26. **M.Hukuhara,** On the zeros of solutions of linear ordinary differential equations, Súgaku 15(1963), 108-109.

27. **Y.P.Kobelkov** and **I.I.Kobyakov,** The existence and uniqueness of solutions of boundary value problems for differential equations with deviating arguments, Differential Equations 10(1974), 610-614.

28. **J.R.Kuttler,** A remark on the paper, 'A maximum principle for fourth order ordinary differential equations' by Chow, Dunninger and Lasota, J. Differential Equations 17(1975), 44-45.

29. **A.Ju.Levin,** Some problems bearing on the oscillation of solutions of linear differential equations, Soviet Math. Dokl. 4(1963), 121-124.

30. **A.Ju.Levin,** Distribution of the zeros of solutions of a linear differential equation, Dokl. Akad. Nauk SSSR 156(1964), 1281-1284.

31. **T.M.MacRobert,** Spherical Harmonics, Pergamon Press, Oxford, Third Edn. (1967).

32. **G.V.Milovanovic** and **J.E.Pécarić,** On an application of Hermite's interpolation polynomial and some related results, Numerical Methods and Approximation Theory, Niš, September 26-28(1984), 93-98.

33. **D.S.Mitrinoviċ,** Analytic Inequalities, Springer - Verlag, Berlin and New York, 1970.

34. **Z.Nehari,** On an inequality of P.R.Beesak, Pacific J. Math. 14(1964), 261-263.

35. **G.M.Phillips** and **P.J.Taylor,** Theory and Applications of Numerical Analysis, Academic Press, Inc., New York, 1974.

36. **Yu.V.Pokornyi,** Some estimates of the Green's function of a multipoint boundary value problem, Math. Notes 4(1968), 810-814.

37. **M.H.Protter** and **H.F.Weinberger,** Maximum Principles in Differential Equations, Prentice-Hall, Englewood Cliffs, 1967.

38. **V.Šeda,** Two remarks on boundary value problems for ordinary differential equations, J. Differential Equations 26(1976), 278-290.

39. **Q.Sheng** and **R.P.Agarwal,** Generalized maximum principle for higher order ordinary differential equations, J. Math. Anal. Appl. to appear.

40. **G.Szegö,** Orthogonal Polynomials, Am. Math. Soc. Colloq. Publ. 23(1939).

41. **A.Torchinsky,** Solution of the problem E 2155, Amer. Math. Monthly 76(1969), 1142-1143.

42. **M.Tumura,** Kôkai zyôbibunhôteisiki ni tuite, Kansû Hôteisiki 30(1941), 20-35.

43. **P.J.Y.Wong** and **R.P.Agarwal,** A new representation for the error function for the Hermite interpolation and sharper pointwise and uniform error bounds for the derivatives, Nonlinear Analysis 19(1992), 769-786.

44. **S.Yakowitz** and **F.Szidarovszky,** An Introduction to Numerical Computations, MacMillan Pub. Co., New York, 1986.

45. **S.Zaidman,** Problem E 2622, Amer. Math. Monthly 83(1976), 740-741.

CHAPTER 3

ABEL - GONTSCHAROFF INTERPOLATION

3.1 INTRODUCTION

Let $-\infty < a < b < \infty$, and $a \leq a_1 \leq a_2 \leq \cdots \leq a_n \leq b$ be the given points. It is easily seen that the *Abel - Gontscharoff interpolating polynomial* $P_{(3.1.1)}(t)$ of degree $(n-1)$ satisfying the *Abel - Gontscharoff conditions*

$$P^{(i)}_{(3.1.1)}(a_{i+1}) = A_{i+1},\ 0 \leq i \leq n-1 \tag{3.1.1}$$

exists uniquely [10,15]. These conditions in particular include the
(i) $(k_1, ..., k_r)$ *right focal point conditions*

$$P^{(i)}_{(3.1.2)}(a_j) = A_{i,j};\ s_{j-1} \leq i \leq s_j - 1,\ s_0 = 0,\ s_j = \sum_{i=1}^{j} k_i\ (k_i \geq 1), \tag{3.1.2}$$

$$1 \leq j \leq r(\geq 2),\ \sum_{i=1}^{r} k_i = n, a \leq a_1 < a_2 < \cdots < a_r \leq b$$

(ii) *two - point right focal conditions*

$$\begin{aligned} &P^{(i)}_{(3.1.3)}(a_1) = A_i,\ 0 \leq i \leq \alpha \\ &P^{(i)}_{(3.1.3)}(a_2) = A_i;\ \alpha+1 \leq i \leq n-1,\ a \leq a_1 < a_2 \leq b. \end{aligned} \tag{3.1.3}$$

The plan of this chapter is as follows : In Section 3.2 we shall provide an explicit representation of the Abel - Gontscharoff interpolating polynomial $P_{(3.1.1)}(t)$. From this the two - point right focal interpolating polynomial $P_{(3.1.3)}(t)$ is deduced which is in a very compact form. In Section 3.3 we shall obtain two different forms of the error function $e_{(3.1.1)}(t) = x(t) - P_{(3.1.1)}(t)$, where $x(t) \in C^{(n)}[a,b]$ and $P_{(3.1.1)}(t)$ is the Abel - Gontscharoff interpolating polynomial of the function $x(t)$, i.e., it satisfies the conditions $P^{(i)}_{(3.1.1)}(a_{i+1}) = x^{(i)}(a_{i+1}),\ 0 \le i \le n-1$. Best possible uniform bounds for $| e^{(k)}_{(3.1.1)}(t) |,\ 0 \le k \le n-1$ are obtained in Section 3.4. Finally, in Section 3.5 we shall indicate some applications of the results presented in Sections 3.2 - 3.4.

3.2 INTERPOLATING POLYNOMIAL REPRESENTATIONS

Theorem 3.2.1. The Abel - Gontscharoff interpolating polynomial $P_{(3.1.1)}(t)$ can be expressed as

$$P_{(3.1.1)}(t) = \sum_{i=0}^{n-1} T_i(t) A_{i+1}, \tag{3.2.1}$$

where $T_0(t) = 1$, and $T_i(t),\ 1 \le i \le n-1$ is the unique polynomial of degree i satisfying

$$\begin{aligned} T_i^{(k)}(a_{k+1}) &= 0,\ 0 \le k \le i-1 \\ T_i^{(i)}(a_{i+1}) &= 1 \end{aligned} \tag{3.2.2}$$

and it can be written as

$$T_i(t) = \frac{1}{1!2!\cdots i!} \begin{vmatrix} 1 & a_1 & a_1^2 & \ldots & a_1^{i-1} & a_1^i \\ 0 & 1 & 2a_2 & \ldots & (i-1)a_2^{i-2} & ia_2^{i-1} \\ \ldots & \ldots & \ldots & \ldots & \ldots & \ldots \\ 0 & 0 & 0 & \ldots & (i-1)! & i!a_i \\ 1 & t & t^2 & \ldots & t^{i-1} & t^i \end{vmatrix} \tag{3.2.3}$$

$$= \int_{a_1}^{t} \int_{a_2}^{t_1} \cdots \int_{a_i}^{t_{i-1}} dt_i dt_{i-1} \cdots dt_1, \quad (t_0 = t). \tag{3.2.4}$$

Proof. It is clear that $T_i(t)$ is a polynomial of degree i. Further, $T_i(t)$ defined in (3.2.3) satisfies (3.2.2) follows from the usual properties of the determinants. Similarly, $T_i(t)$ defined in (3.2.4) satisfies (3.2.2) follows by the successive differentiation. ∎

In particular, we have

$$T_0(t) = 1$$

$$T_1(t) = t - a_1$$

$$T_2(t) = \frac{1}{2}\left[t^2 - 2a_2 t + a_1(2a_2 - a_1)\right].$$

Corollary 3.2.2. The two - point right focal interpolating polynomial $P_{(3.1.3)}(t)$ can be written as

$$P_{(3.1.3)}(t) = \sum_{i=0}^{\alpha} \frac{(t-a_1)^i}{i!} A_i + \sum_{j=0}^{n-\alpha-2}\left[\sum_{i=0}^{j} \frac{(t-a_1)^{\alpha+1+i}(a_1-a_2)^{j-i}}{(\alpha+1+i)!(j-i)!}\right] A_{\alpha+1+j}. \tag{3.2.5}$$

Proof. Since $a_1 = a_2 = \cdots = a_{\alpha+1}$ from (3.2.4) it is clear that $T_i(t) = (t-a_1)^i/i!$, $0 \le i \le \alpha+1$. Further, since $(a_2 =)a_{\alpha+2} = a_{\alpha+3} = \cdots = a_n$ once again from (3.2.4), we have

$$\begin{aligned}
T_{\alpha+1+j}(t) &= \int_{a_1}^{t}\int_{a_1}^{t_1}\cdots\int_{a_1}^{t_\alpha}\int_{a_2}^{t_{\alpha+1}}\cdots\int_{a_2}^{t_{\alpha+1+j-1}} dt_{\alpha+1+j}\cdots dt_1 \\
&= \int_{a_1}^{t}\int_{a_1}^{t_1}\cdots\int_{a_1}^{t_\alpha} \frac{(t_{\alpha+1}-a_2)^j}{j!} dt_{\alpha+1}\cdots dt_1 \\
&= \int_{a_1}^{t}\int_{a_1}^{t_1}\cdots\int_{a_1}^{t_\alpha}\left[\sum_{i=0}^{j} \frac{(t_{\alpha+1}-a_1)^i(a_1-a_2)^{j-i}}{i!(j-i)!}\right] dt_{\alpha+1}\cdots dt_1 \\
&= \sum_{i=0}^{j} \frac{(t-a_1)^{\alpha+1+i}(a_1-a_2)^{j-i}}{(\alpha+1+i)!(j-i)!}, \quad 1 \le j \le n-\alpha-2.
\end{aligned}$$

∎

3.3 ERROR REPRESENTATIONS

Let $A_{i+1} = x^{(i)}(a_{i+1}),\ 0 \le i \le n-1$ where the function $x(t)$ is assumed to be n times continuously differentiable on $[a,b]$. In such a case $P_{(3.1.1)}(t)$ is called the Abel - Gontscharoff interpolating polynomial of the function $x(t)$. For the associated error $e_{(3.1.1)}(t) = x(t) - P_{(3.1.1)}(t)$ here we shall provide two different representations. In particular for $e_{(3.1.3)}(t)$ we shall also obtain its Cauchy's representation.

Repeated Integrals Representation

Theorem 3.3.1. In terms of repeated integrals the error function $e_{(3.1.1)}(t)$ can be written as

$$e_{(3.1.1)}(t) = \int_{a_1}^{t}\int_{a_2}^{t_1}\cdots\int_{a_n}^{t_{n-1}} x^{(n)}(t_n)dt_n\cdots dt_1. \tag{3.3.1}$$

Proof. It suffices to note that

$$e^{(k)}_{(3.1.1)}(t) = \int_{a_{k+1}}^{t}\int_{a_{k+2}}^{t_1}\cdots\int_{a_n}^{t_{n-k-1}} x^{(n)}(t_{n-k})dt_{n-k}\cdots dt_1, \quad 0 \le k \le n-1. \tag{3.3.2}$$

∎

Peano's Representation

Theorem 3.3.2. The error function $e_{(3.1.1)}(t)$ can be written as

$$e_{(3.1.1)}(t) = \int_a^b g_{(3.1.1)}(t,s)x^{(n)}(s)\, ds \tag{3.3.3}$$

where $g_{(3.1.1)}(t,s)$ is Green's function of the boundary value problem

$$\begin{aligned} &z^{(n)} = 0 \\ &z^{(i)}(a_{i+1}) = 0,\ 0 \le i \le n-1 \end{aligned} \tag{3.3.4}$$

and appears as

(3.3.5) $$g_{(3.1.1)}(t,s) = \begin{cases} \displaystyle\sum_{i=0}^{k-1} \frac{T_i(t)}{(n-i-1)!}(a_{i+1}-s)^{n-i-1}, \ a_k \le s \le t \\ \displaystyle -\sum_{i=k}^{n-1} \frac{T_i(t)}{(n-i-1)!}(a_{i+1}-s)^{n-i-1}, \ t \le s \le a_{k+1} \\ \qquad k = 0,\ 1,\ \ldots,\ n \ (a_0 = a,\ a_{n+1} = b). \end{cases}$$

Proof. It is clear that

$$e_{(3.1.1)}^{(n)}(t) = x^{(n)}(t)$$

$$e_{(3.1.1)}^{(i)}(a_{i+1}) = 0, \ 0 \le i \le n-1$$

and therefore

$$e_{(3.1.1)}(t) = \sum_{i=0}^{n-1} T_i(t)\alpha_{i+1} + \frac{1}{(n-1)!}\int_a^t (t-s)^{n-1} x^{(n)}(s)\, ds,$$

where

$$\alpha_{i+1} = -\frac{1}{(n-i-1)!}\int_a^{a_{i+1}} (a_{i+1}-s)^{n-i-1} x^{(n)}(s)\, ds, \ 0 \le i \le n-1.$$

Thus, it follows that

$$\begin{aligned} e_{(3.1.1)}(t) &= -\sum_{i=0}^{n-1} \frac{T_i(t)}{(n-i-1)!}\int_a^{a_{i+1}} (a_{i+1}-s)^{n-i-1} x^{(n)}(s)\, ds \\ &\quad + \frac{1}{(n-1)!}\int_a^t (t-s)^{n-1} x^{(n)}(s)\, ds \\ &= -\sum_{k=0}^{n-1}\int_{a_k}^{a_{k+1}} \sum_{i=k}^{n-1} \frac{T_i(t)}{(n-i-1)!}(a_{i+1}-s)^{n-i-1} x^{(n)}(s)\, ds \qquad (3.3.6) \\ &\quad + \frac{1}{(n-1)!}\int_a^t (t-s)^{n-1} x^{(n)}(s)\, ds. \end{aligned}$$

However, from (3.3.1) it is obvious that

(3.3.7) $$\frac{(t-s)^{n-1}}{(n-1)!} = \sum_{i=0}^{n-1} T_i(t)\frac{(a_{i+1}-s)^{n-i-1}}{(n-i-1)!}$$

and hence (3.3.6) is the same as (3.3.3) follows from the definition of $g_{(3.1.1)}(t,s)$ in (3.3.5). ∎

Corollary 3.3.3. Corresponding to the two - point right focal conditions (3.1.3) Green's function $g_{(3.1.3)}(t,s)$ of the boundary value problem

$$z^{(n)} = 0$$

$$z^{(i)}(a_1) = 0, \; 0 \le i \le \alpha \tag{3.3.8}$$

$$z^{(i)}(a_2) = 0, \; \alpha + 1 \le i \le n-1$$

is given by

(3.3.9) $g_{(3.1.3)}(t,s)$

$$= \frac{1}{(n-1)!} \begin{cases} \sum_{i=0}^{\alpha} \binom{n-1}{i} (t-a_1)^i (a_1 - s)^{n-i-1}, & a \le s \le t \\ -\sum_{i=\alpha+1}^{n-1} \binom{n-1}{i} (t-a_1)^i (a_1-s)^{n-i-1}, & t \le s \le b. \end{cases}$$

Further, for $a_1 \le s, t \le a_2$ the following inequalities hold

$$(-1)^{n-\alpha-1} \frac{\partial^i g_{(3.1.3)}(t,s)}{\partial t^i} \ge 0, \; 0 \le i \le \alpha \tag{3.3.10}$$

$$(-1)^{n-i} \frac{\partial^i g_{(3.1.3)}(t,s)}{\partial t^i} \ge 0, \; \alpha+1 \le i \le n-1. \tag{3.3.11}$$ ∎

Cauchy's Representation

Theorem 3.3.4. In the interval $[a_1, a_2]$ the error function $e_{(3.1.3)}(t)$ can be written as

$$e_{(3.1.3)}(t) = T_n(t)\; x^{(n)}(\xi), \tag{3.3.12}$$

where $\xi \in (a_1, a_2)$, and

$$T_n(t) = \frac{1}{n!} \sum_{i=\alpha+1}^{n} \binom{n}{i} (t-a_1)^i (a_1-a_2)^{n-i}.$$

Proof. In view of (3.3.10), $(-1)^{n-\alpha-1}g_{(3.1.3)}(t,s) \geq 0,\ a_1 \leq s,t \leq a_2$, also since

$$\int_{a_1}^{a_2} g_{(3.1.3)}(t,s)\, ds = \frac{1}{n!}\sum_{i=\alpha+1}^{n}\binom{n}{i}(t-a_1)^i(a_1-a_2)^{n-i}$$

equality (3.3.12) follows on applying the mean value theorem in

$$e_{(3.1.3)}(t) = \int_{a_1}^{a_2} g_{(3.1.3)}(t,s)x^{(n)}(s)\, ds. \qquad (3.3.13) \quad ∎$$

A direct proof of Theorem 3.3.4 similar to that of in Remark 1.4.1 for the Lidstone interpolation, or of Theorem 2.3.1 for the Hermite interpolation can be modelled rather easily.

3.4 ERROR ESTIMATES

We shall wisely use the error representation (3.3.1) to obtain the best possible uniform error bounds for $|\ e^{(k)}_{(3.1.1)}(t)\ |,\ 0 \leq k \leq n-1$.

Lemma 3.4.1. For each $0 \leq k \leq n-1$ the following holds

$$|\ e^{(k)}_{(3.1.1)}(t)\ | \leq M\frac{(b-t)^{n-k}}{(n-k)!},\ a \leq t \leq a_{k+1} \qquad (3.4.1)$$

where $M = \max_{a\leq t\leq b} |\ x^{(n)}(t)\ |$.

Proof. In view of (3.3.2) it immediately follows that

$$\begin{aligned} |\ e^{(k)}_{(3.1.1)}(t)\ | &\leq \int_t^b\int_{t_1}^b\cdots\int_{t_{n-k-1}}^b |\ x^{(n)}(t_{n-k})\ |\, dt_{n-k}\cdots dt_1 \\ &\leq M\frac{(b-t)^{n-k}}{(n-k)!}. \end{aligned} \quad ∎$$

Theorem 3.4.2. For each $0 \leq k \leq n-1$ the following holds

$$\max_{a\leq t\leq b} |\ e^{(k)}_{(3.1.1)}(t)\ | \leq M\frac{(b-a)^{n-k}}{(n-k)!}\binom{n-k-1}{\left[\frac{n-k-1}{2}\right]} \qquad (3.4.2)$$

$$= M\frac{(b-a)^{n-k}}{(n-k)\left[\frac{n-k-1}{2}\right]!\left[\frac{n-k}{2}\right]!}. \qquad (3.4.3)$$

Proof. Since the function $g(t) = e^{(k)}_{(3.1.1)}(t)$ satisfies $g(a_{k+1}) = g'(a_{k+2}) = \cdots = g^{(n-k-1)}(a_n) = 0$, $g^{(n-k)}(t) = x^{(n)}(t)$, and $a \le a_{k+1} \le \cdots \le a_n \le b$ it suffices to prove (3.4.2) for $k = 0$ only. For this, we note that for $a \le t \le a_1$, (3.4.1) implies that

$$|\, e_{(3.1.1)}(t) \,| \le M\frac{(b-a)^n}{n!}. \tag{3.4.4}$$

Further, since

$$e_{(3.1.1)}(t) = \int_{a_1}^{t}\int_{a_2}^{t_1}\cdots\int_{a_j}^{t_{j-1}} e^{(j)}_{(3.1.1)}(t_j)dt_j dt_{j-1}\cdots dt_1 \tag{3.4.5}$$

for $t \ge a_j$, we obtain

$$|\, e_{(3.1.1)}(t) \,| \le \int_a^t\int_a^{t_1}\cdots\int_a^{t_{j-1}} |\, e^{(j)}_{(3.1.1)}(t_j) \,|\, dt_j dt_{j-1}\cdots dt_1$$

$$= \int_a^t \frac{(t-s)^{j-1}}{(j-1)!} |\, e^{(j)}_{(3.1.1)}(s) \,|\, ds. \tag{3.4.6}$$

Thus, in particular

$$|\, e_{(3.1.1)}(t) \,| \le M\int_a^b \frac{(b-s)^{n-1}}{(n-1)!}ds = M\frac{(b-a)^n}{n!} \quad \text{for } a_n \le t \le b.$$

Combining (3.4.1) and (3.4.6), it follows that for $a_j \le t \le a_{j+1}$

$$\begin{aligned}
|\, e_{(3.1.1)}(t) \,| &\le M\int_a^t \frac{(b-s)^{n-j}}{(n-j)!}\frac{(t-s)^{j-1}}{(j-1)!}ds \\
&= M\int_a^b \frac{(b-s)^{n-1}}{(n-j)!(j-1)!}ds \\
&= M\frac{(b-a)^n}{n(n-j)!(j-1)!} \\
&= M\binom{n-1}{n-j}\frac{(b-a)^n}{n!}.
\end{aligned}$$

When j runs through the values 1, 2, ..., $n-1$ the Binomial coefficient $\binom{n-1}{n-j}$ takes its maximum value when $j = \left[\frac{n+1}{2}\right]$, and since $\binom{n-1}{n-\left[\frac{n+1}{2}\right]} = \binom{n-1}{\left[\frac{n-1}{2}\right]}$, (3.4.2) follows for $k = 0$. ∎

For each $0 \le k \le n-1$ the inequality (3.4.2) is the best possible in the sense that equality holds if and only if $x^{(n)}(t) = M$. However, if $a_1 = a$ and $a_n = b$, then (3.4.2) can be improved. Indeed we shall prove the following :

Theorem 3.4.3. Let α and β be integers such that

$$(3.4.7) \quad a = a_1 = \cdots = a_{\alpha+1} < a_{\alpha+2} \le \cdots \le a_{n-\beta} \le a_{n-\beta+1} = \cdots = a_n = b.$$

Then, for each $0 \le k \le n-1$ and $a \le t \le b$ the following holds

$$(3.4.8) \quad | \, e^{(k)}_{(3.1.1)}(t) \, | \le M\frac{(b-a)^{n-k}}{(n-k)!}\begin{cases} \begin{pmatrix} n-k-1 \\ r^* \end{pmatrix}, & \text{if } 0 \le k \le n-\beta-1 \\ 1, & \text{if } n-\beta \le k \le n-1 \end{cases}$$

where $r^* = \max\left\{\alpha - k, \beta, \left[\frac{n-k-1}{2}\right]\right\}$.

Proof. From Lemma 3.4.1 it is clear that for each $0 \le k \le n-1$ and $a_1 \le t \le a_{k+1}$, we have

$$(3.4.9) \qquad | \, e^{(k)}_{(3.1.1)}(t) \, | \le M\frac{(b-a)^{n-k}}{(n-k)!}.$$

Therefore, if $n-\beta \le k \le n-1$, then in view of (3.4.7) inequality (3.4.9) also holds for $a \le t \le b$.

Now we shall show that for each $0 \le k \le n-\beta-1$ and $a_{k+1} \le t \le a_n$ the following holds

$$(3.4.10) \quad | \, e^{(k)}_{(3.1.1)}(t) \, | \le M\frac{(b-a)^{n-k}}{(n-k)!}\min\left\{\begin{pmatrix} n-k-1 \\ m^* \end{pmatrix}, \begin{pmatrix} n-k-1 \\ \ell^* \end{pmatrix}\right\},$$

where

$$m^* = \max\left\{\left[\frac{n-k-1}{2}\right], \beta\right\} \quad \text{and} \quad \ell^* = \max\left\{\left[\frac{n-k-1}{2}\right], \alpha - k\right\}.$$

For this, first we shall prove that for $a_{k+1} \le t \le a_n$

$$(3.4.11) \qquad | \, e^{(k)}_{(3.1.1)}(t) \, | \le M\int_a^t \int_a^{t_1} \cdots \int_a^{t_{\ell-1}} \int_{t_\ell}^b \cdots \int_{t_{n-k-1}}^b dt_{n-k} \cdots dt_1,$$

where ℓ is an integer such that $1 \le \ell \le n-k-1$. Indeed, from (3.3.2) we have

$$(3.4.12) \qquad | \, e^{(k)}_{(3.1.1)}(t) \, | \le M\int_a^t \left| \int_{a_{k+2}}^{t_1} \right| \cdots \left| \int_{a_{n-1}}^{t_{n-k-2}} \right| \int_{t_{n-k-1}}^b dt_{n-k} \cdots dt_1.$$

If $t_1 \le a_{k+2}$, then $\left|\int_{a_{k+2}}^{t_1}\right| \le \int_{t_1}^{b}$. Further, since t_2 lies between a_{k+2} and t_1, we find that $t_1 \le t_2 \le a_{k+2} \le a_{k+3}$ and hence $\left|\int_{a_{k+3}}^{t_2}\right| \le \int_{t_2}^{b}$. Continuing in this way, (3.4.12) leads to (3.4.11) with $\ell = 1$.

If $t_1 \ge a_{k+2}$, then $\left|\int_{a_{k+2}}^{t_1}\right| \le \int_{a}^{t_1}$. Further, since t_2 lies between a_{k+2} and t_1, it is clear that $a_{k+2} \le t_2 \le t_1$. Now there are two possibilities : (i) $t_2 \le a_{k+3}$, then $\left|\int_{a_{k+3}}^{t_2}\right| \le \int_{t_2}^{b}$, and as before we have $\left|\int_{a_{k+4}}^{t_3}\right| \le \int_{t_3}^{b}$, and so on, and from (3.4.12) we get (3.4.11) with $\ell = 2$; (ii) $t_2 \ge a_{k+3}$, then $\left|\int_{a_{k+3}}^{t_2}\right| \le \int_{a}^{t_2}$, and (3.4.12) becomes

$$(3.4.13)\quad |\, e^{(k)}_{(3.1.1)}(t)\,| \le M\int_a^t\int_a^{t_1}\int_a^{t_2}\left|\int_{a_{k+4}}^{t_3}\right|\cdots\left|\int_{a_{n-1}}^{t_{n-k-2}}\right|\int_{t_{n-k-1}}^{b} dt_{n-k}\cdots dt_1.$$

Since t_3 is between a_{k+3} and t_2, we have $a_{k+3} \le t_3 \le t_2$. Now once again we have two possibilities, either $t_3 \le a_{k+4}$ in such a case (3.4.13) leads to (3.4.11) with $\ell = 3$; or $t_3 \ge a_{k+4}$ for which (3.4.13) becomes

$$|\, e^{(k)}_{(3.1.1)}(t)\,| \le M\int_a^t\int_a^{t_1}\int_a^{t_2}\int_a^{t_3}\left|\int_{a_{k+5}}^{t_4}\right|\cdots\left|\int_{a_{n-1}}^{t_{n-k-2}}\right|\int_{t_{n-k-1}}^{b} dt_{n-k}\cdots dt_1.$$

Continuing in this way, we find that (3.4.11) holds.

Next, since the right side of (3.4.11) can be explicitly integrated, we find that

$$(3.4.14)\quad |\, e^{(k)}_{(3.1.1)}(t)\,| \le \frac{M}{(n-k)!}\left\{(-1)^{\ell}(b-t)^{n-k} + \sum_{j=0}^{\ell-1}(-1)^{\ell-j+1}(b-a)^{n-k-j}\binom{n-k}{j}(t-a)^j\right\}.$$

The right side of (3.4.11) attains its maximum at $t = b$, thus from (3.4.14), we obtain

$$(3.4.15)\quad |\, e^{(k)}_{(3.1.1)}(t)\,| \le \frac{M}{(n-k)!}\sum_{j=0}^{\ell-1}\binom{n-k}{j}(-1)^{\ell-j+1}(b-a)^{n-k}.$$

However, since

$$\sum_{j=0}^{\ell-1}\binom{n-k}{j}(-1)^{\ell-j+1} = \binom{n-k-1}{\ell-1},\quad 1 \le \ell \le n-k$$

inequality (3.4.15) is the same as

$$(3.4.16)\quad |\, e^{(k)}_{(3.1.1)}(t)\,| \le \frac{M}{(n-k)!}\binom{n-k-1}{\ell-1}(b-a)^{n-k}.$$

We note that if $0 \le k \le \alpha$, then $\ell - 1 \ge \alpha - k$, and further $\max_{0 \le \ell-1 \le n-k-2} \binom{n-k-1}{\ell-1} \le \binom{n-k-1}{\left[\frac{n-k-1}{2}\right]}$. Therefore, from (3.4.16) we find

$$(3.4.17) \quad | e^{(k)}_{(3.1.1)}(t) | \le \frac{M}{(n-k)!} \binom{n-k-1}{\ell^*} (b-a)^{n-k}, \; 0 \le k \le n-\beta-1.$$

Since $\binom{n-k-1}{\ell-1} = \binom{n-k-1}{n-k-\ell}$, from a previous argument, and the fact that if $0 \le k \le n-\beta-1$ then $n-k-\ell \ge \beta$, (3.4.16) also gives

$$(3.4.18) \quad | e^{(k)}_{(3.1.1)}(t) | \le \frac{M}{(n-k)!} \binom{n-k-1}{m^*} (b-a)^{n-k}, \; 0 \le k \le n-\beta-1.$$

Combining (3.4.17) and (3.4.18), we get the inequality (3.4.10).

We also require to show that

$$(3.4.19) \qquad \min\left\{\binom{n-k-1}{m^*}, \binom{n-k-1}{\ell^*}\right\} = \binom{n-k-1}{r^*},$$

$$0 \le k \le n-\beta-1.$$

For this, since $(\alpha - k) + \beta \le n-k-1$, at least one of $(\alpha-k)$, $\beta \le \left[\frac{n-k-1}{2}\right]$, i.e., at least one of m^*, ℓ^* is $\left[\frac{n-k-1}{2}\right]$. We need to consider three cases :

Case 1 $m^* = \beta$, $\ell^* = \left[\frac{n-k-1}{2}\right]$

In this case the left side of (3.4.19) is $\binom{n-k-1}{m^*} = \binom{n-k-1}{\beta}$. Since $m^* = \beta$ and $\ell^* = \left[\frac{n-k-1}{2}\right]$ imply $\beta \ge \left[\frac{n-k-1}{2}\right] \ge \alpha - k$, we have $r^* = \beta$ and hence the right side of (3.4.19) is $\binom{n-k-1}{\beta}$.

Case 2 $m^* = \left[\frac{n-k-1}{2}\right]$, $\ell^* = \alpha - k$

Obviously the left side of (3.4.19) is $\binom{n-k-1}{\ell^*} = \binom{n-k-1}{\alpha-k}$. Further, $m^* = \left[\frac{n-k-1}{2}\right]$ and $\ell^* = \alpha - k$ provide $\alpha - k \ge \left[\frac{n-k-1}{2}\right] \ge \beta$. Thus, it follows that $r^* = \alpha - k$ and the right side of (3.4.19) is $\binom{n-k-1}{\alpha-k}$.

Case 3 $m^* = \ell^* = \left[\frac{n-k-1}{2}\right]$

The left side of (3.4.19) is $\binom{n-k-1}{\left[\frac{n-k-1}{2}\right]}$. It remains to show that $r^* = \left[\frac{n-k-1}{2}\right]$. For this, from $m^* = \ell^* = \left[\frac{n-k-1}{2}\right]$, it is clear that $\left[\frac{n-k-1}{2}\right] \geq \max\{\alpha - k, \beta\}$ and hence $r^* = \left[\frac{n-k-1}{2}\right]$.

Thus, for each $0 \leq k \leq n-\beta-1$ and $a_{k+1} \leq t \leq a_n$ the inequality (3.4.10) can be written as

$$| e^{(k)}_{(3.1.1)}(t) | \leq M\frac{(b-a)^{n-k}}{(n-k)!}\binom{n-k-1}{r^*}. \tag{3.4.20}$$

Finally, combining (3.4.9), (3.4.20) and using the fact that $\binom{n-k-1}{r^*} \geq 1$, we obtain the required inequalities (3.4.8). ∎

Remark 3.4.1. If $\alpha+1+\beta = n$, i.e., $\beta = n-\alpha-1$ then since $\binom{n-k-1}{\alpha-k} = \binom{n-k-1}{n-\alpha-1}$, $r^* = \max\{\alpha-k, \beta\}$. Thus, for the two - point right focal conditions (3.1.3) with $a_1 = a$ and $a_2 = b$ the inequalities (3.4.8) reduce to

$$| e^{(k)}_{(3.1.1)}(t) | \leq M\frac{(b-a)^{n-k}}{(n-k)!}\begin{cases}\binom{n-k-1}{\alpha-k}, & \text{if } 0 \leq k \leq \alpha\\ 1, & \text{if } \alpha+1 \leq k \leq n-1.\end{cases} \tag{3.4.21}$$

These inequalities have also been obtained in [1], however, by using the representation (3.3.13). ∎

Remark 3.4.2. For each $0 \leq k \leq n-1$, (3.4.8) is sharper than the corresponding (3.4.2). For this, it suffices to note that

$$\binom{n-k-1}{\left[\frac{n-k-1}{2}\right]} \geq \binom{n-k-1}{r^*}, \quad \text{if } 0 \leq k \leq n-\beta-1 \tag{3.4.22}$$

$$\geq \quad 1, \quad \text{if } n-\beta \leq k \leq n-1. \tag{3.4.23}$$

Obviously, in (3.4.22) the strict inequality holds if

$$\left[\frac{n-k-1}{2}\right] < \begin{cases}\max\{\alpha-k, \beta\}, & (n-k-1) \text{ is even}\\ \max\{\alpha-k, \beta\}-1, & (n-k-1) \text{ is odd}\end{cases}$$

and in (3.4.23) provided $n-\beta \le k \le n-3$. ∎

Remark 3.4.3. Inequalities (3.4.8) are the best possible. For this, for a fixed $k,\ 0 \le k \le n-1$ we define an integer $n(k),\ 1 \le n(k) \le n-1$ as follows

$$n(k) = \begin{cases} \left[\dfrac{n+k+1}{2}\right], & \text{if } r^* = \left[\dfrac{n-k-1}{2}\right] \\ n-\beta, & \text{if } r^* = \beta \\ \alpha+1, & \text{if } r^* = \alpha - k. \end{cases} \tag{3.4.24}$$

The n-th degree polynomial

$$x_{n(k)}(t) = \frac{1}{n!}\sum_{j=n(k)}^{n}(-1)^j t^j \binom{n}{j},\ 0 \le t \le 1 \tag{3.4.25}$$

obviously satisfies $\mid x_{n(k)}^{(n)}(t) \mid = 1$. Further, from

$$\begin{aligned} x_{n(k)}^{(i)}(t) &= \frac{1}{n!}\sum_{j=\max\{n(k),i\}}^{n}(-1)^j\binom{n}{j} j(j-1)\cdots(j-i+1)t^{j-i} \\ &= \frac{i!}{n!}\sum_{j=\max\{n(k),i\}}^{n}(-1)^j\binom{n}{j}\binom{j}{i}t^{j-i},\ 0 \le i \le n-1 \end{aligned}$$

we have

$$x_{n(k)}^{(i)}(0) = 0,\ 0 \le i \le n(k)-1 \tag{3.4.26}$$

and

$$\begin{aligned} x_{n(k)}^{(i)}(1) &= \frac{i!}{n!}\sum_{j=i}^{n}(-1)^j\binom{n}{j}\binom{j}{i} \\ &= \frac{i!}{n!}(-1)^i\ \delta_{ni} \\ &= 0,\ n(k) \le i \le n-1. \end{aligned} \tag{3.4.27}$$

For the function $x_{n(k)}(t)$ for which $e_{(3.1.1)}(t) \equiv x_{n(k)}(t)$, it is easy to obtain

$$\max_{0\le t\le 1} \mid x_{n(k)}^{(k)}(t) \mid = \begin{cases} \mid x_{n(k)}^{(k)}(1) \mid = \dfrac{1}{(n-k)!}\dbinom{n-k-1}{n(k)-k-1}, \\ \qquad\qquad 0 \le k \le n(k)-1 \\ \mid x_{n(k)}^{(k)}(0) \mid = \dfrac{1}{(n-k)!},\ n(k) \le k \le n-1. \end{cases} \tag{3.4.28}$$

Clearly,

$$(3.4.29)\quad \binom{n-k-1}{n(k)-k-1} = \begin{cases} \binom{n-k-1}{\left[\frac{n-k-1}{2}\right]}, & \text{if } r^* = \left[\dfrac{n-k-1}{2}\right] \\ \binom{n-k-1}{\beta}, & \text{if } r^* = \beta \\ \binom{n-k-1}{\alpha-k}, & \text{if } r^* = \alpha - k \end{cases}$$

$$= \binom{n-k-1}{r^*}.$$

It can easily be shown that if $0 \le k \le n-\beta-1$ then $k \le n(k)-1$, and if $n-\beta \le k \le n-1$ then $r^* = \beta$ and so $n(k) = n-\beta$. Using all these and (3.4.29) in (3.4.28) lead to

$$(3.4.30)\quad \max_{0\le t\le 1} \mid x_{n(k)}^{(k)}(t) \mid = \frac{1}{(n-k)!}\begin{cases} \binom{n-k-1}{r^*}, & \text{if } 0 \le k \le n-\beta-1 \\ 1, & \text{if } n-\beta \le k \le n-1. \end{cases}$$

Hence, for this function $x_{n(k)}(t)$ in (3.4.8) the equality holds. ■

Remark 3.4.4. In the following table for $b-a=1=M$ and $2 \le n \le 6$ we have computed the right side constants of (3.4.8). For the comparison purpose by $*$ we also indicate the right side constants of (3.4.2). ■

Table 3.4.1.

n	2	3		4			5				6				
k \ α,β	0,1	0,2	1,1	0,3	1,2	2,1	0,4	1,3	2,2	3,1	0,5	1,4	2,3	3,2	4,1
0	$\frac{1}{2}^*$	$\frac{1}{6}$	$\frac{1}{3}^*$	$\frac{1}{24}$	$\frac{1}{8}$	$\frac{1}{8}^*$	$\frac{1}{120}$	$\frac{1}{30}$	$\frac{1}{20}^*$	$\frac{1}{30}$	$\frac{1}{720}$	$\frac{1}{144}$	$\frac{1}{72}$	$\frac{1}{72}^*$	$\frac{1}{144}$
1	1^*	$\frac{1}{2}$	$\frac{1}{2}^*$	$\frac{1}{6}$	$\frac{1}{6}$	$\frac{1}{3}^*$	$\frac{1}{24}$	$\frac{1}{24}$	$\frac{1}{8}^*$	$\frac{1}{8}$	$\frac{1}{120}$	$\frac{1}{120}$	$\frac{1}{30}$	$\frac{1}{20}^*$	$\frac{1}{30}$
2		1	1^*	$\frac{1}{2}$	$\frac{1}{2}$	$\frac{1}{2}^*$	$\frac{1}{6}$	$\frac{1}{6}$	$\frac{1}{6}$	$\frac{1}{3}^*$	$\frac{1}{24}$	$\frac{1}{24}$	$\frac{1}{24}$	$\frac{1}{8}^*$	$\frac{1}{8}$
3				1	1	1^*	$\frac{1}{2}$	$\frac{1}{2}$	$\frac{1}{2}$	$\frac{1}{2}^*$	$\frac{1}{6}$	$\frac{1}{6}$	$\frac{1}{6}$	$\frac{1}{6}$	$\frac{1}{3}^*$
4							1	1	1	1^*	$\frac{1}{2}$	$\frac{1}{2}$	$\frac{1}{2}$	$\frac{1}{2}$	$\frac{1}{2}^*$
5											1	1	1	1	1^*

3.5 SOME APPLICATIONS

The n th order differential equation (2.5.8) together with Abel - Gontscharoff boundary conditions

$$x^{(i)}(a_{i+1}) = A_{i+1}, \ 0 \le i \le n-1 \tag{3.5.1}$$

has been a subject matter of several recent investigations [1-6,8,11-14,16,17,19-26]. As an application of the inequalities (3.4.2) in the interval $[a,b]$, or of (3.4.7) in the interval $[a_1, a_n]$ we note that almost all the results similar to that for Hermite boundary value problem (2.5.8), (2.5.9) in Section 2.5 can be stated for the Abel - Gontscharoff boundary value problem (2.5.8), (3.5.1). To show further importance of these inequalities following [1,9,18] we shall provide (α,β) - right disfocality test for the linear differential equation (2.5.63).

The linear differential equation (2.5.63) is said to be (α,β) - *right disfocal* in $[a,b]$ if and only if the only solution of (2.5.63) satisfying

$$x^{(i)}(a_{i+1}) = 0, \ 0 \le i \le n-1 \tag{3.5.2}$$

where $a \le a_1 = \cdots = a_{\alpha+1} < a_{\alpha+2} \le \cdots \le a_{n-\beta} \le a_{n-\beta+1} = \cdots = a_n \le b$, is the trivial solution. Further, we say that (2.5.63) is *right disfocal* in $[a,b]$ if and only if the only solution of (2.5.63) satisfying (3.5.2) where $a \le a_1 \le \cdots \le a_n \le b$, is the trivial solution.

Theorem 3.5.1. Suppose that the functions $a_i(t)$, $0 \le i \le n-1$ are continuous on $[a,b]$ and $|\ a_i(t)\ | \le L_i$, $0 \le i \le n-1$ for all $t \in [a,b]$. Then, the differential equation (2.5.63) is (α,β) - right disfocal in $[a,b]$ provided $\phi(b-a) \le 1$, where

$$\phi(h) = \sum_{i=0}^{n-\beta-1} \frac{1}{(n-i)!} \binom{n-i-1}{r^*} L_i h^{n-i} + \sum_{i=n-\beta}^{n-1} \frac{1}{(n-i)!} L_i h^{n-i}, \tag{3.5.3}$$

and $r^* = \max\left\{\alpha - i, \beta, \left[\frac{n-i-1}{2}\right]\right\}$.

Proof. Suppose on the contrary that (2.5.63) has a nontrivial solution $x(t)$ satisfying (3.5.2). Applying Theorem 3.4.3, we obtain

$$(3.5.4)\ \ | x^{(i)}(t) | \leq M_n \frac{(a_n - a_1)^{n-i}}{(n-i)!} \begin{cases} \binom{n-i-1}{r^*}, & \text{if } 0 \leq i \leq n-\beta-1 \\ 1, & \text{if } n-\beta \leq i \leq n-1 \end{cases}$$

where $M_n = \max_{a_1 \leq t \leq a_n} | x^{(n)}(t) | = | x^{(n)}(\tau) |$, and $\tau \in [a_1, a_n]$. Therefore, it follows that

$$\begin{aligned} M_n = | x^{(n)}(\tau) | &= | a_0(\tau)x(\tau) + \cdots + a_{n-1}(\tau)x^{(n-1)}(\tau) | \\ &\leq \sum_{i=0}^{n-1} L_i | x^{(i)}(\tau) | \\ &\leq \phi(a_n - a_1)M_n \leq \phi(b-a)M_n. \end{aligned}$$

Clearly, $M_n > 0$, since otherwise $x(t)$ would coincide in $[a_1, a_n]$ with a polynomial of degree $m < n$ and $x^{(m)}(t)$ would not vanish in $[a_1, a_n]$. Hence $\phi(b-a) \geq 1$. It only remains to exclude the possibility of equality. At least one of the numbers L_i, $0 \leq i \leq n-1$ is different from zero, since otherwise $x(t)$ would be a polynomial of degree less than n and cannot satisfy (3.5.2). Thus, if $\phi(b-a) = \phi(a_n - a_1) = 1$ then equality must hold in (3.5.4) for at least one value of i. In view of Remark 3.4.3 this is possible only if $x(t)$ coincides in $[a_1, a_n]$ with a polynomial of degree n. But we can then take τ to be any point in $[a_1, a_n]$, and $| x^{(i)}(\tau) |$ is not constant in $[a_1, a_n]$ for any $0 \leq i \leq n-1$. Therefore, in this case also we have $\phi(b-a) > 1$. ∎

Corollary 3.5.2. Let the conditions of Theorem 3.5.1 be satisfied with

$$(3.5.5) \qquad \phi(h) = \sum_{i=0}^{n-1} \frac{1}{(n-i)!} \binom{n-i-1}{\left[\frac{n-i-1}{2}\right]} L_i h^{n-i}.$$

Then, the differential equation (2.5.63) is right disfocal in $[a, b]$ if $\phi(b-a) \leq 1$. ∎

Remark 3.5.1. By Rolle's theorem the right disfocality of (2.5.63) in $[a, b]$ implies the disconjugacy of (2.5.63) in $[a, b]$. ∎

Lemma 3.5.3. Suppose $x(t) \in C^{(n-1)}[a,b]$ has at least n zeros in $[a,b]$. Then, we can find points a_i, $1 \le i \le 2n-1$ such that

$$a \le a_1 \le a_2 \le \cdots \le a_{2n-1} \le b$$

and

$$\begin{aligned} 0 = x(a_1) &= x'(a_2) = \cdots = x^{(n-2)}(a_{n-1}) = x^{(n-1)}(a_n) \\ &= x^{(n-2)}(a_{n+1}) = \cdots = x'(a_{2n-2}) = x(a_{2n-1}). \end{aligned}$$

Proof. Let $a_1^{(0)}$, ..., $a_n^{(0)}$ be n zeros of $x(t)$ such that $a \le a_1^{(0)} \le \cdots \le a_n^{(0)} \le b$. By Rolle's theorem we can find $(n-1)$ zeros $a_1^{(1)}$, ..., $a_{n-1}^{(1)}$ of $x'(t)$ such that $a_i^{(0)} \le a_i^{(1)} \le a_{i+1}^{(0)}$, $1 \le i \le n-1$. Repeating this process we obtain finally a zero $a_1^{(n-1)}$ of $x^{(n-1)}(t)$ between two zeros $a_1^{(n-2)}$, $a_2^{(n-2)}$ of $x^{(n-2)}(t)$. The points

$$a_1^{(0)},\ a_1^{(1)},\ \ldots,\ a_1^{(n-2)},\ a_1^{(n-1)},\ a_2^{(n-2)},\ \ldots,\ a_{n-1}^{(1)},\ a_n^{(0)}$$

satisfy the requirements of the lemma. ∎

Theorem 3.5.4. Suppose that the functions $a_i(t)$, $0 \le i \le n-1$ are continuous on $[a,b]$ and $|\ a_i(t)\ | \le L_i$, $0 \le i \le n-1$ for all $t \in [a,b]$. Then, the differential equation (2.5.63) is disconjugate in $[a,b]$ if $\phi\left(\frac{b-a}{2}\right) \le 1$ where the function ϕ is defined in (3.5.5).

Proof. It follows from Lemma 3.5.3, Corollary 3.5.2 and Remark 3.5.1. ∎

Finally, we note that the disconjugacy criterion (2.5.65) $(\alpha = 0)$ gives either stronger or weaker results than the inequality $\phi\left(\frac{b-a}{2}\right) \le 1$ in Theorem 3.5.4.

REFERENCES

1. **R.P.Agarwal,** On the right focal point boundary value problems for linear ordinary differential equations, Atti Accad. Naz. Lincei Cl. Sci. Fis. Mat. Natur. 79(1985), 172-177.

2. **R.P.Agarwal,** Boundary Value Problems for Higher Order Differential Equations, World Scientific, Singapore • Philadelphia, 1986.

3. **R.P.Agarwal** and **R.A.Usmani,** Iterative methods for solving right focal point boundary value problems, J. Comp. Appl. Math. 14(1986), 371-390.

4. **R.P.Agarwal** and **R.A.Usmani,** On the right focal point boundary value problems for integro - differential equations, J. Math. Anal. Appl. 126(1987), 51-69.

5. **R.P.Agarwal** and **R.A.Usmani,** Monotone convergence of iterative methods for right focal point boundary value problems, J. Math. Anal. Appl. 130(1988), 451-459.

6. **R.P.Agarwal,** Existence - uniqueness and iterative methods for right focal point boundary value problems for differential equations with deviating arguments, Annales Polonici Mathematici 52(1991), 211-230.

7. **R.P.Agarwal,** Sharp inequalities in polynomial interpolation, in General Inequalities 6, Ed. W.Walter, International Series of Numerical Mathematics, Birkhäuser Verlag 103(1992), 73-92.

8. **R.P.Agarwal, Qin Sheng** and **P.J.Y.Wong,** On Abel - Gontscharoff boundary value problems, Mathl. Comput. Modelling, to appear.

9. **W.A.Coppel,** Disconjugacy, Lecture Notes in Mathematics 220, Springer-Verlag, New York, 1971.

10. **P.J.Davis,** Interpolation and Approximation, Blaisdell Publishing Co., Boston, 1961.

11. **J.Ehme** and **D.Hankerson,** Existence of solutions for right focal boundary value problems, Nonlinear Analysis, 18(1992), 191-197.

12. **U.Elias,** Focal points for a linear differential equation whose coefficients are of constant sign, Trans. Amer. Math. Soc. 249(1979), 187-202.

13. **U.Elias,** Green's function for a non - disconjugate differential operator, J. Differential Equations 37(1980), 318-350.

14. **P.W.Eloe,** Sign properties of Green's functions for two classes of boundary value problems, Canad. Math. Bull. 30(1987), 28-35.

15. **V.L.Gontscharoff,** Theory of Interpolation and Approximation of Functions, Gostekhizdat, Moscow, 1954.

16. **J.Henderson,** Uniqueness of solutions of right focal point boundary value problems for ordinary differential equations, J. Differential Equations 41(1981), 218-227.

17. **J.Henderson,** Existence of solutions of right focal point boundary value problems for ordinary differential equations, Nonlinear Analysis 5(1981), 989-1002.

18. **A.Ju.Levin,** A bound for a function with monotonely distributed zeros of successive derivatives, Mat. Sb. 64(106)(1964), 396-409.

19. **J.Muldowney,** A necessary and sufficient condition for disfocality, Proc. Amer. Math. Soc. 74(1979), 49-55.

20. **Z.Nehari,** Disconjugate linear differential operators, Trans. Amer. Math. Soc. 129(1967), 500-516.

21. **A.C.Peterson,** Green's function for focal type boundary value problems, Rocky Mountain J. Math. 9(1979), 721-732.

22. **A.C.Peterson,** Existence - uniqueness for focal - point boundary value problems, SIAM J. Math. Anal. 12(1981), 173-185.

23. **S.Umamaheswaram** and **M.Venkata Rama,** Green's function for k - point focal boundary value problems, J. Math. Anal. Appl. 148(1990), 350-359.

24. **S.Umamaheswaram** and **M.Venkata Rama,** Existence theorems for some special type of boundary value problems, Nonlinear Analysis 16(1991), 663-668.

25. **S.Umamaheswaram** and **M.Venkata Rama,** Existence theorems for focal boundary value problems, Differential and Integral Equations 4(1991), 883-889.

26. **S.Umamaheswaram** and **M.Venkata Rama,** Focal subfunctions and second order differential inequalities, Rocky Mountain J. Math. 21(1991), 1127-1142.

27. **P.J.Y.Wong** and **R.P.Agarwal,** Abel-Gontscharoff interpolation error bounds for derivatives, Proc. Royal Soc. Edinburgh, 119A(1991), 367-372.

CHAPTER 4

MISCELLANEOUS INTERPOLATION

4.1 INTRODUCTION

Results of Chapters 1 - 3 are used here to obtain the best possible / sharp error bounds for the derivatives of several other interpolating polynomials. Some of these interpolating polynomials satisfy : (i) (n,p) and (p,n) conditions, which arise in determining the intervals of nonoscillation for the linear differential equations, (ii) particular cases of two - point Birkhoff's conditions, (iii) two - point Abel - Gontscharoff - Hermite conditions, and (iv) two - point Abel - Gontscharoff - Lidstone conditions.

4.2 (n,p) AND (p,n) INTERPOLATION

Theorem 4.2.1. The unique (n,p) *interpolating polynomial* $P_{(4.2.1)}(t)$ of degree $(n-1)$ satisfying the (n,p) *conditions*

$$(4.2.1)\qquad \begin{aligned} P^{(i)}_{(4.2.1)}(a_1) &= A_i,\ 0 \le i \le n-2 \\ P^{(p)}_{(4.2.1)}(a_2) &= B,\ 0 \le p \le n-1 \text{ but fixed} \end{aligned}$$

can be explicitly expressed as

$$(4.2.2)\qquad P_{(4.2.1)}(t) = \sum_{i=0}^{n-2} \frac{(t-a_1)^i}{i!} A_i + \left[B - \sum_{i=0}^{n-p-2} \frac{(a_2-a_1)^i}{i!} A_{p+i} \right] \times \frac{(n-p-1)!}{(n-1)!} \frac{(t-a_1)^{n-1}}{(a_2-a_1)^{n-p-1}}.$$

Proof. It is clear that $P_{(4.2.1)}(t)$ in (4.2.2) is a polynomial of degree at most $(n-1)$. Further, since

$$P^{(j)}_{(4.2.1)}(t) = \sum_{i=j}^{n-2} \frac{(t-a_1)^{i-j}}{(i-j)!} A_i + \left[B - \sum_{i=0}^{n-p-2} \frac{(a_2-a_1)^i}{i!} A_{p+i} \right] \times \frac{(n-p-1)!}{(n-j-1)!} \frac{(t-a_1)^{n-j-1}}{(a_2-a_1)^{n-p-1}},\quad 0 \le j \le n-1$$

it is immediate that this $P_{(4.2.1)}(t)$ satisfies the conditions (4.2.1) also. ∎

Theorem 4.2.2. The unique (p,n) *interpolating polynomial* $P_{(4.2.3)}(t)$ of degree $(n-1)$ satisfying the (p,n) *conditions*

$$(4.2.3)\qquad \begin{aligned} P^{(p)}_{(4.2.3)}(a_1) &= A,\ 0 \le p \le n-1 \text{ but fixed} \\ P^{(i)}_{(4.2.3)}(a_2) &= B_i,\ 0 \le i \le n-2 \end{aligned}$$

can be written as

$$(4.2.4)\qquad P_{(4.2.3)}(t) = \sum_{i=0}^{n-2} (-1)^i \frac{(a_2-t)^i}{i!} B_i + \left[A - \sum_{i=0}^{n-p-2} (-1)^i \frac{(a_2-a_1)^i}{i!} B_{p+i} \right] \times (-1)^p \frac{(n-p-1)!}{(n-1)!} \frac{(a_2-t)^{n-1}}{(a_2-a_1)^{n-p-1}}.$$ ∎

Let $A_i = x^{(i)}(a_1)$, $0 \le i \le n-2$, $B = x^{(p)}(a_2)$, where the function $x(t)$ is assumed to be n times continuously differentiable on $[a_1, a_2]$. In such a case we shall call $P_{(4.2.1)}(t)$ the (n,p) interpolating polynomial of the function $x(t)$. The (p,n) interpolating polynomial $P_{(4.2.3)}(t)$ of the function $x(t)$ is defined analogously. For each of the associated errors $e_{(4.2.1)}(t) = x(t) - P_{(4.2.1)}(t)$ and $e_{(4.2.3)}(t) = x(t) - P_{(4.2.3)}(t)$ we shall provide Peano's and Cauchy's representations.

Lemma 4.2.3. Corresponding to the (n,p) conditions (4.2.1) Green's function $g_{(4.2.1)}(t,s)$ of the boundary value problem

$$
\begin{aligned}
& z^{(n)} = 0 \\
(4.2.5) \qquad & z^{(i)}(a_1) = 0, \ 0 \le i \le n-2 \\
& z^{(p)}(a_2) = 0, \ 0 \le p \le n-1 \text{ but fixed}
\end{aligned}
$$

is given by

$$
(4.2.6) \qquad g_{(4.2.1)}(t,s) = -\frac{1}{(n-1)!}\times
\begin{cases}
(t-a_1)^{n-1}\left(\dfrac{a_2-s}{a_2-a_1}\right)^{n-p-1} - (t-s)^{n-1}, \\
\qquad\qquad a_1 \le s \le t \le a_2 \\
(t-a_1)^{n-1}\left(\dfrac{a_2-s}{a_2-a_1}\right)^{n-p-1}, \quad a_1 \le t \le s \le a_2.
\end{cases}
$$

Further, for $0 \le k \le p$ the following hold

$$
(4.2.7) \qquad -\frac{\partial^k g_{(4.2.1)}(t,s)}{\partial t^k} \ge 0, \quad a_1 \le s,t \le a_2,
$$

and

$$
(4.2.8) \quad \int_{a_1}^{a_2} \left| \frac{\partial^k g_{(4.2.1)}(t,s)}{\partial t^k} \right| ds = \frac{1}{(n-k-1)!}(t-a_1)^{n-k-1}\left[\frac{a_2-a_1}{n-p} - \frac{t-a_1}{n-k}\right].
$$

Proof. On differentiating (4.2.6) k times, it is clear that inequalities (4.2.7) hold if and only if

$$
(4.2.9) \qquad (t-a_1)^{n-k-1}\left(\frac{a_2-s}{a_2-a_1}\right)^{n-p-1} \ge (t-s)^{n-k-1}, \ 0 \le k \le p.
$$

However, since $(t-a_1)(a_2-s) \ge (t-s)(a_2-a_1)$, we have

$$
\left(\frac{t-a_1}{t-s}\right)^{n-k-1} \ge \left(\frac{a_2-a_1}{a_2-s}\right)^{n-k-1} \ge \left(\frac{a_2-a_1}{a_2-s}\right)^{n-p-1}, \ 0 \le k \le p
$$

which is the same as (4.2.9).

Now we use (4.2.7), to obtain

$$\int_{a_1}^{a_2}\left|\frac{\partial^k g_{(4.2.1)}(t,s)}{\partial t^k}\right| ds$$

$$= \frac{1}{(n-k-1)!}\int_{a_1}^{t}\left[(t-a_1)^{n-k-1}\left(\frac{a_2-s}{a_2-a_1}\right)^{n-p-1} - (t-s)^{n-k-1}\right] ds$$

$$+\frac{1}{(n-k-1)!}\int_{t}^{a_2}(t-a_1)^{n-k-1}\left(\frac{a_2-s}{a_2-a_1}\right)^{n-p-1} ds$$

$$= \frac{1}{(n-k-1)!}(t-a_1)^{n-k-1}\left[\frac{a_2-a_1}{n-p} - \frac{t-a_1}{n-k}\right],\ 0 \le k \le p$$

which proves (4.2.8). ∎

Theorem 4.2.4. In terms of Green's function $g_{(4.2.1)}(t,s)$ the error function $e_{(4.2.1)}(t)$ can be written as

$$e_{(4.2.1)}(t) = \int_{a_1}^{a_2} g_{(4.2.1)}(t,s)x^{(n)}(s)\, ds. \quad \blacksquare \tag{4.2.10}$$

Theorem 4.2.5. If $x(t) \in C^{(n)}[a_1,a_2]$, then

$$e_{(4.2.1)}(t) = \frac{1}{(n-1)!}(t-a_1)^{n-1}\left[\frac{t-a_1}{n} - \frac{a_2-a_1}{n-p}\right] x^{(n)}(\xi), \tag{4.2.11}$$

where $\xi \in (a_1,a_2)$.

Proof. Since $-g_{(4.2.1)}(t,s) \ge 0$, (4.2.11) follows from (4.2.10) as an application of the mean value theorem and the equality (4.2.8). ∎

Lemma 4.2.6. Corresponding to the (p,n) conditions (4.2.3) Green's function $g_{(4.2.3)}(t,s)$ of the boundary value problem

$$\begin{aligned} &z^{(n)} = 0 \\ &z^{(p)}(a_1) = 0,\ 0 \le p \le n-1 \text{ but fixed} \\ &z^{(i)}(a_2) = 0,\ 0 \le i \le n-2 \end{aligned} \tag{4.2.12}$$

is given by

(4.2.13) $$g_{(4.2.3)}(t,s) = \frac{(-1)^{n+1}}{(n-1)!}\times$$

$$\begin{cases} (a_2-t)^{n-1}\left(\dfrac{s-a_1}{a_2-a_1}\right)^{n-p-1}, & a_1 \le s \le t \le a_2 \\ (a_2-t)^{n-1}\left(\dfrac{s-a_1}{a_2-a_1}\right)^{n-p-1} - (s-t)^{n-1}, & \\ & a_1 \le t \le s \le a_2. \end{cases}$$

Further, for $0 \le k \le p$ the following hold

(4.2.14) $$(-1)^{n+k+1}\,\frac{\partial^k g_{(4.2.3)}(t,s)}{\partial t^k} \ge 0,\ a_1 \le s,t \le a_2,$$

and

(4.2.15) $$\int_{a_1}^{a_2}\left|\frac{\partial^k g_{(4.2.3)}(t,s)}{\partial t^k}\right| ds = \frac{1}{(n-k-1)!}(a_2-t)^{n-k-1}\times$$

$$\left[\frac{a_2-a_1}{n-p} - \frac{a_2-t}{n-k}\right]. \quad \blacksquare$$

Theorem 4.2.7. In terms of Green's function $g_{(4.2.3)}(t,s)$ the error function $e_{(4.2.3)}(t)$ can be written as

(4.2.16) $$e_{(4.2.3)}(t) = \int_{a_1}^{a_2} g_{(4.2.3)}(t,s)x^{(n)}(s)\,ds. \quad \blacksquare$$

Theorem 4.2.8. If $x(t) \in C^{(n)}[a_1,a_2]$, then

(4.2.17) $$e_{(4.2.3)}(t) = \frac{(-1)^n}{(n-1)!}(a_2-t)^{n-1}\left[\frac{a_2-t}{n} - \frac{a_2-a_1}{n-p}\right]x^{(n)}(\xi),$$

where $\xi \in (a_1,a_2)$. $\blacksquare$

The above results and Theorem 2.4.4 will be used now to obtain the best possible uniform bounds for $|\ e^{(k)}_{(4.2.i)}(t)\ |;\ i=1,3,\ 0 \le k \le n-1$.

Theorem 4.2.9. If $x(t) \in C^{(n)}[a_1,a_2]$, then the following hold

(4.2.18) $$|\ e^{(k)}_{(4.2.i)}(t)\ | \le \alpha_{n,k}(a_2-a_1)^{n-k}M_n;\ i=1,3,\ 0 \le k \le n-1$$

where $M_n = \max_{a_1 \le t \le a_2} | x^{(n)}(t) |$, and

$$(4.2.19) \qquad \alpha_{n,k} = \frac{1}{(n-k)!} \begin{cases} \dfrac{p-k}{n-p}, & 0 \le k \le p-1 \\[2ex] \dfrac{(n-k-1)^{n-k-1}}{(n-p)^{n-k}}, & k = p \\[2ex] \dfrac{k-p}{n-p}, & p+1 \le k \le n-1. \end{cases}$$

Proof. From (4.2.10) and (4.2.8) it is clear that

$$(4.2.20) \qquad | e^{(k)}_{(4.2.1)}(t) | \le \frac{1}{(n-k-1)!} \left(\frac{v^{n-k-1}}{n-p} - \frac{v^{n-k}}{n-k} \right) (a_2 - a_1)^{n-k} M_n$$
$$= \phi_k(v) M_n; \ 0 \le k \le p, \ 0 \le v \le 1.$$

By maximizing $\phi_k(v)$ directly we obtain the constants $\alpha_{n,k}$, $0 \le k \le p-1$ when $v = 1$, and $\alpha_{n,p}$ when $v = (n-k-1)/(n-p)$.

For $p+1 \le k \le n-1$, $\partial^k g_{(4.2.1)}(t,s) \big/ \partial t^k$ as a function of s changes sign in (a_1, a_2), and hence the above procedure cannot be used. However, this case can be deduced from Theorem 2.4.4 by noting that the function $y(t) = e^{(p)}_{(4.2.1)}(t) \in C^{(n-p)}[a_1, a_2]$ satisfies $y^{(i)}(a_1) = 0$, $0 \le i \le n-p-2$ and $y(a_2) = 0$.

A similar proof holds for the function $e_{(4.2.3)}(t)$. ∎

Remark 4.2.1. In (4.2.18) the constants $\alpha_{n,k}$ are the best possible as they are exact for the functions

$$x_1(t) = (t - a_1)^{n-1} \left[\frac{a_2 - a_1}{n-p} - \frac{t - a_1}{n} \right] \text{ for (4.2.1)}$$

and

$$x_2(t) = (a_2 - t)^{n-1} \left[\frac{a_2 - a_1}{n-p} - \frac{a_2 - t}{n} \right] \text{ for (4.2.3)}$$

and only for these functions up to a constant factor. ∎

Remark 4.2.2. For $0 \le p \le n-2$ as in the proof of Theorem 4.2.9 once again let $y(t) = e^{(p)}_{(4.2.1)}(t)$, so that in view of (2.3.11) it can be written as

$$e^{(p)}_{(4.2.1)}(t) = y(t) = \int_{a_1}^{a_2} g^*(t,s) y^{(n-p)}(s) \, ds$$

where $g^*(t,s)$ is the same as $g_{(2.1.4)}(t,s)$ with $n = n-p$ and $m = n-p-1$. Thus, from (2.3.13) it follows that

$$| e^{(p)}_{(4.2.1)}(t) | \le \frac{1}{(n-p)!}(t-a_1)^{n-p-1}(a_2-t)M_n.$$

Now on using the fact that $e^{(i)}_{(4.2.1)}(a_1) = 0,\ 0 \le i \le p-1$ we immediately get

$$| e^{(k)}_{(4.2.1)}(t) | \le \frac{1}{(p-k-1)!}\int_{a_1}^{t}(t-s)^{p-k-1}\frac{1}{(n-p)!}(s-a_1)^{n-p-1}(a_2-s)M_n\, ds,$$

$$0 \le k \le p-1$$

and hence

$$\begin{aligned} | e^{(k)}_{(4.2.1)}(t) | &\le \frac{1}{(p-k-1)!(n-p)!}(a_2-a_1)^{n-k}M_n\int_0^1 u^{n-p-1}(1-u)^{p-k}du \\ &= \frac{p-k}{(n-p)\ (n-k)!}(a_2-a_1)^{n-k}M_n,\ 0 \le k \le p-1. \end{aligned}$$ ∎

Remark 4.2.3. For the linear differential equation (2.5.63), (n,p) and (p,n) boundary conditions arise in determining the intervals of nonoscillation [5,6]. Further, the boundary value problems consisting of the differential equation (2.5.8) and the boundary conditions (4.2.1) or (4.2.3) have been studied extensively in [1,2,4,7]. ∎

4.3 $(0,0;\ m,n-m)$ INTERPOLATION

Theorem 4.3.1. The unique $(0,0;\ m,n-m)$ *interpolating polynomial* $P_{(4.3.1)}(t)$ *of degree* $(n+1)$ *satisfying the* $(0,0;\ m,n-m)$ *conditions*

$$P_{(4.3.1)}(a_1) = \bar{\alpha}_0,\quad P_{(4.3.1)}(a_2) = \bar{\beta}_0$$

$$P^{(i+2)}_{(4.3.1)}(a_1) = A_i;\ 0 \le i \le m-1,\ 1 \le m \le n-1 \text{ but fixed} \tag{4.3.1}$$

$$P^{(i+2)}_{(4.3.1)}(a_2) = B_i,\ 0 \le i \le n-m-1$$

can be expressed as

$$P_{(4.3.1)}(t) = \frac{a_2-t}{a_2-a_1}\bar{\alpha}_0 + \frac{t-a_1}{a_2-a_1}\bar{\beta}_0 + \int_{a_1}^{a_2} g(t,s)P_{(2.1.4)}(s)\, ds, \tag{4.3.2}$$

where $g(t,s)$ is the same as $g_{(2.1.5)}(t,s)$ with $m=1$. ∎

Let $\bar{\alpha}_0 = x(a_1)$, $\bar{\beta}_0 = x(a_2)$, $A_i = x^{(i+2)}(a_1)$, $0 \le i \le m-1$, $B_i = x^{(i+2)}(a_2)$, $0 \le i \le n-m-1$ where the function $x(t)$ is $(n+2)$ times continuously differentiable on $[a_1, a_2]$. In such a case we shall call $P_{(4.3.1)}(t)$ the $(0,0;\ m, n-m)$ interpolating polynomial of the function $x(t)$. For the associated error $e_{(4.3.1)}(t) = x(t) - P_{(4.3.1)}(t)$ we shall provide Peano's and Cauchy's representations.

Lemma 4.3.2. Corresponding to the $(0,0;\ m, n-m)$ conditions (4.3.1) Green's function $g_{(4.3.1)}(t,s)$ of the boundary value problem

$$
\begin{aligned}
&z^{(n+2)} = 0 \\
&z(a_1) = z(a_2) = 0 \\
&z^{(i+2)}(a_1) = 0;\ 0 \le i \le m-1,\ 1 \le m \le n-1 \text{ but fixed} \\
&z^{(i+2)}(a_2) = 0,\ 0 \le i \le n-m-1
\end{aligned}
\tag{4.3.3}
$$

is given by

$$
g_{(4.3.1)}(t,s) = \int_{a_1}^{a_2} g(t,\tau) g_{(2.1.4)}(\tau,s)\, d\tau. \tag{4.3.4}
$$

Further, the following hold

$$
(-1)^{n-m+1} g_{(4.3.1)}(t,s) \ge 0, \quad a_1 \le s, t \le a_2, \tag{4.3.5}
$$

$$
\int_{a_1}^{a_2} |\, g_{(4.3.1)}(t,s)\,|\, ds
$$

$$
= \frac{(a_2-a_1)^{n+2}}{n!} \left\{ \sum_{i=0}^{n-m} \binom{n-m}{i} \frac{(-1)^i}{(m+i+1)(m+i+2)} \times v(1-v^{m+i+1}) = \psi_1(v) \right\} \tag{4.3.6}
$$

$$
= \frac{(a_2-a_1)^{n+2}}{n!} \Big\{ (1-v) B_v(m+2, n-m+1) + v\, B_{1-v}(n-m+2, m+1) \Big\}; \tag{4.3.7}
$$

and

$$
\int_{a_1}^{a_2} \left| \frac{\partial g_{(4.3.1)}(t,s)}{\partial t} \right| ds = \frac{(a_2-a_1)^{n+1}}{n!} \Big\{ B_v(m+2, n-m+1) \tag{4.3.8}
$$

$$+B_{1-v}(n-m+2,m+1) = \psi_2(v)\Big\},$$

where $v = (t-a_1)/(a_2-a_1)$, $0 \le v \le 1$ and B_v is the incomplete Beta function.

Proof. From Lemma 2.3.3 (i) it is clear that $(-1)^{n-m}g_{(2.1.4)}(t,s) \ge 0$, $a_1 \le s,t \le a_2$. Further, from the definition of $g_{(2.1.5)}(t,s)$ with $m=1$ it is obvious that $-g(t,s) \ge 0$, $a_1 \le s,t \le a_2$. Thus, (4.3.5) follows from the representation (4.3.4).

To show (4.3.6), we use (2.3.13) to obtain

$$\int_{a_1}^{a_2} | \, g_{(4.3.1)}(t,s) \, | \, ds$$

$$= \int_{a_1}^{a_2}\int_{a_1}^{a_2} (-1)^{n-m+1} g(t,\tau) g_{(2.1.4)}(\tau,s) \, d\tau \, ds$$

$$= \int_{a_1}^{a_2} -g(t,\tau)\left\{\int_{a_1}^{a_2}(-1)^{n-m}g_{(2.1.4)}(\tau,s)\,ds\right\}d\tau$$

$$= \int_{a_1}^{a_2} -g(t,\tau)\left\{\frac{1}{n!}(\tau-a_1)^m(a_2-\tau)^{n-m}\right\}d\tau$$

$$(4.3.9) \quad = \frac{(a_2-a_1)^{n+2}}{n!}\left\{(1-v)\int_0^v u^{m+1}(1-u)^{n-m}du \right.$$

$$\left. +v\int_v^1 u^m(1-u)^{n-m+1}du\right\}$$

$$= \frac{(a_2-a_1)^{n+2}}{n!}\left\{(1-v)\sum_{i=0}^{n-m}\binom{n-m}{i}(-1)^i\frac{v^{m+i+2}}{m+i+2}\right.$$

$$\left. +v\sum_{i=0}^{n-m+1}\binom{n-m+1}{i}(-1)^i\frac{1}{m+i+1}\left[1-v^{m+i+1}\right]\right\}$$

$$= \frac{(a_2-a_1)^{n+2}}{n!}\psi_1(v).$$

Since $\int_v^1 u^m(1-u)^{n-m+1}du = \int_0^{1-v} u^{n-m+1}(1-u)^m du$, (4.3.7) follows from (4.3.9).

Finally, to prove (4.3.8) it suffices to note that

$$\int_{a_1}^{a_2}\left|\frac{\partial g_{(4.3.1)}(t,s)}{\partial t}\right| ds$$

$$= \int_{a_1}^{a_2}\int_{a_1}^{a_2}\left|\frac{\partial g(t,\tau)}{\partial t}\right|(-1)^{n-m}g_{(2.1.4)}(\tau,s)\,d\tau\,ds$$

$$= \frac{(a_2-a_1)^{n+1}}{n!}\left\{\int_0^v u^{m+1}(1-u)^{n-m}du + \int_v^1 u^m(1-u)^{n-m+1}du\right\}. \qquad \blacksquare$$

Theorem 4.3.3. In terms of Green's function $g_{(4.3.1)}(t,s)$ the error function $e_{(4.3.1)}(t)$ can be written as

$$e_{(4.3.1)}(t) = \int_{a_1}^{a_2} g_{(4.3.1)}(t,s)x^{(n+2)}(s)\,ds. \qquad \blacksquare \tag{4.3.10}$$

Theorem 4.3.4. If $x(t) \in C^{(n+2)}[a_1,a_2]$, then

$$e_{(4.3.1)}(t) = \frac{(a_2-a_1)^{n+2}}{n!}(-1)^{n-m+1}\psi_1(v)x^{(n+2)}(\xi), \tag{4.3.11}$$

where $\xi \in (a_1,a_2)$ and $v = (t-a_1)/(a_2-a_1)$.

Proof. Since $(-1)^{n-m+1}g_{(4.3.1)}(t,s) \geq 0$, (4.3.11) follows from (4.3.10) as an application of the mean value theorem and the equality (4.3.6). ∎

Theorem 4.3.5. If $x(t) \in C^{(n+2)}[a_1,a_2]$, then the following hold

$$|\, e^{(k)}_{(4.3.1)}(t)\,| \leq \beta_{n+2,k}(a_2-a_1)^{n+2-k}M_{n+2};\ k=0,1 \tag{4.3.12}$$

where $M_{n+2} = \max\limits_{a_1\leq t\leq a_2} |\, x^{(n+2)}(t)\,|$, and

$$\beta_{n+2,0} = \frac{1}{n!}B_{v_1}(m+2,n-m+1), \tag{4.3.13}$$

where v_1 is the root of the equation $B_v(m+2,n-m+1) = B_{1-v}(n-m+2,m+1)$,

$$\beta_{n+2,1} = \frac{m!(n-m)!}{n!(n+2)!}\max\{n+1-m,m+1\}; \tag{4.3.14}$$

and

$$|\, e^{(k+2)}_{(4.3.1)}(t)\,| = \left|\int_{a_1}^{a_2}\frac{\partial^k g_{(2.1.4)}(t,s)}{\partial t^k}\,y^{(n)}(s)\,ds\right| = |\, e^{(k)}_{(2.1.4)}(t)\,|, \tag{4.3.15}$$

$$0 \leq k \leq n-1$$

where $y(t) = x''(t)$.

Proof. From (4.3.10) and (4.3.6) it is clear that

$$|\, e_{(4.3.1)}(t)\,| \leq \frac{(a_2-a_1)^{n+2}}{n!}\max_{0\leq v\leq 1}\psi_1(v)\ M_{n+2}.$$

Thus, to show (4.3.12) for $k = 0$ it suffices to observe that $\psi_1'(v) = -B_v(m + 2, n - m + 1) + B_{1-v}(n - m + 2, m + 1)$, $\psi_1'(0) > 0$, $\psi_1'(1) < 0$ and $\psi_1''(v) = -v^m(1-v)^{n-m} \le 0$, $0 \le v \le 1$.

Next, from (4.3.10) and (4.3.8) we have

$$| e'_{(4.3.1)}(t) | \le \frac{(a_2 - a_1)^{n+1}}{n!} \max_{0 \le v \le 1} \psi_2(v)\, M_{n+2}.$$

Since $\psi_2'(v) = v^m(1-v)^{n-m}(2v-1)$, it is clear that $\psi_2(v)$ attains its maximum at 0 or 1. Thus,

$$\begin{aligned} \max_{0 \le v \le 1} \psi_2(v) &= \max\left\{\frac{m!(n-m+1)!}{(n+2)!}, \frac{(m+1)!(n-m)!}{(n+2)!}\right\} \\ &= \frac{m!(n-m)!}{(n+2)!} \max\{n-m+1, m+1\}. \end{aligned}$$

This completes the proof of (4.3.12) for $k = 1$.

To show (4.3.15), in view of (4.3.4) and (4.3.10), it suffices to note that

$$e^{(k+2)}_{(4.3.1)}(t) = \int_{a_1}^{a_2} \frac{\partial^k g_{(2.1.4)}(t,s)}{\partial t^k}\, x^{(n+2)}(s)\, ds, \; 0 \le k \le n-1. \quad \blacksquare$$

Remark 4.3.1. In inequalities (4.3.12) the constants $\beta_{n+2,k}$; $k = 0, 1$ are the best possible as the equality holds for the function

$$x_3(t) = \frac{1}{n!} \sum_{i=0}^{n-m} \binom{n-m}{i} \frac{(-1)^i}{(m+i+1)(m+i+2)} \left(\frac{t-a_1}{a_2-a_1}\right) \times \left[1 - \left(\frac{t-a_1}{a_2-a_1}\right)^{m+i+1}\right]$$

and only for this function up to a constant factor. For this, it is clear that $| x_3^{(n+2)}(t) | = 1/(a_2 - a_1)^{n+2}$, and $\max_{a_1 \le t \le a_2} | x_3(t) | = \beta_{n+2,0}$ is included in the proof. Further, since

$$x_3'(t) = \frac{1}{n!}\frac{1}{(a_2-a_1)}\left[-B_v(m+2, n-m+1) + B_{1-v}(n-m+2, m+1)\right]$$

and

$$x_3''(t) = -\frac{1}{n!}\frac{1}{(a_2-a_1)^2}\, v^m(1-v)^{n-m}$$

where $v = (t - a_1)/(a_2 - a_1)$, it follows that

$$\max_{a_1 \le t \le a_2} | x_3'(t) | = \frac{1}{n!} \frac{1}{(a_2 - a_1)} \max\{\psi_2(0), \psi_2(1)\} = \frac{1}{(a_2 - a_1)} \beta_{n+2,1}. \quad \blacksquare$$

Remark 4.3.2. The estimates obtained in Chapter 2 for $| e^{(k)}_{(2.1.4)}(t) |$, $0 \le k \le n-1$ can be used in (4.3.15) to obtain best possible / sharp bounds for $| e^{(k+2)}_{(4.3.1)}(t) |$, $0 \le k \le n-1$, i.e., in (4.3.12) the constants $\beta_{n+2,k}$, $2 \le k \le n+1$ can be determined. ∎

Remark 4.3.3. In case $n = 2m$ and $m = m$, the equation $B_v(m+2, n-m+1) = B_{1-v}(n-m+2, m+1)$ reduces to $B_v(m+2, m+1) = B_{1-v}(m+2, m+1)$. Since for this reduced equation $v_1 = \frac{1}{2}$ is the only root, we find that

$$(4.3.16) \qquad \beta_{2m+2,0} = \frac{1}{(2m)!} B_{\frac{1}{2}}(m+2, m+1)$$

$$= \frac{m!(m+1)!}{(2m)!(2m+2)!} \left[1 - \frac{1}{2^{2m+2}} \sum_{i=0}^{m+1} \binom{2m+2}{i} \right].$$

Further, for $n = 2m$ we deduce that

$$(4.3.17) \qquad \beta_{2m+2,1} = \frac{m!(m+1)!}{(2m)!(2m+2)!}.$$

Remark 4.3.4. From Remarks 4.3.3, 4.3.2 and Theorem 2.4.13 we obtain $\beta_{6,0} = \frac{11}{64}\frac{1}{6!}$, $\beta_{6,1} = \frac{1}{2}\frac{1}{6!}$, $\beta_{6,2} = \frac{1}{384}$, $\beta_{6,3} = \frac{1}{72\sqrt{3}}$, $\beta_{6,4} = \frac{1}{12}$, $\beta_{6,5} = \frac{1}{2}$; and $\beta_{8,0} = \frac{93}{1280}\frac{1}{8!}$, $\beta_{8,1} = \frac{1}{5}\frac{1}{8!}$, $\beta_{8,2} = \frac{1}{46080}$, $\beta_{8,3} = \frac{\sqrt{5}}{30000}$, $\beta_{8,4} = \frac{1}{1920}$, $\beta_{8,5} = \frac{1}{120}$, $\beta_{8,6} = \frac{1}{10}$, $\beta_{8,7} = \frac{1}{2}$. These constants are the best possible and have earlier been obtained by Varma and Howell [8]. ∎

4.4 (0; $m, n-m$) INTERPOLATION

Theorem 4.4.1. The unique (0; $m, n-m$) *interpolating polynomial* $P_{(4.4.1)}(t)$ of degree n satisfying the (0; $m, n-m$) *conditions*

$$\begin{aligned} &P_{(4.4.1)}(\eta) = \bar{\alpha}_0, \ a_1 \le \eta \le a_2 \\ (4.4.1) \qquad &P^{(i+1)}_{(4.4.1)}(a_1) = A_i; \ 0 \le i \le m-1, \ 1 \le m \le n-1 \text{ but fixed} \\ &P^{(i+1)}_{(4.4.1)}(a_2) = B_i, \ 0 \le i \le n-m-1 \end{aligned}$$

can be expressed as

$$(4.4.2) \qquad P_{(4.4.1)}(t) = \bar{\alpha}_0 + \int_\eta^t P_{(2.1.4)}(s)\, ds. \qquad \blacksquare$$

Let $\bar{\alpha}_0 = x(\eta)$, $A_i = x^{(i+1)}(a_1)$, $0 \le i \le m-1$, $B_i = x^{(i+1)}(a_2)$, $0 \le i \le n-m-1$ where the function $x(t)$ is $(n+1)$ times continuously differentiable on $[a_1, a_2]$. In such a case we shall call $P_{(4.4.1)}(t)$ the $(0;\ m, n-m)$ interpolating polynomial of the function $x(t)$. For the associated error $e_{(4.4.1)}(t) = x(t) - P_{(4.4.1)}(t)$ we shall provide Peano's and Cauchy's representations.

Lemma 4.4.2. Corresponding to the $(0;\ m, n-m)$ conditions (4.4.1) Green's function $g_{(4.4.1)}(t,s)$ of the boundary value problem

$$\begin{aligned} &z^{(n+1)} = 0 \\ &z(\eta) = 0 \\ (4.4.3) \qquad &z^{(i+1)}(a_1) = 0; \ 0 \le i \le m-1, \ 1 \le m \le n-1 \text{ but fixed} \\ &z^{(i+1)}(a_2) = 0, \ 0 \le i \le n-m-1 \end{aligned}$$

is given by

$$(4.4.4) \qquad g_{(4.4.1)}(t,s) = \int_\eta^t g_{(2.1.4)}(\tau, s)\, d\tau.$$

Further, the following hold

$$(4.4.5) \qquad (-1)^{n-m}(t-\eta) g_{(4.4.1)}(t,s) \ge 0, \ a_1 \le s, t \le a_2;$$

and

$$(4.4.6) \quad \int_{a_1}^{a_2} | \, g_{(4.4.1)}(t,s) \, | \, ds = \frac{1}{n!}(a_2 - a_1)^{n+1} \left\{ \sum_{i=0}^{n-m} \binom{n-m}{i} \frac{(-1)^i}{m+i+1} \times \right.$$

$$[\text{sgn}\ (v-\bar{\eta})]\,(v^{m+i+1}-\bar{\eta}^{m+i+1}) = \psi_3(v)\Big\}$$

where $v = (t-a_1)/(a_2-a_1)$, $\bar{\eta} = (\eta-a_1)/(a_2-a_1)$, $0 \le v, \bar{\eta} \le 1$.

Proof. As in Lemma 4.3.2 the relation (4.4.4) immediately leads to (4.4.5). To show (4.4.6), once again we follow as in Lemma 4.3.2 to obtain

$$\begin{aligned}\int_{a_1}^{a_2} |\ g_{(4.4.1)}(t,s)\ |\ ds &= \int_{a_1}^{a_2}\left|\int_{\eta}^{t} g_{(2.1.4)}(\tau,s)\ d\tau\right| ds \\ &= \int_{a_1}^{a_2}\int_{\min\{\eta,t\}}^{\max\{\eta,t\}} (-1)^{n-m} g_{(2.1.4)}(\tau,s)\ d\tau\ ds \\ &= \int_{\min\{\eta,t\}}^{\max\{\eta,t\}}\left\{\int_{a_1}^{a_2}(-1)^{n-m} g_{(2.1.4)}(\tau,s)\ ds\right\} d\tau \\ &= \int_{\min\{\eta,t\}}^{\max\{\eta,t\}}\left\{\frac{1}{n!}(\tau-a_1)^m(a_2-\tau)^{n-m}\right\} d\tau \\ &= \frac{1}{n!}(a_2-a_1)^{n+1}\int_{\min\{\bar{\eta},v\}}^{\max\{\bar{\eta},v\}} u^m(1-u)^{n-m}du \\ &= \frac{1}{n!}(a_2-a_1)^{n+1}\psi_3(v). \quad \blacksquare\end{aligned}$$

Theorem 4.4.3. In terms of Green's function $g_{(4.4.1)}(t,s)$ the error function $e_{(4.4.1)}(t)$ can be written as

$$e_{(4.4.1)}(t) = \int_{a_1}^{a_2} g_{(4.4.1)}(t,s)x^{(n+1)}(s)\ ds. \quad \blacksquare \tag{4.4.7}$$

Theorem 4.4.4. If $x(t) \in C^{(n+1)}[a_1,a_2]$, then

$$e_{(4.4.1)}(t) = \frac{(a_2-a_1)^{n+1}}{n!}(-1)^{n-m}[\text{sgn}\ (t-\eta)]\psi_3(v)x^{(n+1)}(\xi), \tag{4.4.8}$$

where $\xi \in (a_1,a_2)$. $\blacksquare$

Theorem 4.4.5. If $x(t) \in C^{(n+1)}[a_1,a_2]$, then the following hold

$$|\ e_{(4.4.1)}(t)\ | \le \gamma_{n+1,0}(a_2-a_1)^{n+1}M_{n+1}, \tag{4.4.9}$$

where $M_{n+1} = \max_{a_1\le t\le a_2}\ |\ x^{(n+1)}(t)\ |$, and

$$\gamma_{n+1,0} = \frac{1}{n!}\max\left\{B_{\bar{\eta}}(m+1,n-m+1), B_{1-\bar{\eta}}(n-m+1,m+1)\right\}; \tag{4.4.10}$$

and

$$(4.4.11) \qquad |\, e^{(k+1)}_{(4.4.1)}(t)\,| = \left| \int_{a_1}^{a_2} \frac{\partial^k g_{(2.1.4)}(t,s)}{\partial t^k}\, y^{(n)}(s)\, ds \right| = |\, e^{(k)}_{(2.1.4)}(t)\,|,$$

$$0 \le k \le n-1$$

where $y(t) = x'(t)$.

Proof. From (4.4.7) and (4.4.6) it is clear that

$$|\, e_{(4.4.1)}(t)\,| \le \frac{(a_2 - a_1)^{n+1}}{n!} \max_{0 \le v \le 1} \psi_3(v)\, M_{n+1}.$$

Now since $\psi_3'(v) = [\mathrm{sgn}\ (v - \bar{\eta})]v^m(1-v)^{n-m} \ge 0,\ 0 \le v \le 1$, it follows that $\psi_3(v) \le \max\{\psi_3(0), \psi_3(1)\}$.

The proof of (4.4.11) is obvious from (4.4.4) and (4.4.7). ∎

Remark 4.4.1. In (4.4.9) the constant $\gamma_{n+1,0}$ is the best possible as the equality holds for the function

$$x_4(t) = \frac{1}{n!} \sum_{i=0}^{n-m} \binom{n-m}{i} \frac{(-1)^i}{m+i+1} [\mathrm{sgn}\ (t - \eta)] \times$$

$$\left[\left(\frac{t - a_1}{a_2 - a_1} \right)^{m+i+1} - \left(\frac{\eta - a_1}{a_2 - a_1} \right)^{m+i+1} \right]$$

and only for this function up to a constant factor. ∎

Remark 4.4.2. A remark similar to that of Remark 4.3.2 for (4.4.11) holds. ∎

Remark 4.4.3. If $\eta = a_1$ or a_2 then

$$(4.4.12) \qquad \gamma_{n+1,0} = \frac{m!(n-m)!}{n!(n+1)!}.$$

Also, if $n = 2m,\ m = m$ and $\eta = (a_1 + a_2)/2$ then

$$\gamma_{2m+1,0} = \frac{1}{2} \frac{(m!)^2}{(2m)!(2m+1)!}. \qquad ∎$$

4.5 $(0, 2, 0;\ m, n-m)$ INTERPOLATION

Theorem 4.5.1. Let $x(t) \in C^{(n+3)}[a_1, a_2]$ and $P_{(4.5.1)}(t)$ be the $(0, 2, 0;\ m, n-m)$ *interpolating polynomial* of degree $(n+2)$ of the function $x(t)$, i.e., it satisfies

$$P_{(4.5.1)}(a_1) = x(a_1) = \bar{\alpha}_0, \quad P_{(4.5.1)}(a_2) = x(a_2) = \bar{\beta}_0$$

$$P''_{(4.5.1)}(a_1) = x''(a_1) = \bar{\alpha}_2$$

$$P^{(i+3)}_{(4.5.1)}(a_1) = x^{(i+3)}(a_1) = A_i; \ 0 \le i \le m-1, \qquad 1 \le m \le n-1 \text{ but fixed} \tag{4.5.1}$$

$$P^{(i+3)}_{(4.5.1)}(a_2) = x^{(i+3)}(a_2) = B_i, \ 0 \le i \le n-m-1.$$

Then, for the error function $e_{(4.5.1)}(t) = x(t) - P_{(4.5.1)}(t)$ the following hold

$$| e^{(k)}_{(4.5.1)}(t) | \le \zeta_{n+3,k}(a_2 - a_1)^{n+3-k} M_{n+3}; \ k = 0,1,2 \tag{4.5.2}$$

where $M_{n+3} = \max_{a_1 \le t \le a_2} | x^{(n+3)}(t) |$, and

$$\zeta_{n+3,0} = \frac{1}{n!}\sum_{i=0}^{n-m} \binom{n-m}{i} \frac{(-1)^i}{(m+i+1)(m+i+2)(m+i+3)} \times v_1\left(1 - v_1^{m+i+2}\right), \tag{4.5.3}$$

where v_1 is the root of the equation

$$\int_0^v u \ B_u(m+1, n-m+1)du = \int_v^1 (1-u) B_u(m+1, n-m+1)du,$$

$$\zeta_{n+3,1} = \frac{1}{n!}\sum_{i=0}^{n-m} \binom{n-m}{i} \frac{(-1)^i}{(m+i+1)(m+i+3)}, \tag{4.5.4}$$

$$\zeta_{n+3,2} = \frac{m!(n-m)!}{n!(n+1)!}; \tag{4.5.5}$$

and

$$| e^{(k+3)}_{(4.5.1)}(t) | = \left| \int_{a_1}^{a_2} \frac{\partial^k g_{(2.1.4)}(t,s)}{\partial t^k} \ y^{(n)}(s) \ ds \right| = | e^{(k)}_{(2.1.4)}(t) |, \qquad 0 \le k \le n-1 \tag{4.5.6}$$

where $y(t) = x'''(t)$. ∎

Remark 4.5.1. The constants $\zeta_{n+3,k}$; $k = 0,1,2$ in (4.5.2) are the best possible as the equality holds for the function

$$x_5(t) = \frac{(a_2-a_1)^{n+3}}{n!} \sum_{i=0}^{n-m} \binom{n-m}{i} \frac{(-1)^i}{(m+i+1)(m+i+2)(m+i+3)} \times v(1-v^{m+i+2}),$$

where $v = (t-a_1)/(a_2-a_1)$, $0 \le v \le 1$ and only for this function up to a constant factor. ∎

Remark 4.5.2. A remark similar to that of Remark 4.3.2 for (4.5.6) holds. ∎

4.6 $(0:\ell-1, \ell:\ell+j-1;\ m, n-m)$ INTERPOLATION

Theorem 4.6.1. Let $x(t) \in C^{(n+\ell+j)}[a_1,a_2]$ and $P_{(4.6.1)}(t)$ be the $(0:\ell-1,\ell:\ell+j-1;\ m,n-m)$ *interpolating polynomial* of degree $(n+\ell+j-1)$ of the function $x(t)$, i.e., it satisfies

$$P^{(i)}_{(4.6.1)}(a_1) = x^{(i)}(a_1) = \bar{\alpha}_i;\ 0 \le i \le \ell-1,\ \ell \ge 1$$

$$P^{(\ell+i)}_{(4.6.1)}(a_2) = x^{(\ell+i)}(a_2) = \bar{\beta}_i;\ 0 \le i \le j-1,\ j \ge 1$$

$$P^{(\ell+j+i)}_{(4.6.1)}(a_1) = x^{(\ell+j+i)}(a_1) = A_i;\ 0 \le i \le m-1, \quad 1 \le m \le n-1 \text{ but fixed} \tag{4.6.1}$$

$$P^{(\ell+j+i)}_{(4.6.1)}(a_2) = x^{(\ell+j+i)}(a_2) = B_i,\ 0 \le i \le n-m-1.$$

Then, for the error function $e_{(4.6.1)}(t) = x(t) - P_{(4.6.1)}(t)$ the following hold

$$|\, e^{(k)}_{(4.6.1)}(t)\,| \le \xi_{n+\ell+j,k}(a_2-a_1)^{n+\ell+j-k} M_{n+\ell+j},\ 0 \le k \le \ell+j-1 \tag{4.6.2}$$

where $M_{n+\ell+j} = \max_{a_1 \le t \le a_2} |\, x^{(n+\ell+j)}(t)\,|$, and

$$\xi_{n+\ell+j,k} = \frac{1}{n!(\ell-k-1)!} \sum_{i=0}^{m} \binom{m}{i} \times \frac{(-1)^i (n-m+i)!}{(n-m+i+j+\ell-k)\ (n-m+i+j)!},\ 0 \le k \le \ell-1 \tag{4.6.3}$$

$$= \frac{(n-m)!(m+j+\ell-k-1)!}{n!(j+\ell-k-1)!(n+j+\ell-k)!},\ \ell \le k \le \ell+j-1; \tag{4.6.4}$$

and

$$(4.6.5) \qquad |\, e^{(\ell+j+k)}_{(4.6.1)}(t)\,| = \left| \int_{a_1}^{a_2} \frac{\partial^k g_{(2.1.4)}(t,s)}{\partial t^k}\, y^{(n)}(s)\, ds \right| = |\, e^{(k)}_{(2.1.4)}(t)\,|,$$

$$0 \le k \le n-1$$

where $y(t) = x^{(\ell+j)}(t)$. ∎

Remark 4.6.1. The constants $\xi_{n+\ell+j,k}$, $0 \le k \le \ell+j-1$ in (4.6.2) are the best possible as the equality holds for the function

$$x_6(t) = \frac{(a_2-a_1)^{n+\ell+j}}{n!(\ell-1)!(j-1)!} \int_0^v (v-u)^{\ell-1} \left[\int_u^1 (u-\tau)^{j-1}(1-\tau)^{n-m}\tau^m d\tau \right] du,$$

where $v = (t-a_1)/(a_2-a_1)$, $0 \le v \le 1$ and only for this function up to a constant factor. ∎

Remark 4.6.2. A remark similar to that of Remark 4.3.2 for (4.6.5) holds. ∎

Remark 4.6.3. For $m = n = 0$ the constants $\xi_{n+\ell+j,k}$, $0 \le k \le \ell+j-1$ reduce to

$$\xi_{\ell+j,k} = \frac{1}{(j+\ell-k)!} \begin{cases} \begin{pmatrix} j+\ell-k-1 \\ \ell-k-1 \end{pmatrix}, & 0 \le k \le \ell-1 \\ 1, & \ell \le k \le \ell+j-1 \end{cases}$$

which are exactly the same as in Remark 3.4.1. ∎

4.7 (0; *Lidstone*) INTERPOLATION

Theorem 4.7.1. Let $x(t) \in C^{(2m+1)}[0,1]$ and $P_{(4.7.1)}(t)$ be the (0; *Lidstone*) *interpolating polynomial* of degree $2m$ of the function $x(t)$, i.e., it satisfies

$$P_{(4.7.1)}(\eta) = x(\eta) = \bar{\alpha}_0, \ 0 \le \eta \le 1$$

$$(4.7.1) \qquad P^{(2i+1)}_{(4.7.1)}(0) = x^{(2i+1)}(0) = \alpha_i, \ 0 \le i \le m-1$$

$$P^{(2i+1)}_{(4.7.1)}(1) = x^{(2i+1)}(1) = \beta_i, \ 0 \le i \le m-1.$$

Then, for the error function $e_{(4.7.1)}(t) = x(t) - P_{(4.7.1)}(t)$ the following hold

$$(4.7.2) \qquad |\, e_{(4.7.1)}(t)\,| \le \mu_{2m+1,0}\, M_{2m+1},$$

where $M_{2m+1} = \max_{0\le t\le 1} | x^{(2m+1)}(t) |$, and

(4.7.3)
$$\mu_{2m+1,0} = (-1)^m \begin{cases} E_{2m+1}(1) - E_{2m+1}(\eta), & 0 \le \eta \le \frac{1}{2} \\ E_{2m+1}(\eta) - E_{2m+1}(0), & \frac{1}{2} \le \eta \le 1 \end{cases}$$
$$\le (-1)^m \frac{4(2^{2m+2}-1)}{(2m+2)!} B_{2m+2};$$

and

(4.7.4)
$$| e^{(k+1)}_{(4.7.1)}(t) | = \left| \int_0^1 \frac{\partial^k g_m(t,s)}{\partial t^k} y^{(2m)}(s)\, ds \right| = | e^{(k)}_{(1.1.1)}(t) |,$$
$$0 \le k \le 2m-1$$

where $y(t) = x'(t)$. ∎

Remark 4.7.1. In (4.7.2) the constant $\mu_{2m+1,0}$ given in (4.7.3) is the best possible as the equality holds for the function

$$x_7(t) = (-1)^m [\mathrm{sgn}\ (t-\eta)]\, [E_{2m+1}(t) - E_{2m+1}(\eta)]$$

and only for this function up to a constant factor. ∎

Remark 4.7.2. The estimates obtained in Chapter 1 for $| e^{(k)}_{(1.1.1)}(t) |$, $0 \le k \le 2m-1$ can be used in (4.7.4) to obtain the best possible bounds for $| e^{(k+1)}_{(4.7.1)}(t) |$, $0 \le k \le 2m-1$. ∎

4.8 (0, 2, 0; *Lidstone*) INTERPOLATION

Theorem 4.8.1. Let $x(t) \in C^{(2m+3)}[0,1]$ and $P_{(4.8.1)}(t)$ be the (0, 2, 0; *Lidstone*) *interpolating polynomial* of degree $(2m+2)$ of the function $x(t)$, i.e., it satisfies

(4.8.1)
$$P_{(4.8.1)}(0) = x(0) = \bar{\alpha}_0, \quad P_{(4.8.1)}(1) = x(1) = \bar{\beta}_0$$
$$P''_{(4.8.1)}(0) = x''(0) = \bar{\alpha}_2$$
$$P^{(2i+3)}_{(4.8.1)}(0) = x^{(2i+3)}(0) = \alpha_i, \ 0 \le i \le m-1$$
$$P^{(2i+3)}_{(4.8.1)}(1) = x^{(2i+3)}(1) = \beta_i, \ 0 \le i \le m-1.$$

Then, for the error function $e_{(4.8.1)}(t) = x(t) - P_{(4.8.1)}(t)$ the following hold

(4.8.2) $$| e^{(k)}_{(4.8.1)}(t) | \le \nu_{2m+3,k}\ M_{2m+3};\ k = 0,1,2$$

where $M_{2m+3} = \max\limits_{0\le t\le 1} | x^{(2m+3)}(t) |$, and

(4.8.3) $$\nu_{2m+3,0} = (-1)^{m+1}\Big[\frac{1}{2}t_1(1-t_1)E_{2m+1}(0) - (1-2t_1)E_{2m+3}(0) + E_{2m+3}(t_1)\Big],$$

where t_1 is the root of the equation $\left(t - \frac{1}{2}\right) E_{2m+1}(0) - 2E_{2m+3}(0) = E_{2m+2}(t)$,

(4.8.4) $$\nu_{2m+3,1} = (-1)^m \left[2E_{2m+3}(0) - \frac{1}{2}E_{2m+1}(0)\right],$$

(4.8.5) $$\nu_{2m+3,2} = 2(-1)^m E_{2m+1}(1);$$

and

(4.8.6) $$| e^{(k+3)}_{(4.8.1)}(t) | = \left|\int_0^1 \frac{\partial^k g_m(t,s)}{\partial t^k}\ y^{(2m)}(s)\ ds\right| = | e^{(k)}_{(1.1.1)}(t) |,$$

$$0 \le k \le 2m-1$$

where $y(t) = x'''(t)$. ∎

Remark 4.8.1. In (4.8.2) the constants $\nu_{2m+3,k};\ k = 0,1,2$ are the best possible as the equality holds for the function

$$x_8(t) = (-1)^{m+1}\left[\frac{1}{2}t(1-t)E_{2m+1}(0) - (1-2t)E_{2m+3}(0) + E_{2m+3}(t)\right]$$

and only for this function up to a constant factor. ∎

Remark 4.8.2. A remark similar to that of Remark 4.7.2 for (4.8.6) holds. ∎

4.9 (1, 3, 0, 1; *Lidstone*) INTERPOLATION

Theorem 4.9.1. Let $x(t) \in C^{(2m+4)}[0,1]$ and $P_{(4.9.1)}(t)$ be the $(1,3,0,1;$ *Lidstone*$)$ *interpolating polynomial* of degree $(2m+3)$ of the function $x(t)$, i.e., it satisfies

$$P_{(4.9.1)}(\eta) = x(\eta) = \bar{\alpha}_0,\ 0 \le \eta \le 1$$

$$P'_{(4.9.1)}(0) = x'(0) = \bar{\alpha}_1,\ \ P'_{(4.9.1)}(1) = x'(1) = \bar{\beta}_1$$

$$(4.9.1) \qquad P'''_{(4.9.1)}(0) = x'''(0) = \bar{\alpha}_3$$

$$P^{(2i+4)}_{(4.9.1)}(0) = x^{(2i+4)}(0) = \alpha_i,\ 0 \le i \le m-1$$

$$P^{(2i+4)}_{(4.9.1)}(1) = x^{(2i+4)}(1) = \beta_i,\ 0 \le i \le m-1.$$

Then, for the error function $e_{(4.9.1)}(t) = x(t) - P_{(4.9.1)}(t)$ the following hold

$$(4.9.2) \qquad |\ e^{(k)}_{(4.9.1)}(t)\ | \le \sigma_{2m+4,k}\ M_{2m+4};\ k = 0,1,2,3$$

where $M_{2m+4} = \max\limits_{0\le t\le 1} |\ x^{(2m+4)}(t)\ |$, and

$$(4.9.3)\ \sigma_{2m+4,0} = (-1)^m \times$$

$$\max\left\{\frac{1}{12}(3\eta^2 - 2\eta^3 - 1)E_{2m+1}(0) - \eta(1-\eta)E_{2m+3}(0) + E_{2m+4}(\eta),\right.$$

$$\left.\frac{1}{12}\eta^2(2\eta - 3)E_{2m+1}(0) + \eta(1-\eta)E_{2m+3}(0) - E_{2m+4}(\eta)\right\},$$

$$(4.9.4) \qquad \sigma_{2m+4,k} = \nu_{2m+3,k-1},\ k = 1,2,3;$$

and

$$(4.9.5) \qquad |\ e^{(k+4)}_{(4.9.1)}(t)\ | = \left|\int_0^1 \frac{\partial^k g_m(t,s)}{\partial t^k}\ y^{(2m)}(s)\ ds\right| = |\ e^{(k)}_{(1.1.1)}(t)\ |,$$

$$0 \le k \le 2m-1$$

where $y(t) = x''''(t)$. ∎

Remark 4.9.1. The constants $\sigma_{2m+4,k};\ k = 0,1,2,3$ in (4.9.2) are the best possible as the equality holds for the function

$$x_9(t) = (-1)^{m+1}[\mathrm{sgn}\ (t-\eta)]\int_\eta^t \left[\frac{1}{2}s(1-s)E_{2m+1}(0) - (1-2s)E_{2m+3}(0)\right.$$

$$\left. + E_{2m+3}(s)\right] ds$$

and only for this function up to a constant factor. ∎

Remark 4.9.2. A remark similar to that of Remark 4.7.2 for (4.9.5) holds. ∎

4.10 $(0 : \ell - 1, \ell : \ell + j - 1;$ *Lidstone*$)$ INTERPOLATION

Theorem 4.10.1. Let $x(t) \in C^{(2m+\ell+j)}[0,1]$ and $P_{(4.10.1)}(t)$ be the $(0 : \ell-1, \ell : \ell+j-1;$ *Lidstone*$)$ *interpolating polynomial* of degree $(2m+\ell+j-1)$ of the function $x(t)$, i.e., it satisfies

$$(4.10.1)\qquad \begin{aligned} &P^{(i)}_{(4.10.1)}(0) = x^{(i)}(0) = \bar{\alpha}_i;\ 0 \le i \le \ell-1,\ \ell \ge 1 \\ &P^{(\ell+i)}_{(4.10.1)}(1) = x^{(\ell+i)}(1) = \bar{\beta}_i;\ 0 \le i \le j-1,\ j \ge 1 \\ &P^{(\ell+j+2i)}_{(4.10.1)}(0) = x^{(\ell+j+2i)}(0) = \alpha_i,\ 0 \le i \le m-1 \\ &P^{(\ell+j+2i)}_{(4.10.1)}(1) = x^{(\ell+j+2i)}(1) = \beta_i,\ 0 \le i \le m-1. \end{aligned}$$

Then, for the error function $e_{(4.10.1)}(t) = x(t) - P_{(4.10.1)}(t)$ the following hold

$$(4.10.2)\qquad |\, e^{(k)}_{(4.10.1)}(t)\, | \le \kappa_{2m+\ell+j,k}\, M_{2m+\ell+j},\ 0 \le k \le \ell+j-1$$

where $M_{2m+\ell+j} = \max_{0\le t\le 1} |\, x^{(2m+\ell+j)}(t)\, |$, and

(4.10.3) $\kappa_{2m+\ell+j,k}$

$$= (-1)^m \left\{ \frac{1}{(\ell-k-1)!} \sum_{i=0}^{j-1} \frac{(-1)^i}{(j+\ell-k-i-1)\,(j-i-1)!} E_{2m+i+1}(1) \right.$$

$$+(-1)^{j+1} \sum_{i=0}^{\ell-k-1} \frac{1}{(\ell-k-i-1)!} E_{2m+j+i+1}(0)$$

$$\left. +(-1)^j E_{2m+j+\ell-k}(1) \right\},\ 0 \le k \le \ell-1$$

$$= (-1)^m \left\{ \sum_{i=0}^{j+\ell-k-1} \frac{(-1)^i}{(j+\ell-k-i-1)!} E_{2m+i+1}(1) \right.$$

$$\left. +(-1)^{j+\ell-k} E_{2m+j+\ell-k}(0) \right\},\ \ell \le k \le \ell+j-1;$$

and

$$(4.10.4) \qquad |\, e^{(\ell+j+k)}_{(4.10.1)}(t)\,| = \left| \int_0^1 \frac{\partial^k g_m(t,s)}{\partial t^k}\; y^{(2m)}(s)\; ds \right| = |\, e^{(k)}_{(1.1.1)}(t)\,|,$$

$$0 \le k \le 2m-1$$

where $y(t) = x^{(\ell+j)}(t)$. ∎

Remark 4.10.1. The constants $\kappa_{2m+\ell+j,k}$, $0 \le k \le \ell+j-1$ in (4.10.2) are the best possible as the equality holds for the function

$$x_{10}(t) = \frac{1}{(\ell-1)!(j-1)!}\int_0^t (t-s)^{\ell-1}\left[\int_s^1 (s-\tau)^{j-1}(-1)^m E_{2m}(\tau)\; d\tau\right] ds$$

and only for this function up to a constant factor. ∎

Remark 4.10.2. A remark similar to that of Remark 4.7.2 for (4.10.4) holds. ∎

4.11 (0, 2, 1; *Lidstone*) INTERPOLATION

Theorem 4.11.1. Let $x(t) \in C^{(2m+3)}[0,1]$ and $P_{(4.11.1)}(t)$ be the (0, 2, 1; *Lidstone*) *interpolating polynomial* of degree $(2m+2)$ of the function $x(t)$, i.e., it satisfies

$$(4.11.1) \qquad \begin{aligned} P_{(4.11.1)}(0) &= x(0) = \bar{\alpha}_0 \\ P'_{(4.11.1)}(1) &= x'(1) = \bar{\beta}_1 \\ P''_{(4.11.1)}(0) &= x''(0) = \bar{\alpha}_2 \\ P^{(2i+3)}_{(4.11.1)}(0) &= x^{(2i+3)}(0) = \alpha_i,\ 0 \le i \le m-1 \\ P^{(2i+3)}_{(4.11.1)}(1) &= x^{(2i+3)}(1) = \beta_i,\ 0 \le i \le m-1. \end{aligned}$$

Then, for the error function $e_{(4.11.1)}(t) = x(t) - P_{(4.11.1)}(t)$ the following hold

$$(4.11.2) \qquad |\, e^{(k)}_{(4.11.1)}(t)\,| \le \vartheta_{2m+3,k}\; M_{2m+3};\ k = 0, 1, 2$$

where $M_{2m+3} = \max_{0 \le t \le 1} |\, x^{(2m+3)}(t)\,|$, and

$$(4.11.3) \qquad \vartheta_{2m+3,0} = \nu_{2m+3,1},$$

(4.11.4) $$\vartheta_{2m+3,1} = (-1)^{m+1} E_{2m+1}(0),$$

(4.11.5) $$\vartheta_{2m+3,2} = \nu_{2m+3,2};$$

and

(4.11.6) $$| e^{(k+3)}_{(4.11.1)}(t) | = \left| \int_0^1 \frac{\partial^k g_m(t,s)}{\partial t^k} \, y^{(2m)}(s) \, ds \right| = | e^{(k)}_{(1.1.1)}(t) |,$$

$$0 \le k \le 2m-1$$

where $y(t) = x'''(t)$. ∎

Remark 4.11.1. The constants $\vartheta_{2m+3,k}$; $k = 0,1,2$ in (4.11.2) are the best possible as the equality holds for the function

$$x_{11}(t) = (-1)^m \left[E_{2m+3}(0) - E_{2m+3}(t) - \frac{1}{2} t(2-t) E_{2m+1}(0) \right]$$

and only for this function up to a constant factor. ∎

Remark 4.11.2. A remark similar to that of Remark 4.7.2 for (4.11.6) holds. ∎

REFERENCES

1. **R.P.Agarwal,** Some new results on two - point problems for higher order differential equations, Funkcialaj Ekvacioj 29(1986), 197-212.

2. **R.P.Agarwal,** Boundary Value Problems for Higher Order Differential Equations, World Scientific, Singapore • Philadelphia, 1986.

3. **R.P.Agarwal,** Sharp inequalities in polynomial interpolation, in General Inequalities 6, Ed. W.Walter, International Series of Numerical Mathematics, Birkhäuser Verlag 103(1992), 73-92.

4. **L.Erbe,** Boundary value problems for ordinary differential equations, Rocky Mountain J. Math. 4(1971), 709-729.

5. **A.Ju.Levin,** Some problems bearing on the oscillation of solutions of linear differential equations, Soviet Math. Dokl. 4(1963), 121-124.

6. **A.Ju.Levin,** Distribution of the zeros of solutions of a linear differential equation, Dokl. Akad. Nauk SSSR 156(1964), 1281-1284.

7. **A.C.Peterson,** Existence - uniqueness for two - point boundary value problems for nth order nonlinear differential equations, Rocky Mountain J. Math. 2(1977), 103-109.

8. **A.K.Varma** and **G.Howell,** Best error bounds for derivatives in two point Birkhoff interpolation problem, J. Approximation Theory 38(1983), 258-268.

9. **P.J.Y.Wong** and **R.P.Agarwal,** Optimal error bounds for the derivatives of two - point mixed interpolation, to appear.

CHAPTER 5

PIECEWISE - POLYNOMIAL INTERPOLATION

5.1 INTRODUCTION

Although polynomials have several attractive features, polynomial interpolation of a given function often has the drawback of producing approximations that may be wildly oscillatory. To overcome this difficulty we divide the interval of interest into small subintervals and in each subinterval consider polynomials of relatively low degree and finally 'piece together' these polynomials. This subject has steadily developed over the past fifty years, and at present there are thousands of research papers on piecewise - polynomial interpolation and their applications. The plan of this chapter is as follows : In Section 5.2 we collect some results from analysis which will be used repeatedly throughout the remaining monograph. In Section 5.3 we shall follow the general treatment of Birkhoff, Ciarlet, Schultz and Varga [8,10,11] to provide an explicit representation of piecewise - Hermite interpolates. This representation is then used to obtain error bounds for the derivatives of piecewise - Hermite interpolates in L_∞ and L_2 - norms. These bounds improve on those given by Schultz [21], and extend to cases not considered by him. Section 5.4 contains an explicit representation of piecewise - Lidstone interpolates and the error bounds for its derivatives in L_∞ and L_2 - norms. Results of Sections 5.3 and 5.4 are extended to two variables in Sections 5.5 and 5.6, respectively.

5.2 PRELIMINARIES

Let $-\infty < a < b < \infty$ and $-\infty < c < d < \infty$. For the intervals $[a,b]$ and $[c,d]$ we let $\Delta : a = t_0 < t_1 < \cdots < t_{N+1} = b$, and $\Delta' : c = u_0 < u_1 < \cdots < u_{M+1} = d$ denote uniform partitions of $[a,b]$ and $[c,d]$ with stepsizes $h = (b-a)/(N+1)$ and $\ell = (d-c)/(M+1)$, respectively. Further, we let $\rho = \Delta \times \Delta'$ be a rectangular partition of $[a,b]\times[c,d]$. For the functions $x(t)$ and $f(t,u)$ and each positive integer r, we shall denote by $D^r x = d^r x/dt^r$, $D_t^r f = \partial^r f/\partial t^r$ and $D_u^r f = \partial^r f/\partial u^r$. For each nonnegative integer r and for each p, $1 \le p \le \infty$, we will let $PC^{r,p}[a,b]$ be the set of all real - valued functions $x(t)$ such that : (i) $x(t)$ is $(r-1)$ times continuously differentiable on $[a,b]$, (ii) there exist s_i, $0 \le i \le L+1$ with $a = s_0 < s_1 < \cdots < s_{L+1} = b$ such that on each open subinterval (s_i, s_{i+1}), $0 \le i \le L$, $D^{r-1}x$ is continuously differentiable, and (iii) the L_p - norm of $D^r x$ is finite, i.e.,

$$\| D^r x \|_p = \left(\sum_{i=0}^{L} \int_{s_i}^{s_{i+1}} | D^r x(t) |^p \, dt \right)^{\frac{1}{p}} < \infty.$$

Of course, for the case $p = \infty$ it reduces to

$$\| D^r x \|_\infty = \max_{0 \le i \le L} \sup_{t \in (s_i, s_{i+1})} | D^r x(t) | < \infty.$$

Similarly, for each nonnegative integer r and for each p, $1 \le p \le \infty$, we will let $PC^{r,p}([a,b] \times [c,d])$ be the set of all real - valued functions $f(t,u)$ such that : (i) $f(t,u)$ is $(r-1)$ times continuously differentiable, i.e., $D_t^\mu D_u^\nu f$, $0 \le \mu + \nu \le r-1$ exists and continuous on $[a,b] \times [c,d]$, (ii) there exist s_i, $0 \le i \le L+1$ and v_j, $0 \le j \le R+1$ with $a = s_0 < s_1 < \cdots < s_{L+1} = b$

and $c = v_0 < v_1 < \cdots < v_{R+1} = d$ such that on each open subrectangle $(s_i, s_{i+1}) \times (v_j, v_{j+1})$; $0 \le i \le L$, $0 \le j \le R$ and for all $0 \le \mu \le r-1$, $0 \le \nu \le r-1$ such that $\mu + \nu = r-1$, $D_t^\mu D_u^\nu f$ is continuously differentiable, and (iii) for all $0 \le \mu \le r$, $0 \le \nu \le r$ such that $\mu + \nu = r$ the L_p - norm of $D_t^\mu D_u^\nu f$ is finite, i.e.,

$$\| D_t^\mu D_u^\nu f \|_p = \left(\sum_{i=0}^{L} \sum_{j=0}^{R} \int_{s_i}^{s_{i+1}} \int_{v_j}^{v_{j+1}} | D_t^\mu D_u^\nu f(t,u) |^p \, du \, dt \right)^{\frac{1}{p}} < \infty.$$

For the particular case $p = \infty$ it reduces to

$$\| D_t^\mu D_u^\nu f \|_\infty = \max_{\substack{0 \le i \le L \\ 0 \le j \le R}} \sup_{(t,u) \in (s_i, s_{i+1}) \times (v_j, v_{j+1})} | D_t^\mu D_u^\nu f(t,u) | < \infty.$$

We will also need the set $PC^{r_1, r_2, p}([a,b] \times [c,d])$ of all real - valued functions $f(t,u)$ such that : (i) $D_t^\mu D_u^\nu f$; $0 \le \mu \le r_1 - 1$, $0 \le \nu \le r_2 - 1$ exists and continuous on $[a,b] \times [c,d]$, (ii) on each open subrectangle $(s_i, s_{i+1}) \times (v_j, v_{j+1})$; $0 \le i \le L$, $0 \le j \le R$ and for all $0 \le \mu \le r_1$, $0 \le \nu \le r_2$, $D_t^\mu D_u^\nu f$ exists and continuous, and (iii) for all $0 \le \mu \le r_1$, $0 \le \nu \le r_2$ the L_p - norm of $D_t^\mu D_u^\nu f$ is finite.

Theorem 5.2.1. (Peano's Kernel Theorem) If E is a linear functional on $PC^{n,1}[a,b]$, $n \ge 1$ and $E(p(t)) = 0$ for all polynomials $p(t)$ of degree $(n-1)$, then for all $x(t) \in PC^{n,1}[a,b]$

$$E(x(t)) = \frac{1}{(n-1)!} E_t \left[\int_a^b (t-s)_+^{n-1} D^n x(s) \, ds \right], \tag{5.2.1}$$

where

$$(t-s)_+^{n-1} = \begin{cases} (t-s)^{n-1}, & t \ge s \\ 0, & t < s \end{cases}$$

and E_t means the linear functional E applied to the expression

$$\int_a^b (t-s)_+^{n-1} D^n x(s) \, ds$$

considered as a function of t.

Proof. Taylor's formula with the exact remainder can be written as

$$x(t) = \sum_{i=0}^{n-1} \frac{(t-a)^i}{i!} D^i x(a) + \frac{1}{(n-1)!} \int_a^b (t-s)_+^{n-1} D^n x(s) \, ds. \tag{5.2.2}$$

The result now follows by applying E to both sides of (5.2.2) and by using the linearity of E and the fact that E vanishes on all polynomials of degree $(n-1)$. ∎

Theorem 5.2.2. (Wirtinger's Inequality [6,17]) If $x(t) \in PC^{1,2}[a,b]$ and $x(a) = x(b) = 0$, then

$$\int_a^b x^2(t)\, dt \leq \frac{(b-a)^2}{\pi^2} \int_a^b [Dx(t)]^2\, dt. \tag{5.2.3}$$

Moreover, in (5.2.3) the equality holds if and only if $x(t) = c \sin\left(\frac{\pi(x-a)}{b-a}\right)$, where c is an arbitrary constant.

Proof. Expanding $x(t)$ and $Dx(t)$ in their respective Fourier series, we have

$$x(t) \sim \sum_{n=1}^{\infty} a_n \sin\left(\frac{n\pi(x-a)}{b-a}\right)$$

and

$$Dx(t) \sim \sum_{n=1}^{\infty} a_n \frac{n\pi}{(b-a)} \cos\left(\frac{n\pi(x-a)}{b-a}\right).$$

Therefore, by Parseval's equality it follows that

$$\int_a^b x^2(t)\, dt = \sum_{n=1}^{\infty} a_n^2$$

and

$$\int_a^b [Dx(t)]^2\, dt = \sum_{n=1}^{\infty} a_n^2 \frac{n^2\pi^2}{(b-a)^2}.$$

From the above relations the inequality (5.2.3) is immediate. ∎

Corollary 5.2.3. If $x(t) \in PC^{2,2}[a,b]$ and $x(a) = x(b) = 0$, then

$$\int_a^b x^2(t)\, dt \leq \frac{(b-a)^4}{\pi^4} \int_a^b [D^2x(t)]^2\, dt. \tag{5.2.4}$$

Moreover, in (5.2.4) the equality holds if and only if $x(t) = c \sin\left(\frac{\pi(x-a)}{b-a}\right)$, where c is an arbitrary constant.

Proof. since

$$\begin{aligned}\int_a^b [Dx(t)]^2\, dt &= x(t)\, Dx(t)\Big|_a^b - \int_a^b x(t)\, D^2x(t)\, dt \\ &= -\int_a^b x(t)\, D^2x(t)\, dt\end{aligned}$$

by the Cauchy - Schwarz inequality it follows that

$$(5.2.5) \qquad \int_a^b [Dx(t)]^2\, dt \le \left\{\int_a^b x^2(t)\, dt\right\}^{\frac{1}{2}} \left\{\int_a^b \left[D^2x(t)\right]^2 dt\right\}^{\frac{1}{2}}.$$

A combination of (5.2.3) and (5.2.5) immediately gives (5.2.4). ∎

Next result is a generalization of Theorem 5.2.2.

Theorem 5.2.4. (Cimmino's Inequality [12]) If $x(t) \in PC^{n,2}[a,b]$ and $x^{(i)}(a) = x^{(i)}(b) = 0,\ 0 \le i \le n-1$ then

$$(5.2.6) \qquad \int_a^b [D^k x(t)]^2\, dt \le \left(\frac{b-a}{\lambda_{n,k}}\right)^{2n-2k} \int_a^b [D^n x(t)]^2\, dt,\ 0 \le k \le n$$

where $\lambda_{n,n} = 1$ and for $0 \le k \le n-1$, $\lambda_{n,k}$ is the least positive zero of the Wronskian of n linearly independent solutions of the differential equation

$$(5.2.7) \qquad y^{(2n)} - (-1)^{n+k}\, y^{(2k)} = 0$$

satisfying the partial initial conditions $y^{(i)}(0) = 0,\ 0 \le i \le n-1$. Equivalently, in (5.2.6), $\lambda_{n,k}^{2n-2k}$ is the first eigenvalue of the boundary value problem

$$(5.2.8) \qquad \begin{aligned} &y^{(2n)} - \lambda(-1)^{n+k}\, y^{(2k)} = 0,\quad \lambda > 0 \\ &y^{(i)}(0) = y^{(i)}(1) = 0,\quad 0 \le i \le n-1. \end{aligned}$$

In (5.2.6) equality holds if and only if $x(t)$ is the first eigenfunction of the boundary value problem $y^{(2n)} - \lambda(-1)^{n+k}\, y^{(2k)} = 0,\ \lambda > 0;\ y^{(i)}(a) = y^{(i)}(b) = 0,\ 0 \le i \le n-1$.

Proof. We shall indicate the proof for the case $k = 0$. Let K_{2n} be the subspace of $PC^{2n,2}[0,1]$ such that every $y(t) \in K_{2n}$ satisfies the conditions $y^{(i)}(0) = y^{(i)}(1) = 0,\ 0 \le i \le n-1$. Let $T : K_{2n} \to PC^{2n,2}[0,1]$ be an operator defined by $Ty = (-1)^n D^n y^{(n)}$. It is well known that the functional $G : K_{2n} \to \Re$ given by $G(y) = \| y^{(n)} \|_2^2$ attains its minimum in the set of all functions of K_{2n} satisfying the condition $\| y \|_2 = 1$ only for the solutions of $Ty = \lambda y$ (which is equivalent to (5.2.8)). It is also easily seen that for $y(t) \in K_{2n}$ we have $G(y) = (Ty, y) = (y, Ty)$, where $(\cdot,\cdot)$ is the usual inner product. Hence, by $Ty = \lambda y$ it follows that the minimum of the functional G taken over all the functions of K_{2n} satisfying the condition $\| y \|_2 = 1$ is equal to the smallest positive eigenvalue of the operator T (i.e., smallest positive eigenvalue of (5.2.8)). Denoting it by λ_1, we get $G\left(\frac{y}{\|y\|_2}\right) \geq \lambda_1$ for any $y(t) \in K_{2n}$ such that $\| y \|_2 > 0$. ∎

The first few $\lambda_{n,k}$ are given in the following table.

Table 5.2.1.

k \ n	1	2	3	4	5
0	π	4.730041	6.283185	7.818707	9.343298
1		2π	7.853205	9.427056	10.995831
2			8.986819	10.535315	12.111801
3				11.526918	13.059858
4					13.975860

Remark 5.2.1. For the first eigenvalue λ_1 of (5.2.8) with $k = 0$ the following sharp lower bound is known [13,15]

$$\lambda_1 = \lambda_{n,0}^{2n} \geq \frac{(4n-1)![(n-1)!]^2}{(2n-1)!(2n-2)!}. \tag{5.2.9}$$ ∎

Corollary 5.2.5. If $x(t) \in PC^{2n-j,2}[a,b]$, $0 \leq j \leq n$ and $x^{(i)}(a) = x^{(i)}(b) = 0$, $0 \leq i \leq n-1$ then

$$\int_a^b [D^k x(t)]^2 \, dt \leq \left(\frac{b-a}{\lambda_{n,k}}\right)^{2n-2k} \left(\frac{b-a}{\lambda_{n,j}}\right)^{2n-2j} \int_a^b [D^{2n-j} x(t)]^2 \, dt; \tag{5.2.10}$$

$$0 \leq k \leq n, \; 0 \leq j \leq n.$$

Proof. Since

$$\begin{aligned}\int_a^b [D^n x(t)]^2\, dt &= D^{n-1}x(t)\ D^n x(t)\ \Big|_a^b - \int_a^b D^{n-1}x(t)\ D^{n+1}x(t)\ dt \\ &= -\int_a^b D^{n-1}x(t)\ D^{n+1}x(t)\ dt \\ &= (-1)^2 \int_a^b D^{n-2}x(t)\ D^{n+2}x(t)\ dt \\ &\quad \cdots \\ &= (-1)^{n-j} \int_a^b D^j x(t)\ D^{2n-j}x(t)\ dt\end{aligned}$$

by the Cauchy - Schwarz inequality it follows that

$$\int_a^b [D^n x(t)]^2\, dt \le \left\{\int_a^b [D^j x(t)]^2\, dt\right\}^{\frac{1}{2}} \left\{\int_a^b [D^{2n-j}x(t)]^2\, dt\right\}^{\frac{1}{2}}.$$

Therefore, in view of (5.2.6), we have

$$\int_a^b [D^n x(t)]^2\, dt \le \left(\frac{b-a}{\lambda_{n,j}}\right)^{n-j} \left\{\int_a^b [D^n x(t)]^2\, dt\right\}^{\frac{1}{2}} \left\{\int_a^b [D^{2n-j}x(t)]^2\, dt\right\}^{\frac{1}{2}},$$

which implies that

$$\int_a^b [D^n x(t)]^2\, dt \le \left(\frac{b-a}{\lambda_{n,j}}\right)^{2n-2j} \int_a^b [D^{2n-j}x(t)]^2\, dt. \tag{5.2.11}$$

A combination of (5.2.6) and (5.2.11) immediately gives (5.2.10). ∎

It is clear that Corollary 5.2.3 is a particular case of Corollary 5.2.5.

Remark 5.2.2. From Corollary 5.2.5 it is immediate that

$$\lambda_{n+1,k}^{2n+2-2k} \ge \lambda_{n,k}^{2n-2k}\lambda_{n,n-1}^2,\ \ 0 \le k \le n. \tag{5.2.12}$$

Thus, the constants provided in Table 5.2.1 can be used recursively to obtain lower bounds of $\lambda_{n,k}$, $0 \le k \le n-1$ for $n \ge 6$. ∎

Theorem 5.2.6. (Schmidt's Inequality [20]) If $p_n(t)$ is a polynomial of degree n, then

$$\int_a^b [Dp_n(t)]^2\, dt \le 4k_n(b-a)^{-2} \int_a^b [p_n(t)]^2\, dt, \tag{5.2.13}$$

where k_n is the largest eigenvalue of the matrix

$$A_n = (a_{ij})_{0\le i,j\le n} = \left(\int_{-1}^{1} L_i'(t)\ L_j'(t)\ dt\right)_{0\le i,j\le n},$$

and $(L_i(t))_{0\le i\le n}$ is the orthonormal system of Legendre polynomials.

Proof. Following Dörfler [14] it suffices to consider the special case of $a = -1$ and $b = 1$, for which (5.2.13) reduces to

$$\int_{-1}^{1} [Dp_n(t)]^2\ dt \le k_n \int_{-1}^{1} [p_n(t)]^2\ dt. \tag{5.2.14}$$

From the Fourier - Legendre expansion we have

$$p_n(t) = \sum_{i=0}^{n} \beta_i\ L_i(t), \tag{5.2.15}$$

where $\beta_i = \int_{-1}^{1} p_n(t)\ L_i(t)\ dt,\ 0 \le i \le n.$

Thus, in view of (5.2.14) and (5.2.15) it follows that

$$k_n = \sup_{\bar\beta\ne 0} \frac{\int_{-1}^{1}[Dp_n(t)]^2\ dt}{\int_{-1}^{1}[p_n(t)]^2\ dt} = \sup_{\bar\beta\ne 0} \frac{\bar\beta^T A_n\ \bar\beta}{\bar\beta^T\bar\beta} = \sup_{\bar\beta\ne 0}\ R[\bar\beta].$$

Obviously, the $(n+1)\times(n+1)$ matrix A_n is symmetric, nonnegative definite and $R[\bar\beta]$ is the Rayleigh quotient of A_n. Thus, k_n is the same as the maximum eigenvalue of A_n follows from the variational characterization of the eigenvalues of symmetric matrices in terms of Rayleigh quotient. ∎

Remark 5.2.3. In (5.2.13) the constants k_n are obviously the best possible. ∎

Remark 5.2.4. From the explicit representation

$$L_0(t) = \sqrt{\frac{1}{2}}, \qquad L_1(t) = \sqrt{\frac{3}{2}}\ t$$

$$L_2(t) = \sqrt{\frac{5}{8}}(3t^2 - 1), \qquad L_3(t) = \sqrt{\frac{7}{8}}(5t^3 - 3t)$$

$$L_4(t) = \frac{3}{8\sqrt{2}}(35t^4 - 30t^2 + 3), \quad L_5(t) = \frac{1}{8}\sqrt{\frac{11}{2}}(63t^5 - 70t^3 + 15t)$$

...

it follows that

$$A_1 = \begin{bmatrix} 0 & 0 \\ 0 & 3 \end{bmatrix}, \quad A_2 = \begin{bmatrix} 0 & 0 & 0 \\ 0 & 3 & 0 \\ 0 & 0 & 15 \end{bmatrix}, \quad A_3 = \begin{bmatrix} 0 & 0 & 0 & 0 \\ 0 & 3 & 0 & \sqrt{21} \\ 0 & 0 & 15 & 0 \\ 0 & \sqrt{21} & 0 & 42 \end{bmatrix}, \quad \ldots$$

and hence $k_1 = 3$, $k_2 = 15$, $k_3 = \frac{1}{2}(45 + \sqrt{1605})$, $k_4 = \frac{1}{2}(105 + \sqrt{7245})$. The computation of k_n for $n > 4$ seems to be possible only by a computer. Results up to $n \leq 6$ are as follows

$$\begin{aligned} k_1 &= 3, & k_2 &= 15 \\ k_3 &= 42.53122562, & k_4 &= 95.05878288 \\ k_5 &= 184.7262344, & k_6 &= 326.1507671. \end{aligned}$$ ∎

Remark 5.2.5. Although, Goetgheluck [16] contains the computed values of k_n up to $n \leq 65$, we shall show that

$$k_n \leq \frac{1}{8}\, n(n+1)(n+2)(n+3). \tag{5.2.16}$$

This bound is sharper than those given in [7,18]. For this, in view of (5.2.15) and the orthonormality relations of $L_i(t)$ it follows that

$$\int_{-1}^{1} [p_n(t)]^2 \, dt = \sum_{i=0}^{n} \beta_i^2. \tag{5.2.17}$$

On the other hand from the recurrence relation [1]

$$P'_{n+1}(t) = (2n+1)P_n(t) - P_{n-1}(t) \quad \text{and} \quad \int_{-1}^{1} [P_n(t)]^2 \, dt = \frac{2}{2n+1},$$

where $P_n(t)$ are the classical Legendre polynomials, we find that

$$L'_{2i}(t) = \sqrt{4i+1} \sum_{k=0}^{i-1} \sqrt{4k+3}\; L_{2k+1}(t)$$

and

$$L'_{2i+1}(t) = \sqrt{4i+3} \sum_{k=0}^{i} \sqrt{4k+1}\; L_{2k}(t).$$

Thus, for the case $n = 2m+1$ from (5.2.15) it follows that

$$\begin{aligned}
p'_{2m+1}(t) &= \sum_{i=1}^{2m+1} \beta_i\, L'_i(t) = \sum_{i=0}^{m} \beta_{2i+1}\, L'_{2i+1}(t) + \sum_{i=1}^{m} \beta_{2i}\, L'_{2i}(t) \\
&= \sum_{i=0}^{m} \beta_{2i+1} \left(\sqrt{4i+3} \sum_{k=0}^{i} \sqrt{4k+1}\, L_{2k}(t) \right) \\
&\qquad + \sum_{i=1}^{m} \beta_{2i} \left(\sqrt{4i+1} \sum_{k=0}^{i-1} \sqrt{4k+3}\, L_{2k+1}(t) \right) \\
&= \sum_{i=0}^{m} \sqrt{4i+1}\, L_{2i}(t) \left(\sum_{k=i}^{m} \sqrt{4k+3}\, \beta_{2k+1} \right) \\
&\qquad + \sum_{i=0}^{m-1} \sqrt{4i+3}\, L_{2i+1}(t) \left(\sum_{k=i+1}^{m} \sqrt{4k+1}\, \beta_{2k} \right),
\end{aligned}$$

and hence

$$\begin{aligned}
\text{(5.2.18)} \qquad \int_{-1}^{1} [p'_{2m+1}(t)]^2\, dt &= \sum_{i=0}^{m} (4i+1) \left(\sum_{k=i}^{m} \sqrt{4k+3}\, \beta_{2k+1} \right)^2 \\
&\qquad + \sum_{i=0}^{m-1} (4i+3) \left(\sum_{k=i+1}^{m} \sqrt{4k+1}\, \beta_{2k} \right)^2 .
\end{aligned}$$

Next, since the first sum is

$$\begin{aligned}
&\sum_{i=0}^{m} (4i+1) \left(\sum_{k=i}^{m} \sqrt{4k+3}\, \beta_{2k+1} \right)^2 \\
&\quad \le \sum_{i=0}^{m} (4i+1) \left(\sum_{k=i}^{m} (4k+3) \right) \left(\sum_{k=i}^{m} \beta_{2k+1}^2 \right) \\
&\quad \le \sum_{i=0}^{m} (4i+1)(m+1-i)(2m+3+2i) \left(\sum_{k=0}^{m} \beta_{2k+1}^2 \right) \\
&\quad = \left\{ \sum_{i=0}^{m} \left[(m+1)(2m+3) + (8m^2+20m+11)i - 6i^2 - 8i^3 \right] \right\} \left(\sum_{k=0}^{m} \beta_{2k+1}^2 \right) \\
&\quad = (m+1)(m+2)(2m+1) \left(m + \frac{3}{2} \right) \left(\sum_{k=0}^{m} \beta_{2k+1}^2 \right)
\end{aligned}$$

and the same process yields the same estimate for the second term, we find that

$$\int_{-1}^{1} [p'_{2m+1}(t)]^2\, dt \le (m+1)(m+2)(2m+1)\left(m+\frac{3}{2}\right)\sum_{i=0}^{2m+1} \beta_i^2$$

$$= (m+1)(m+2)(2m+1)\left(m+\frac{3}{2}\right)\int_{-1}^{1} [p_{2m+1}(t)]^2\, dt.$$

This proves (5.2.16) for the case $n = 2m+1$. The proof for $n = 2m$ is similar. ∎

Remark 5.2.6. Theorem 5.2.6 can be repeatedly used to obtain inequalities of the form

$$\int_a^b [D^\ell p_n(t)]^2\, dt \le \bar{k}_{n,\ell}\, [4(b-a)^{-2}]^\ell \int_a^b [p_n(t)]^2\, dt, \quad 1 \le \ell \le n \tag{5.2.19}$$

where

$$\bar{k}_{n,\ell} = k_{n-\ell+1} \cdots k_{n-1} k_n. \tag{5.2.20}$$

However, these constants are too large. In fact, from an extension of Theorem 5.2.6 it is clear that for each ℓ in (5.2.19) the constant $\bar{k}_{n,\ell}$ can be replaced by the optimal constant $\tilde{k}_{n,\ell}$, which is the largest eigenvalue of the matrix $\left(\int_{-1}^{1} L_i^{(\ell)}(t)\, L_j^{(\ell)}(t)\, dt\right)_{0\le i,j\le n}$.

From a direct computation it follows that

$$\tilde{k}_{n,n} = (2n+1)\left(\frac{(2n)!}{2^n\, n!}\right)^2,$$

$$\tilde{k}_{n,n-1} = \frac{1}{3}(2n+1)\left(\frac{(2n)!}{2^n\, n!}\right)^2,$$

$$\tilde{k}_{n,n-2} = \frac{1}{30}(2n-1)(2n-3)\left(\frac{(2n-4)!}{2^{n-2}(n-2)!}\right)^2 \Big\{(6n^3 - 19n^2 + 16n + 6) + (n-2)\sqrt{36n^4 - 84n^3 + 73n^2 + 92n + 24}\Big\},$$

$$\tilde{k}_{n,n-3} = \frac{1}{2}\left[(y+r) + \sqrt{(y+r)^2 - 4B}\right]$$

where

$$B = \frac{xyzr - q^2 zx - p^2 yr + p^2 q^2}{xz - p^2}$$

and

$$x = (2n-5)\left(\frac{(2n-6)!}{2^{n-3}(n-3)!}\right)^2, \quad y = \frac{1}{3}(2n-3)\left(\frac{(2n-4)!}{2^{n-2}(n-2)!}\right)^2$$

$$z = \frac{1}{15}(2n-1)(3n^2-14n+18)\left(\frac{(2n-4)!}{2^{n-2}(n-2)!}\right)^2$$

$$p = \frac{1}{3}(2n-6)\frac{(2n-6)!(2n-4)!}{2^{2n-4}(n-3)!(n-2)!}\sqrt{(2n-5)(2n-1)}$$

$$q = \frac{1}{30}(n-3)\frac{(2n-2)!(2n-4)!}{2^{2n-4}(n-2)!(n-1)!}\sqrt{(2n-3)(2n+1)}$$

$$r = \frac{1}{315}(2n+1)(5n^2-26n+38)\left(\frac{(2n-2)!}{2^{n-1}(n-1)!}\right)^2.$$

In particular, we note that

$$\tilde{k}_{3,2} = 525 < \bar{k}_{3,2} = k_2k_3 = 15 \times \frac{1}{2}(45+\sqrt{1605}) = 637.968\cdots.$$

Further, $\tilde{k}_{n,1} = \bar{k}_{n,1} = k_n$. ∎

5.3 PIECEWISE HERMITE INTERPOLATION

For a fixed Δ, we define the set $H_m(\Delta) = \{h(t) \in C^{(m-1)}[a,b] : h(t)$ is a polynomial of degree at most $(2m-1)$ in each subinterval $[t_i, t_{i+1}]$, $0 \le i \le N\}$. It is clear that $H_m(\Delta)$ is of dimension $m(N+2)$.

Definition 5.3.1. For a given $x(t) \in C^{(m-1)}[a,b]$ we say $H_m^\Delta x(t)$ is the $H_m(\Delta)$ - *interpolate of* $x(t)$, also known as *Hermite interpolate of* $x(t)$ if $H_m^\Delta x(t) \in H_m(\Delta)$ with $D^k H_m^\Delta x(t_i) = x^{(k)}(t_i) = x_i^{(k)}$; $0 \le i \le N+1$, $0 \le k \le m-1$.

For $x(t) \in C^{(m-1)}[a,b]$ it is trivial to show that $H_m^\Delta x(t)$ uniquely exists and can be explicitly expressed as

$$H_m^\Delta x(t) = \sum_{i=0}^{N+1}\sum_{j=0}^{m-1} h_{m,i,j}(t)\, x_i^{(j)}, \tag{5.3.1}$$

where $h_{m,i,j}(t)$, $0 \le i \le N+1$ are the basic elements of $H_m(\Delta)$ satisfying

$$D^{\nu} h_{m,i,j}(t_{\mu}) = \delta_{i\mu}\, \delta_{\nu j}\ ;\ 0 \le \nu \le m-1,\ 0 \le \mu \le N+1. \tag{5.3.2}$$

It is clear that in the above Definition 5.3.1 as well as in the representation (5.3.1) the function $x(t)$ need not be in $C^{(m-1)}[a,b]$, rather it is sufficient (which we shall assume throughout) that for the function $x(t)$, $x_i^{(k)}$; $0 \le i \le N+1$, $0 \le k \le m-1$ exist.

In particular, we have

$$\begin{aligned}
h_{1,i,0}(t) &= -\frac{1}{h}(t - t_{i+1}), \quad t_i \le t \le t_{i+1},\ 0 \le i \le N \\
&= \frac{1}{h}(t - t_{i-1}), \quad t_{i-1} \le t \le t_i,\ 1 \le i \le N+1 \\
&= 0, \quad \text{otherwise;} \\
h_{2,i,0}(t) &= \frac{(t-t_i)^2}{h^2}\left[\frac{2}{h}(t-t_i) - 3\right] + 1, \quad t_i \le t \le t_{i+1},\ 0 \le i \le N \\
&= \frac{(t-t_{i-1})^2}{h^2}\left[-\frac{2}{h}(t-t_{i-1}) + 3\right], \quad t_{i-1} \le t \le t_i,\ 1 \le i \le N+1 \\
&= 0, \quad \text{otherwise;} \\
h_{2,i,1}(t) &= \frac{1}{h^2}(t-t_{i+1})^2(t-t_i), \quad t_i \le t \le t_{i+1},\ 0 \le i \le N \\
&= \frac{1}{h^2}(t-t_{i-1})^2(t-t_i), \quad t_{i-1} \le t \le t_i,\ 1 \le i \le N+1 \\
&= 0, \quad \text{otherwise;} \\
h_{3,i,0}(t) &= \frac{(t-t_{i+1})^3}{h^3}\left[-\frac{6}{h^2}(t-t_i)^2 - \frac{3}{h}(t-t_i) - 1\right], \\
&\qquad t_i \le t \le t_{i+1},\ 0 \le i \le N \\
&= \frac{(t-t_{i-1})^3}{h^3}\left[\frac{6}{h^2}(t-t_i)^2 - \frac{3}{h}(t-t_i) + 1\right], \\
&\qquad t_{i-1} \le t \le t_i,\ 1 \le i \le N+1 \\
&= 0, \quad \text{otherwise;}
\end{aligned}$$

$$
\begin{aligned}
h_{3,i,1}(t) &= \frac{(t-t_{i+1})^3}{h^3}\left[-\frac{3}{h}(t-t_i)^2-(t-t_i)\right], \quad t_i \le t \le t_{i+1},\ 0 \le i \le N \\
&= \frac{(t-t_{i-1})^3}{h^3}\left[-\frac{3}{h}(t-t_i)^2+(t-t_i)\right], \quad t_{i-1} \le t \le t_i,\ 1 \le i \le N+1 \\
&= 0, \qquad \text{otherwise;} \\
h_{3,i,2}(t) &= -\frac{1}{2h^3}(t-t_{i+1})^3(t-t_i)^2, \quad t_i \le t \le t_{i+1},\ 0 \le i \le N \\
&= \frac{1}{2h^3}(t-t_{i-1})^3(t-t_i)^2, \quad t_{i-1} \le t \le t_i,\ 1 \le i \le N+1 \\
&= 0, \qquad \text{otherwise.}
\end{aligned}
$$

We begin with the following equalities for the basic elements of $H_m(\Delta)$. Some of these results will be used in the next chapter.

Lemma 5.3.1. For $0 \le i \le N$ and $k = 0, 1$ the following equalities hold

$$
\max_{t_i \le t \le t_{i+1}} \left[| D^k h_{1,i,0}(t) | + | D^k h_{1,i+1,0}(t) | \right] = a_{1,0,k}\, h^{-k}, \tag{5.3.3}
$$

where $a_{1,0,0} = 1$ and $a_{1,0,1} = 2$. ∎

Lemma 5.3.2. For $0 \le i \le N$ and $0 \le k \le 3$ the following equalities hold

$$
\max_{t_i \le t \le t_{i+1}} \left[| D^k h_{2,i,j}(t) | + | D^k h_{2,i+1,j}(t) | \right] = a_{2,j,k}\, h^{j-k}; \ j = 0, 1 \tag{5.3.4}
$$

where the constants $a_{2,j,k}$ are given in the following table.

Table 5.3.1.

j \ k	0	1	2	3
0	1	3	12	24
1	$\frac{1}{4}$	1	6	12

Proof. The proof is by direct computation. ∎

Lemma 5.3.3. For $0 \le i \le N$ and $0 \le k \le 3$ the following equalities hold

$$\max_{t_i \le t \le t_{i+1}} \mid D^k h_{2,i,1}(t) \mid = b_{2,1,k}\ h^{1-k}, \tag{5.3.5}$$

where $b_{2,1,0} = \frac{4}{27}$, $b_{2,1,1} = 1$, $b_{2,1,2} = 4$ and $b_{2,1,3} = 6$. ∎

Lemma 5.3.4. For $0 \le i \le N$ and $0 \le k \le 5$ the following equalities hold

$$\max_{t_i \le t \le t_{i+1}} \left[\mid D^k h_{3,i,j}(t) \mid + \mid D^k h_{3,i+1,j}(t) \mid \right] = a_{3,j,k}\ h^{j-k};\ j = 0,1,2 \tag{5.3.6}$$

where the constants $a_{3,j,k}$ are given in the following table.

Table 5.3.2.

j \ k	0	1	2	3	4	5
0	1	$\frac{15}{4}$	$\frac{20}{\sqrt{3}}$	120	720	1440
1	$\frac{5}{16}$	1	$\frac{10}{\sqrt{3}}$	60	360	720
2	$\frac{1}{32}$	$\frac{\sqrt{3}}{18}$	1	12	60	120

Proof. The proof is by direct computation, and we convince by giving the details for the case $j = 1,\ k = 1$. For $t \in [t_i, t_{i+1}],\ 0 \le i \le N$ we have

$$\mid Dh_{3,i,1}(t) \mid = \frac{(t-t_{i+1})^2}{h^4} \mid 3(t-t_i)[3(t-t_i)+h]+(t-t_{i+1})[6(t-t_i)+h] \mid \tag{5.3.7}$$

and

$$\mid Dh_{3,i+1,1}(t) \mid = \frac{(t-t_i)^2}{h^4} \mid 3(t-t_{i+1})[-3(t-t_{i+1})+h] \tag{5.3.8}$$
$$+(t-t_i)[-6(t-t_{i+1})+h] \mid .$$

Rewriting (5.3.7) and (5.3.8) in terms of a new variable $s = t - t_i,\ s \in [0, h]$

$$\mid Dh_{3,i,1}(s+t_i) \mid = \frac{(s-h)^2}{h^4} \mid 15s^2 - 2hs - h^2 \mid$$

and

$$| Dh_{3,i+1,1}(s+t_i) | = \frac{s^2}{h^4} \, | -15s^2 + 28hs - 12h^2 | \, .$$

Thus, it follows that

$$(5.3.9) \quad | Dh_{3,i,1}(s+t_i) | + | Dh_{3,i+1,1}(s+t_i) |$$

$$= \frac{1}{h^3}(4s^3 - 6hs^2 + h^3), \; 0 \le s \le \frac{h}{3}$$

$$= \frac{1}{h^4}(30s^4 - 60hs^3 + 30h^2s^2 - h^4), \; \frac{h}{3} \le s \le \frac{2h}{3}$$

$$= \frac{1}{h^3}(-4s^3 + 6hs^2 - h^3), \; \frac{2h}{3} \le s \le h.$$

Now it is easy to verify that the right side of (5.3.9) attains its maximum at $s = 0$ and h, and its maximum value is indeed 1. ∎

Lemma 5.3.5. For $0 \le i \le N$ and $0 \le k \le 5$ the following equalities hold

$$(5.3.10) \qquad \max_{t_i \le t \le t_{i+1}} | D^k h_{3,i,j}(t) | = b_{3,j,k} \, h^{j-k}; \; j = 1, 2$$

where the constants $b_{3,j,k}$ are given in the following table.

Table 5.3.3.

$j \backslash k$	0	1	2	3	4	5
1	$\frac{16}{81}$	1	$\frac{8(28+19\sqrt{19})}{225}$	36	192	360
2	$\frac{54}{3125}$	$\frac{3(3+8\sqrt{6})}{1000}$	1	9	36	60

∎

Now we shall derive error bounds for the derivatives of the interpolation error $x(t) - H_m^{\Delta}x(t)$ in terms of the derivatives of $x(t)$ in L_∞ and L_2 - norms.

Theorem 5.3.6. Let $x(t) \in PC^{n,\infty}[a,b]$, $n = 1, 2$. Then,

$$(5.3.11) \qquad \| D^k(x - H_1^{\Delta}x) \|_\infty \le c_{1,n,k} \, h^{n-k} \, \| D^n x \|_\infty \, , \; 0 \le k \le n-1$$

where $c_{1,1,0} = \frac{1}{2}$, $c_{1,2,0} = \frac{1}{8}$ and $c_{1,2,1} = \frac{1}{2}$.

Proof. Without loss of generality we assume that $a = 0,\ b = 1$ and $h = 1$. While the proof for the case $n = 2$ is included in Theorem 2.4.13, we shall give the proof for the case $n = 1$. For this, we have

$$H_1^\Delta x(t) = (1-t)\ x(0) + t\ x(1).$$

Therefore, in view of Theorem 5.2.1 it follows that

$$x(t) - H_1^\Delta x(t) = \int_0^1 g_1(t,s) Dx(s)\ ds, \tag{5.3.12}$$

where

$$g_1(t,s) = \begin{cases} 1-t, & s \le t \\ -t, & t \le s. \end{cases} \tag{5.3.13}$$

Hence, we find that

$$\begin{aligned} |\ (x - H_1^\Delta x)(t)\ | &\le \left(\int_0^t (1-t)ds + \int_t^1 t\ ds\right) \|\ Dx\ \|_\infty \\ &= 2t(1-t)\ \|\ Dx\ \|_\infty \\ &\le \frac{1}{2}\ \|\ Dx\ \|_\infty\ . \end{aligned}$$ ∎

Remark 5.3.1. From Theorem 2.4.13 it is clear that the constants $c_{1,2,0} = \frac{1}{8}$ and $c_{1,2,1} = \frac{1}{2}$ are the best possible in the inequalities (5.3.11). To show that $c_{1,1,0} = \frac{1}{2}$ is also the best possible, we consider the function

$$\begin{aligned} x_m(t) &= \frac{1}{4}\left(1 - \frac{2}{m}\right) - t, \quad 0 \le t \le \frac{1}{2} - \frac{1}{m} \\ &= \frac{1}{8}m(2t-1)^2 - \frac{1}{4}, \quad \frac{1}{2} - \frac{1}{m} \le t \le \frac{1}{2} + \frac{1}{m} \\ &= \frac{1}{4}\left(1 - \frac{2}{m}\right) - (1-t), \quad \frac{1}{2} + \frac{1}{m} \le t \le 1. \end{aligned}$$

It is easy to see that $x_m(t) \in C^{(1)}[0,1]$ and $\|\ Dx_m\ \|_\infty = 1$. Further, for $h = 1$ we have

$$\left| x_m\left(\frac{1}{2}\right) - H_1^\Delta x_m\left(\frac{1}{2}\right) \right| = \frac{1}{2} - \frac{1}{2m} \to \frac{1}{2}, \quad \text{as } m \to \infty.$$ ∎

The following best possible result for the case $m = 2$ is due to Varma and Katsifarakis [22].

Theorem 5.3.7. Let $x(t) \in PC^{n,\infty}[a,b]$, $2 \le n \le 4$. Then,

$$\| D^k(x - H_2^{\Delta}x) \|_\infty \le c_{2,n,k}\, h^{n-k} \| D^n x \|_\infty \,, \quad 0 \le k \le n-1 \tag{5.3.14}$$

where the constants $c_{2,n,k}$ are given in the following table.

Table 5.3.4.

k \ n	2	3	4
0	$\frac{1}{16}$	$\frac{1}{96}$	$\frac{1}{384}$
1	$0.251497657+0$	$\frac{13\sqrt{13}-46}{27}$	$\frac{1}{72\sqrt{3}}$
2		$\frac{8}{27}$	$\frac{1}{12}$
3			$\frac{1}{2}$

Proof. Without loss of generality we assume that $a = 0,\ b = 1$ and $h = 1$, so that

$$H_2^{\Delta}x(t) = (1-t)^2(1+2t)x(0) + t^2(3-2t)x(1) + t(1-t)^2x'(0) + t^2(t-1)x'(1).$$

Therefore, in view of Theorem 5.2.1 for $0 \le k \le n-1$ it follows that

$$D^k(x - H_2^{\Delta}x)(t) = \frac{1}{(n-1)!}\int_0^1 g_n^{(k)}(t,s) D^n x(s)\, ds, \quad 2 \le n \le 4 \tag{5.3.15}$$

where $g_n^{(k)}(t,s) = \partial^k g_n(t,s)/\partial t^k$, and

$$g_2(t,s) = t^2[(3-2t)s - (2-t)], \quad t \le s \tag{5.3.16}$$

$$g_3(t,s) = t^2(1-s)[(3-2t)s - 1], \quad t \le s \tag{5.3.17}$$

$$g_4(t,s) = t^2(1-s)^2[(3-2t)s-t], \quad t \le s \tag{5.3.18}$$

and $g_n(t,s)$ for $s \le t$ is the same as $(-1)^n g_n(1-t, 1-s)$ for $t \le s$.

While the proof for the case $n = 4,\ 0 \le k \le 3$ is included in Theorem 2.4.13, we shall give the proof for the case $n = 3$, whereas for $n = 2$ the computation is similar.

From (5.3.15) and (5.3.17), we have

$$| D^k(x - H_2^{\Delta}x)(t) | \le \left(\frac{1}{2}\int_0^1 | g_3^{(k)}(t,s) | \, ds\right) \| D^3x \|_{\infty} \ ; \ k = 0,1,2 \tag{5.3.19}$$

where

$$g_3(t,s) = \begin{cases} t^2(3-2t)(1-s)(s-\alpha), & t \le s \\ (1-t)^2(1+2t)s(s-\beta), & s \le t \end{cases} \tag{5.3.20}$$

and

$$\alpha = \frac{1}{3-2t}, \qquad \beta = \frac{2t}{1+2t}.$$

For $\frac{1}{2} \le t \le 1$ it is easy to note that $0 < \alpha \le 1,\ \alpha \le t$ and $0 < \beta < 1,\ \beta \le t$. Thus, for $\frac{1}{2} \le t \le 1$ we have

$$\begin{aligned} \int_0^1 | g_3(t,s) | \, ds &= (1-t)^2(1+2t)\left(\int_0^{\beta} s(\beta - s)ds + \int_{\beta}^{t} s(s-\beta)ds\right) \\ &\quad + t^2(3-2t)\int_t^1 (1-s)(s-\alpha)ds \\ &= \frac{2t^3(3-3t+4t^3)(1-t)^2}{3(1+2t)^2} + \frac{2}{3}\,t^3(1-t)^3 \end{aligned}$$

$$= \frac{8t^3(1-t)^2}{3(1+2t)^2}. \tag{5.3.21}$$

From the symmetry arguments for $0 \le t \le \frac{1}{2}$ we also have

$$\int_0^1 | g_3(t,s) | \, ds = \frac{8t^2(1-t)^3}{3(3-2t)^2}. \tag{5.3.22}$$

Now from (5.3.21) and (5.3.22) it is easy to see that

$$\int_0^1 | g_3(t,s) | \, ds \le \frac{1}{48}, \ 0 \le t \le 1.$$

Thus, from (5.3.19) we have

$$\| x - H_2^{\Delta} x \|_\infty \leq \frac{1}{96} \| D^3 x \|_\infty .$$

This completes the proof for $k = 0$.

For $k = 1$, from (5.3.20) we have

$$g_3^{(1)}(t,s) = 2 \begin{cases} 3t(1-t)(1-s)(s-\alpha_1), & t \leq s \\ -3t(1-t)s(s-\beta_1), & s < t \end{cases} \tag{5.3.23}$$

where

$$\alpha_1 = \frac{1}{3(1-t)}, \qquad \beta_1 = \frac{3t-1}{3t}.$$

Once again it is easy to note that for $0 \leq t \leq \frac{1}{2}$, $\alpha_1 \geq t$, $\alpha_1 \leq 0$ and for $0 \leq t \leq \frac{1}{3}$, $\beta_1 \leq 0$ whereas for $\frac{1}{3} \leq t \leq \frac{1}{2}$, $\beta_1 \leq t$, $\beta_1 \geq 0$. Hence, we find that

$$\begin{aligned} \int_0^1 | g_3^{(1)}(t,s) | \, ds &= 6t(1-t) \left(\int_0^t s(s-\beta_1) ds + \int_t^{\alpha_1} (1-s)(\alpha_1 - s) ds \right. \\ &\qquad \left. + \int_{\alpha_1}^1 (1-s)(s-\alpha_1) ds \right) \\ &= \frac{2(2-3t)^3 t}{27(1-t)^2}, \quad 0 \leq t \leq \frac{1}{3}; \qquad (5.3.24) \\ &= 6t(1-t) \left(\int_0^{\beta_1} s(\beta_1 - s) ds + \int_{\beta_1}^t s(s-\beta_1) ds \right. \\ &\qquad \left. + \int_t^{\alpha_1} (1-s)(\alpha_1 - s) ds + \int_{\alpha_1}^1 (1-s)(s-\alpha_1) ds \right) \\ &= \frac{2t(1-t)}{27} \left[\frac{(3t-1)^3}{t^3} + \frac{(2-3t)^3}{(1-t)^3} \right], \quad \frac{1}{3} \leq t \leq \frac{1}{2}. \qquad (5.3.25) \end{aligned}$$

Now we maximize the right sides of (5.3.24) and (5.3.25) in their respective intervals, to obtain

$$\int_0^1 | g_3^{(1)}(t,s) | \, ds \leq \frac{26\sqrt{13} - 92}{27}, \quad 0 \leq t \leq \frac{1}{2}.$$

Thus, from the symmetry and (5.3.19) we get

$$\| D(x - H_2^{\Delta} x) \|_\infty \leq \frac{13\sqrt{13} - 46}{27} \| D^3 x \|_\infty .$$

Finally, for $k = 2$ from (5.3.23) we have

$$g_3^{(2)}(t,s) = 2\begin{cases} 3(1-2t)(1-s)(s-\alpha_2), & t \le s \\ -3(1-2t)s(s-\beta_2), & s < t \end{cases} \tag{5.3.26}$$

where

$$\alpha_2 = \frac{1}{3(1-2t)}, \qquad \beta_2 = \frac{2(2-3t)}{3(1-2t)}.$$

For $\frac{1}{2} \le t \le 1$, $\alpha_2 \le 0$ and for $\frac{1}{2} \le t \le \frac{2}{3}$, $\beta_2 \le 0$ whereas for $\frac{2}{3} \le t \le 1$, $0 \le \beta_2 \le t$. Thus, it follows that

$$\begin{aligned} \int_0^1 \mid g_3^{(2)}(t,s) \mid ds &= 6(2t-1)\left(\int_0^t s(s-\beta_2)ds + \int_t^1 (1-s)(s-\alpha_2)ds\right) \\ &= 8t^2(1-t)^2, \ \frac{1}{2} \le t \le \frac{2}{3}; \qquad (5.3.27) \\ &= 6(2t-1)\left(\int_0^{\beta_2} s(\beta_2 - s)ds + \int_{\beta_2}^t s(s-\beta_2)ds \right. \\ &\qquad \left. + \int_t^1 (1-s)(s-\alpha_2)ds\right) \\ &= 8t^2(1-t)^2 + \frac{16}{27}\frac{(3t-2)^3}{(2t-1)^2}, \ \frac{2}{3} \le t \le 1. \qquad (5.3.28) \end{aligned}$$

Maximizing the right sides of (5.3.27) and (5.3.28) in their respective intervals, we get

$$\int_0^1 \mid g_3^{(2)}(t,s) \mid ds \le \frac{16}{27}, \ \frac{1}{2} \le t \le 1.$$

Thus, from the symmetry and (5.3.19) we obtain

$$\| D^2(x - H_2^{\Delta}x) \|_\infty \le \frac{8}{27} \| D^3 x \|_\infty . \tag{5.3.29}$$

■

Remark 5.3.2. From Theorem 2.4.13 it is clear that the constants $c_{2,4,k}$, $0 \le k \le 3$ given in Table 5.3.4 are the best in the inequalities (5.3.14). To show that the other constants $c_{2,n,k}$, $0 \le k \le n-1$ are as well the best possible, we consider the function

$$\begin{aligned} x_m(t) &= \left(1 - \frac{2}{m}\right)t^2 - \frac{4}{3}t^3 + \frac{6m-4}{3m^2}\,t + \frac{4m-3m^2-2}{6m^3}, \ 0 \le t \le \frac{1}{2} - \frac{1}{m} \\ &= \frac{m(2t-1)^4}{48} - \frac{(2t-1)^2}{4} + \frac{1}{12}, \ \frac{1}{2} - \frac{1}{m} \le t \le \frac{1}{2} + \frac{1}{m} \\ &= \left(1 - \frac{2}{m}\right)(1-t)^2 - \frac{4}{3}(1-t)^3 + \frac{6m-4}{3m^2}(1-t) + \frac{4m-3m^2-2}{6m^2}, \end{aligned}$$

$$\frac{1}{2}+\frac{1}{m}\le t\le 1.$$

It is easy to see that $x_m(t)\in C^{(3)}[0,1]$ and $\| D^3x_m \|_\infty = 8$. Further, for $h=1$ we have

$$\frac{1}{8}\left|x_m\left(\frac{1}{2}\right)-H_2^{\Delta}x_m\left(\frac{1}{2}\right)\right| = \frac{1}{96}-\frac{1}{24m^2}+\frac{1}{24m^3}.$$

Thus, in (5.3.14) the constant $c_{2,3,0}=\frac{1}{96}$ is the best possible.

Next, we consider the function

$$\begin{aligned} y_m(t) &= \frac{1}{6}t^3+\frac{1}{2}At^2, \quad 0\le t\le \alpha-\frac{1}{m} \\ &= -\frac{m}{24}(t-\alpha)^4+\frac{1}{2}\left(\alpha-\frac{1}{2m}+A\right)(t-\alpha)^2+B(t-\alpha)+C, \\ &\qquad\qquad \alpha-\frac{1}{m}\le t\le \alpha+\frac{1}{m} \\ &= -\frac{1}{6}t^3+\frac{1}{2}(2\alpha+A)t^2+\left(\frac{1}{2}-2\alpha-A\right)t+D, \quad \alpha+\frac{1}{m}\le t\le 1 \end{aligned}$$

where

$$\alpha=\frac{\sqrt{13}-1}{6},\quad \beta=\frac{5-\sqrt{13}}{6},\quad A=\alpha^2-2\alpha+\frac{1}{2}+\frac{1}{3m^2},$$

$$B=\frac{1}{2}\alpha^2+A\alpha-\frac{1}{6m^2},\quad C=\frac{1}{6}\alpha^3+\frac{1}{2}A\alpha^2-\frac{1}{24m^3},\quad D=\frac{1}{3}\alpha^3+\frac{\alpha}{3m^2}.$$

Again, it is easy to see that $y_m(t)\in C^{(3)}[0,1]$ and $\| D^3y_m \|_\infty = 1$. Further, for $h=1$ we have

$$\lim_{m\to\infty} \mid D(y_m-H_2^{\Delta}y_m)(\beta)\mid = \frac{13\sqrt{13}-46}{27}.$$

Thus, in (5.3.14) the constant $c_{2,3,1}=\frac{13\sqrt{13}-46}{27}$ is the best possible.

Finally, we consider

$$\begin{aligned} z_m(t) &= \frac{1}{6}t^3+\frac{1}{2}\left(-\frac{7}{18}+\frac{1}{2m}\right)t^2+\left(\frac{1}{6m^2}-\frac{1}{3m}\right)t+A_1, \quad 0\le t\le \frac{2}{3}-\frac{1}{m} \\ &= -\frac{m}{24}\left(t-\frac{2}{3}\right)^4+\frac{5}{36}\left(t-\frac{2}{3}\right)^2-\frac{1}{27}\left(t-\frac{2}{3}\right)-\frac{1}{27}, \\ &\qquad\qquad \frac{2}{3}-\frac{1}{m}\le t\le \frac{2}{3}+\frac{1}{m} \\ &= -\frac{1}{6}t^3+\frac{1}{2}\left(\frac{17}{18}+\frac{1}{2m}\right)t^2-\left(\frac{4}{9}+\frac{1}{3m}+\frac{1}{6m^2}\right)t+A_2, \\ &\qquad\qquad \frac{2}{3}+\frac{1}{m}\le t\le 1 \end{aligned}$$

where

$$A_1 = \frac{1}{9m} - \frac{1}{9m^2} + \frac{1}{24m^3}, \quad A_2 = \frac{8}{81} + \frac{1}{9m} + \frac{1}{9m^2} + \frac{1}{24m^3}.$$

Again, $z_m(t) \in C^{(3)}[0,1]$ and $\| D^3 z_m \|_\infty = 1$, also for $h = 1$ we have

$$| D^2(z_m - H_2^{\Delta} z_m)(1) | = \frac{8}{27} - \frac{2}{3m^2}.$$

Therefore, in (5.3.14) the constant $c_{2,3,2} = \frac{8}{27}$ is also the best possible.

In an analogous way we can show that in (5.3.14) the constants $c_{2,2,k}$; $k = 0, 1$ are the best possible. ∎

Theorem 5.3.8. Let $x(t) \in PC^{n,\infty}[a,b]$, $2 \le n \le 6$. Then,

(5.3.30) $$\| D^k(x - H_3^{\Delta} x) \|_\infty \le c_{3,n,k}\, h^{n-k} \| D^n x \|_\infty, \quad 0 \le k \le n-1$$

where the constants $c_{3,n,k}$ are given in the following table.

Table 5.3.5.

k \ n	2	3	4	5	6
0	$\frac{25}{256}$	$\frac{19}{3072}$	$\frac{39+55\sqrt{33}}{524288}$	$\frac{1}{10240}$	$\frac{1}{46080}$
1	0.35863821 + 0	0.22055373 − 1	0.24601405 − 2	0.34045545 − 3	$\frac{\sqrt{5}}{30000}$
2		$\frac{5}{24}$	$\frac{1}{54}$	$\frac{1}{384}$	$\frac{1}{1920}$
3			$\frac{12\sqrt{6}}{125}$	$\frac{108}{3125}$	$\frac{1}{120}$
4				$\frac{32}{81}$	$\frac{1}{10}$
5					$\frac{1}{2}$

Proof. Without loss of generality we assume that $a = 0$, $b = 1$ and $h = 1$, so that

$$\begin{aligned} H_3^{\Delta} x(t) &= (1-t)^3(6t^2+3t+1)x(0) + t^3[6(1-t)^2 + 3(1-t) + 1]x(1) \\ &\quad + (1-t)^3(3t^2+t)x'(0) + t^3[-3(1-t)^2 - (1-t)]x'(1) \\ &\quad + \frac{1}{2}(1-t)^3 t^2 x''(0) + \frac{1}{2} t^3 (1-t)^2 x''(1). \end{aligned}$$

Therefore, in view of Theorem 5.2.1 for $0 \le k \le n-1$ it follows that

$$(5.3.31)\quad D^k(x - H_3^\Delta x)(t) = \frac{1}{(n-1)!}\int_0^1 g_n^{(k)}(t,s)D^n x(s)\,ds - \begin{cases} D^k\left[\frac{1}{2}(1-t)^3t^2x''(0) + \frac{1}{2}t^3(1-t)^2x''(1)\right], \\ \qquad\qquad\qquad n = 2 \\ 0, \quad 3 \le n \le 6 \end{cases}$$

where $g_n^{(k)}(t,s) = \partial^k g_n(t,s)/\partial t^k$, and

$$(5.3.32)\quad g_2(t,s) = t^3[(6t^2 - 15t + 10)s - 3t^2 + 8t - 6], \quad t \le s$$

$$(5.3.33)\quad g_3(t,s) = t^3[(-6t^2+15t-10)s^2+(6t^2-16t+12)s-t^2+3t-3], \quad t \le s$$

$$(5.3.34)\quad g_4(t,s) = t^3(1-s)[(-6t^2+15t-10)s^2 + (3t^2-9t+8)s - 1], \quad t \le s$$

$$(5.3.35)\quad g_5(t,s) = t^3(1-s)^2[(-6t^2+15t-10)s^2 - (2t-4)s - t], \quad t \le s$$

$$(5.3.36)\quad g_6(t,s) = t^3(1-s)^3[(-6t^2+15t-10)s^2 - t(3t-5)s - t^2], \quad t \le s$$

and $g_n(t,s)$ for $s \le t$ is the same as $(-1)^n g_n(1-t, 1-s)$ for $t \le s$.

While the proof for the case $n = 6$, $0 \le k \le 5$ is included in Theorem 2.4.13, we shall give the proof only for the case $n = 2$, whereas for $n = 3, 4$ and 5 the computation is similar, however tedious.

From (5.3.31) and (5.3.32), we have

$$(5.3.37)\quad |\, D^k(x - H_3^\Delta x)(t)\,| \le \left\{\int_0^1 |\, g_2^{(k)}(t,s)\,|\,ds + \left|D^k\left[\frac{1}{2}(1-t)^3t^2\right]\right| + \left|D^k\left[\frac{1}{2}t^3(1-t)^2\right]\right|\right\} \|\, D^2x\,\|_\infty\ ; \ k = 0, 1$$

where

$$(5.3.38)\quad g_2(t,s) = \begin{cases} t^3(6t^2 - 15t + 10)(s - \alpha), & t \le s \\ (t-1)^3(6t^2 + 3t + 1)(s - \beta), & s \le t \end{cases}$$

and

$$\alpha = \frac{3t^2 - 8t + 6}{6t^2 - 15t + 10}, \qquad \beta = \frac{3t^2 + t}{6t^2 + 3t + 1}.$$

It is easy to note that $0 < \alpha, \beta < 1,\ \alpha \ge t$ and $\beta \le t$. Thus, for $0 \le t \le 1$ we have

$$\begin{aligned}\int_0^1 \mid g_2(t,s) \mid ds &= (1-t)^3(6t^2+3t+1)\left(\int_0^\beta (\beta-s)ds + \int_\beta^t (s-\beta)ds\right)\\ &\quad + t^3(6t^2-15t+10)\left(\int_t^\alpha (\alpha-s)ds + \int_\alpha^1 (s-\alpha)ds\right)\\ &= \frac{1}{2}\Big\{(1-t)^3(6t^2+3t+1)[\beta^2+(t-\beta)^2]\\ &\quad + t^3(6t^2-15t+10)[(1-\alpha)^2+(t-\alpha)^2]\Big\}.\end{aligned}$$

The right side of the above inequality attains its maximum at $t = \frac{1}{2}$, and its maximum value is $\frac{17}{256}$. Thus, in view of

$$\left|\frac{1}{2}(1-t)^3t^2\right| + \left|\frac{1}{2}t^3(1-t)^2\right| = \frac{1}{2}t^2(1-t)^2 \le \frac{1}{32},$$

from (5.3.37) we find that

$$\| x - H_3^\Delta x \|_\infty \le \frac{25}{256} \| D^2 x \|_\infty .$$

This completes the proof for $k = 0$.

For $k = 1$, from (5.3.38) we have

$$g_2^{(1)}(t,s) = \begin{cases} 30t^2(1-t)^2(s-\alpha_1), & t \le s \\ 30t^2(1-t)^2(s-\beta_1), & s < t \end{cases} \tag{5.3.39}$$

where

$$\alpha_1 = \frac{15t^2-32t+18}{30(1-t)^2}, \qquad \beta_1 = \frac{15t^2-2t-1}{30t^2}.$$

Once again it is easy to note that $\alpha_1 \ge t,\ \beta_1 \le t,\ \alpha_1 \ge 0$ and $\beta_1 \le 1$ for all $t \in (0,1)$; $\alpha_1 \ge 1$ for $t \in \left[\frac{2}{3}, 1\right)$; and $\beta_1 \le 0$ for $t \in \left(0, \frac{1}{3}\right]$. Hence, we find that

$$\begin{aligned}&\int_0^1 \mid g_2^{(1)}(t,s) \mid ds\\ &= 30t^2(1-t)^2\left(\int_0^t (s-\beta_1)ds + \int_t^{\alpha_1} (\alpha_1-s)ds + \int_{\alpha_1}^1 (s-\alpha_1)ds\right)\end{aligned}$$

$$(5.3.40)\quad = 15t^2(1-t)^2[(t-\beta_1)^2 - \beta_1^2 + (t-\alpha_1)^2 + (1-\alpha_1)^2],\ 0 \le t \le \frac{1}{3};$$

$$= 30t^2(1-t)^2\left(\int_0^{\beta_1}(\beta_1 - s)ds + \int_{\beta_1}^{t}(s-\beta_1)ds + \int_t^{\alpha_1}(\alpha_1 - s)ds + \int_{\alpha_1}^{1}(s-\alpha_1)ds\right)$$

$$(5.3.41)\quad = 15t^2(1-t)^2[(t-\beta_1)^2 + \beta_1^2 + (t-\alpha_1)^2 + (1-\alpha_1)^2],\ \frac{1}{3} \le t \le \frac{2}{3};$$

$$= 30t^2(1-t)^2\left(\int_0^{\beta_1}(\beta_1 - s)ds + \int_{\beta_1}^{t}(s-\beta_1)ds + \int_t^{1}(\alpha_1 - s)ds\right)$$

$$(5.3.42)\quad = 15t^2(1-t)^2[(t-\beta_1)^2 + \beta_1^2 + (t-\alpha_1)^2 - (1-\alpha_1)^2],\ \frac{2}{3} \le t \le 1.$$

Now we maximize the right sides of (5.3.40) - (5.3.42) in their respective intervals with the aid of IBM3081 computer using double precision in order to reduce the roundoff errors to a minimum. The absolute maximum value turns out to be 0.26241317... + 0. Thus, in view of

$$\left|D\left[\frac{1}{2}(1-t)^3t^2\right]\right| + \left|D\left[\frac{1}{2}t^3(1-t)^2\right]\right|$$

$$= \frac{1}{2}t(1-t)\begin{cases} 2(1-2t), & 0 \le t \le \frac{2}{5} \\ 10t - 2 - 10t^2, & \frac{2}{5} \le t \le \frac{3}{5} \\ 2(2t-1), & \frac{3}{5} \le t \le 1 \end{cases}$$

$$\le \frac{\sqrt{3}}{18},$$

from (5.3.37) it follows that

$$\| D(x - H_3^{\Delta}x) \|_\infty \le (0.35863821... + 0) \| D^2 x \|_\infty .$$

This completes the proof for $k = 1$. ∎

Remark 5.3.3. From Theorem 2.4.13 it is clear that the constants $c_{3,6,k}$, $0 \le k \le 5$ given in Table 5.3.5 are the best in the inequalities (5.3.30). The constant $c_{3,2,0} = \frac{25}{256}$ is also the best possible. However, for the remaining constants $c_{3,2,1}$, $c_{3,n,k}$, $3 \le n \le 5$, $0 \le k \le n-1$ the sharpness of the inequalities (5.3.30) remain undecided. In Tables 5.3.6 - 5.3.9 we compute the actual value of $\| D^k(x - H_3^{\Delta}x) \|_\infty$, $2 \le n \le 5$, $0 \le k \le n-1$ for several simple functions and compare these with the corresponding right side bounds in (5.3.30).

Table 5.3.6.

$$x(t) = t^5 \sin \frac{1}{t},\ t \in \left[-\frac{1}{4}, \frac{1}{4}\right],\ n = 2$$

N	7	15	24
$\| x - H_3^{\Delta}x \|_\infty$	0.80901928 − 5	0.12814502 − 5	0.33635119 − 6
Bound	0.10662905 − 3	0.26657262 − 4	0.10918815 − 4
$\| D(x - H_3^{\Delta}x) \|_\infty$	0.49311092 − 3	0.18147165 − 3	0.89342391 − 4
Bound	0.62654466 − 2	0.31327233 − 2	0.20049430 − 2

Table 5.3.7.

$$x(t) = t^3 \mid t \mid,\ t \in \left[-\frac{1}{4}, \frac{1}{4}\right],\ n = 3$$

N	14	24	34	54
$\| x - H_3^{\Delta}x \|_\infty$	0.19051898 − 8	0.24691259 − 9	0.64273374 − 10	0.10540289 − 10
Bound	0.13744213 − 5	0.29687500 − 6	0.10819060 − 6	0.27880823 − 7
$\| D(x - H_3^{\Delta}x) \|_\infty$	0.57870370 − 6	0.12500000 − 6	0.45553936 − 7	0.11739294 − 7
Bound	0.14703582 − 3	0.52932895 − 4	0.27006579 − 4	0.10936549 − 4
$\| D^2(x - H_3^{\Delta}x) \|_\infty$	0.13681268 − 3	0.49252566 − 4	0.25128860 − 4	0.10176150 − 4
Bound	0.41666667 − 1	0.25000000 − 1	0.17857143 − 1	0.11363636 − 1

Table 5.3.8.

$$x(t) = t^4 \mid t \mid,\ t \in \left[-\frac{1}{4}, \frac{1}{4}\right],\ n = 4$$

N	14	24	34	54
$\| x - H_3^{\Delta} x \|_{\infty}$	0.48225309 − 9	0.37500000 − 10	0.69725412 − 11	0.72764218 − 12
Bound	0.25074637 − 7	0.32496730 − 8	0.84591654 − 9	0.13872315 − 9
$\| D(x - H_3^{\Delta} x) \|_{\infty}$	0.50149205 − 7	0.64993370 − 8	0.16918307 − 8	0.27744591 − 9
Bound	0.27334894 − 5	0.59043372 − 6	0.21517264 − 6	0.55450199 − 7
$\| D^2(x - H_3^{\Delta} x) \|_{\infty}$	0.11574074 − 4	0.25000000 − 5	0.91107872 − 6	0.23478588 − 6
Bound	0.61728395 − 3	0.22222222 − 3	0.11337868 − 3	0.45913682 − 4
$\| D^3(x - H_3^{\Delta} x) \|_{\infty}$	0.41666667 − 2	0.15000000 − 2	0.76530612 − 3	0.30991736 − 3
Bound	0.23515102 + 0	0.14109061 + 0	0.10077901 + 0	0.64132095 − 1

Table 5.3.9.

$$x(t) = t^5 \mid t \mid,\ t \in \left[-\frac{1}{4}, \frac{1}{4}\right],\ n = 5$$

N	14	24	34	54
$\| x - H_3^{\Delta} x \|_{\infty}$	0.21433471 − 10	0.10000000 − 11	0.13281045 − 12	0.88201473 − 14
Bound	0.72337963 − 9	0.56250000 − 10	0.10458812 − 10	0.10914633 − 11
$\| D(x - H_3^{\Delta} x) \|_{\infty}$	0.22084537 − 8	0.17172937 − 9	0.31930412 − 10	0.33322181 − 11
Bound	0.75656767 − 7	0.98051170 − 8	0.25523524 − 8	0.41856418 − 9
$\| D^2(x - H_3^{\Delta} x) \|_{\infty}$	0.46296296 − 6	0.60000000 − 7	0.15618492 − 7	0.25613008 − 8
Bound	0.17361111 − 4	0.37500000 − 5	0.13666181 − 5	0.35217881 − 6
$\| D^3(x - H_3^{\Delta} x) \|_{\infty}$	0.22222222 − 3	0.48000000 − 4	0.17492712 − 4	0.45078928 − 5
Bound	0.69120000 − 2	0.24883200 − 2	0.12695510 − 2	0.51411570 − 3
$\| D^4(x - H_3^{\Delta} x) \|_{\infty}$	0.80000000 − 1	0.28800000 − 1	0.14693878 − 1	0.59504162 − 2
Bound	0.23703704 + 1	0.14222222 + 1	0.10158730 + 1	0.64646465 + 0

∎

To obtain error bounds for the derivatives of the interpolation error $x(t) - H_m^{\Delta} x(t)$ in terms of the derivatives of $x(t)$ in L_2 - norm first we shall prove a variational characterization of the Hermite interpolate $H_m^{\Delta} x(t)$. This characterization in particular for $m = 1$ shows that $H_1^{\Delta} x(t)$ is the piecewise linear interpolating function which has the minimum least square variation; and for $m = 2$ indicates that $H_2^{\Delta}x(t)$ is the piecewise cubic interpolating function which has the minimum least square curvature.

Theorem 5.3.9. Let Δ and $\left\{x_i^{(k)}\right\}_{i=0}^{N+1}\,{}_{k=0}^{m-1}$ be given and

$$V \equiv \left\{\omega(t) \in PC^{m,2}[a,b] \;:\; \omega^{(k)}(t_i) = x_i^{(k)};\; 0 \le i \le N+1,\; 0 \le k \le m-1\right\}.$$

The variational problem of finding the function $p(t) \in V$ which minimizes $\| D^m\omega \|_2^2$ over all $\omega(t) \in V$ has the unique solution $H_m^{\Delta}x(t)$.

Proof. First we shall show that $p(t) \in V$ is a solution of the variational problem if and only if the inner product

$$(D^m p, D^m \delta) = 0 \tag{5.3.43}$$

for all functions $\delta(t) \in V_0 = \{\omega(t) \in PC^{m,2}[a,b] : \omega^{(k)}(t_i) = 0;\; 0 \le i \le N+1,\; 0 \le k \le m-1\}$, i.e., $p(t)$ is a solution of the generalized Euler's equation.

For this, if $p(t) \in V$ then it is clear that for all real numbers α and $\delta(t) \in V_0$ we have $p(t) + \alpha\,\delta(t) \in V$. Moreover, if $p(t)$ is a solution of the variational problem, then the function

$$\begin{aligned} F(\alpha) &= \| D^m(p+\alpha\delta) \|_2^2 \\ &= (D^m(p+\alpha\delta), D^m(p+\alpha\delta)) \\ &= (D^m p, D^m p) + 2\alpha(D^m p, D^m\delta) + \alpha^2(D^m\delta, D^m\delta) \end{aligned} \tag{5.3.44}$$

attains its minimum when $\alpha = 0$. This implies that $\left.\frac{dF}{d\alpha}\right|_{\alpha=0} = 0$, which is the same as (5.3.43).

Conversely, if $p(t) \in V$ is a solution of (5.3.43) and $\omega(t) \in V$, then $\omega(t) - p(t) \in V_0$ and hence $(D^m p, D^m(\omega - p)) = 0$. However, then

$$\begin{aligned} \| D^m\omega \|_2^2 &= (D^m\omega, D^m\omega) \\ &= (D^m(\omega-p) + D^m p, D^m(\omega-p) + D^m p) \\ &= (D^m(\omega-p), D^m(\omega-p)) + 2\,(D^m(\omega-p), D^m p) + (D^m p, D^m p) \\ &= \| D^m(\omega-p) \|_2^2 + \| D^m p \|_2^2 . \end{aligned} \tag{5.3.45}$$

From the above inequality it is clear that

$$\| D^m p \|_2^2 \le \| D^m \omega \|_2^2 \tag{5.3.46}$$

for all $\omega(t) \in V$, i.e., $p(t)$ is a solution of the variational problem. Moreover, we have equality in (5.3.46) if and only if $\| D^m(\omega - p) \|_2^2 = 0$, or in view of Theorem 5.2.4 if and only if

$$\lambda_{m,0}^{2m} \| \omega - p \|_2^2 \le (b-a)^{2m} \| D^m(\omega - p) \|_2^2 = 0,$$

or $\omega = p$. Thus, the variational problem and the generalized Euler's equation (5.3.43) have at most one solution.

To complete the proof, we shall show that $H_m^\Delta x(t)$ is a solution of the generalized Euler's equation (5.3.43). For any $\delta(t) \in V_0$, an integration by parts, boundary conditions and the fact that $H_m^\Delta x(t)$ is a polynomial of degree $(2m-1)$ in each subinterval $[t_i, t_{i+1}]$, $0 \le i \le N$ lead to

$$\begin{aligned}
(D^m H_m^\Delta x, D^m \delta) &= \int_a^b D^m H_m^\Delta x(t)\; D^m \delta(t)\, dt \\
&= \sum_{i=0}^N \int_{t_i}^{t_{i+1}} D^m H_m^\Delta x(t)\; D^m \delta(t)\, dt \\
&= \sum_{i=0}^N \Big[D^{m-1}\delta(t)\; D^m H_m^\Delta x(t) \Big|_{t_i}^{t_{i+1}} \\
&\qquad - \int_{t_i}^{t_{i+1}} D^{m-1}\delta(t)\; D^{m+1} H_m^\Delta x(t)\, dt \Big] \\
&= -\sum_{i=0}^N \int_{t_i}^{t_{i+1}} D^{m-1}\delta(t)\; D^{m+1} H_m^\Delta x(t)\, dt \\
&= (-1)^2 \sum_{i=0}^N \int_{t_i}^{t_{i+1}} D^{m-2}\delta(t)\; D^{m+2} H_m^\Delta x(t)\, dt \\
&\qquad \cdots \\
&= (-1)^m \sum_{i=0}^N \int_{t_i}^{t_{i+1}} \delta(t)\; D^{2m} H_m^\Delta x(t)\, dt \\
&= 0. \qquad \blacksquare
\end{aligned}$$

As a particular case of (5.3.45) we obtain the 'First Integral Relation'.

Corollary 5.3.10. If $x(t) \in PC^{m,2}[a,b]$, then

$$\| D^m H^{\Delta}_m x \|_2^2 + \| D^m H^{\Delta}_m x - D^m x \|_2^2 = \| D^m x \|_2^2 . \tag{5.3.47}$$ ∎

By using the same type of integration by parts argument that we used in the proof of Theorem 5.3.9, we obtain the following result.

Theorem 5.3.11. If $y(t) \in PC^{2m,2}[a,b]$, $y^{(k)}(t_i) = x^{(k)}(t_i)$; $0 \le i \le N+1$, $0 \le k \le m-1$ then

$$\| D^m (y - H^{\Delta}_m x) \|_2^2 = (-1)^m (y - H^{\Delta}_m x, D^{2m} y). \tag{5.3.48}$$ ∎

As a corollary of the above result we obtain the 'Second Integral Relation'.

Corollary 5.3.12. If $x(t) \in PC^{2m,2}[a,b]$, then

$$\| D^m (x - H^{\Delta}_m x) \|_2^2 = (-1)^m (x - H^{\Delta}_m x, D^{2m} x). \tag{5.3.49}$$ ∎

Both of these integral relations (5.3.47) and (5.3.49) were introduced by Ahlberg, Nilson and Walsh [4,5].

Theorem 5.3.13. Let $x(t) \in PC^{n,2}[a,b]$, $n = 1, 2$. Then,

$$\| D^k (x - H^{\Delta}_1 x) \|_2 \le \bar{c}_{1,n,k}\, h^{n-k} \| D^n x \|_2 \; ; \; k = 0, 1 \tag{5.3.50}$$

where the constants $\bar{c}_{1,n,k}$ are given in the following table.

Table 5.3.10.

k \ n	1	2
0	$\frac{1}{\pi}$	$\frac{1}{\pi^2}$
1	1	$\frac{1}{\pi}$

Proof. For $n = 1,\ k = 1$ the inequality (5.3.50) follows directly from the first integral relation (5.3.47). For $n = 1,\ k = 0$ we note that $x(t_i) - H_1^{\Delta}x(t_i) = 0$, for all $0 \le i \le N+1$, and hence by Theorem 5.2.2 we have

$$(5.3.51)\quad \int_{t_i}^{t_{i+1}} [x(t) - H_1^{\Delta}x(t)]^2\, dt \le \frac{(t_{i+1} - t_i)^2}{\pi^2} \int_{t_i}^{t_{i+1}} [Dx(t) - DH_1^{\Delta}x(t)]^2\, dt,$$

$$0 \le i \le N.$$

Summing both sides of the inequality (5.3.51) with respect to i from 0 to N and taking the square root of both sides of the resulting inequality, we obtain

$$(5.3.52)\qquad \| x - H_1^{\Delta}x \|_2 \le \frac{h}{\pi} \| D(x - H_1^{\Delta}x) \|_2 \ .$$

Now using (5.3.50) for the case $n = 1,\ k = 1$ to bound the right side of (5.3.52), we obtain the required inequality

$$(5.3.53)\qquad \| x - H_1^{\Delta}x \|_2 \le \frac{h}{\pi} \| Dx \|_2 \ .$$

For the case $n = 2,\ k = 1$ we apply the Cauchy - Schwarz inequality to the second integral relation (5.3.49) for $m = 1$, to get

$$\| D(x - H_1^{\Delta}x) \|_2^2 \le \| D^2x \|_2 \| x - H_1^{\Delta}x \|_2 \ .$$

Combining this with (5.3.52) gives that

$$(5.3.54)\qquad \| D(x - H_1^{\Delta}x) \|_2 \le \frac{h}{\pi} \| D^2x \|_2 \ .$$

Finally, for the case $n = 2,\ k = 0$ we use (5.3.54) to bound the right side of (5.3.52), to find

$$(5.3.55)\qquad \| x - H_1^{\Delta}x \|_2 \le \frac{h^2}{\pi^2} \| D^2x \|_2 \ . \qquad \blacksquare$$

Remark 5.3.4. Inequality (5.3.55) can also be obtained directly from Corollary 5.2.3. Indeed, from (5.2.4), we have

$$\int_{t_i}^{t_{i+1}} [x(t) - H_1^{\Delta}x(t)]^2\, dt \le \frac{(t_{i+1} - t_i)^4}{\pi^4} \int_{t_i}^{t_{i+1}} [D^2x(t) - D^2H_1^{\Delta}x(t)]^2\, dt,$$

$$0 \le i \le N.$$

However, since $H_1^\Delta x(t)$ is a linear function in each interval $[t_i, t_{i+1}]$, $0 \le i \le N$ it follows that

$$\int_{t_i}^{t_{i+1}} [x(t) - H_1^\Delta x(t)]^2 \, dt \le \frac{(t_{i+1} - t_i)^4}{\pi^4} \int_{t_i}^{t_{i+1}} [D^2 x(t)]^2 \, dt, \ 0 \le i \le N.$$

From the above inequality (5.3.55) is immediate. ∎

Theorem 5.3.14. Let $x(t) \in PC^{n,2}[a,b]$, $2 \le n \le 4$. Then,

$$\| D^k(x - H_2^\Delta x) \|_2 \le \bar{c}_{2,n,k} \, h^{n-k} \, \| D^n x \|_2 \, , \ 0 \le k \le 2 \tag{5.3.56}$$

where the constants $\bar{c}_{2,n,k}$ in terms of $\lambda_{n,k}$ (see Table 5.2.1) are given in the following table.

Table 5.3.11.

k \ n	2	3	4
0	$\lambda_{2,0}^{-2}$ $= 0.446962 - 1$	$\lambda_{3,0}^{-3} + 2\sqrt{15}\lambda_{2,0}^{-4}$ $= 0.195060 - 1$	$\lambda_{2,0}^{-4}$ $= 0.199775 - 2$
1	$\lambda_{2,1}^{-1}$ $= \frac{1}{2\pi}$	$\lambda_{3,1}^{-2} + 2\sqrt{15}\lambda_{2,0}^{-2}\lambda_{2,1}^{-1}$ $= 0.713165 - 1$	$\lambda_{2,0}^{-2}\lambda_{2,1}^{-1}$ $= 0.711362 - 2$
2	1	$\lambda_{3,2}^{-1} + 2\sqrt{15}\lambda_{2,0}^{-2}$ $= 0.457490 + 0$	$\lambda_{2,0}^{-2}$ $= 0.446962 - 1$

Proof. For $n = 2$, $k = 2$ the inequality (5.3.56) follows directly from the first integral relation (5.3.47). For $n = 2$, $k = 0$ and 1 we note that $x(t_i) - H_2^\Delta x(t_i) = D(x(t_i) - H_2^\Delta x(t_i)) = 0$, for all $0 \le i \le N+1$, and therefore in view of Theorem 5.2.4, we have

$$\int_{t_i}^{t_{i+1}} [D^k(x(t) - H_2^\Delta x(t))]^2 \, dt \le \frac{(t_{i+1} - t_i)^{4-2k}}{\lambda_{2,k}^{4-2k}} \int_{t_i}^{t_{i+1}} [D^2(x(t) - H_2^\Delta x(t))]^2 \, dt;$$

$$0 \le i \le N, \ k = 0, 1.$$

Summing both sides of the above inequality with respect to i from 0 to N and taking the square root of both sides of the resulting inequality, we obtain

$$\| D^k(x - H_2^\Delta x) \|_2 \le \frac{h^{2-k}}{\lambda_{2,k}^{2-k}} \| D^2(x - H_2^\Delta x) \|_2 \ ; \ k = 0, 1. \tag{5.3.57}$$

In the above inequality we bound the right side by (5.3.56) for $n = 2, \ k = 2$ to obtain

$$\| D^k(x - H_2^\Delta x) \|_2 \le \frac{h^{2-k}}{\lambda_{2,k}^{2-k}} \| D^2 x \|_2 \ ; \ k = 0, 1. \tag{5.3.58}$$

For the case $n = 4, \ k = 2$ we apply the Cauchy - Schwarz inequality to the second integral relation (5.3.49) for $m = 2$, to get

$$\| D^2(x - H_2^\Delta x) \|_2^2 \le \| D^4 x \|_2 \, \| x - H_2^\Delta x \|_2 \ .$$

Combining this inequality with (5.3.57) for $k = 0$ gives that

$$\| D^2(x - H_2^\Delta x) \|_2 \le \frac{h^2}{\lambda_{2,0}^2} \| D^4 x \|_2 \ . \tag{5.3.59}$$

For $n = 4, \ k = 0$ and 1 we use Corollary 5.2.5, to obtain

$$\int_{t_i}^{t_{i+1}} [D^k(x(t) - H_2^\Delta x(t))]^2 \, dt \le \frac{(t_{i+1} - t_i)^{8-2k}}{\lambda_{2,k}^{4-2k} \lambda_{2,0}^4} \int_{t_i}^{t_{i+1}} [D^4(x(t) - H_2^\Delta x(t))]^2 \, dt.$$

However, since $H_2^\Delta x(t)$ is a cubic function in each interval $[t_i, t_{i+1}]$, $0 \le i \le N$ it follows that

$$\| D^k(x - H_2^\Delta x) \|_2 \le \frac{h^{4-k}}{\lambda_{2,k}^{2-k} \lambda_{2,0}^2} \| D^4 x \|_2 \ ; \ k = 0, 1. \tag{5.3.60}$$

For $n = 3, \ 0 \le k \le 2$ let $q(t)$ be the unique quintic polynomial in $[t_i, t_{i+1}]$, $0 \le i \le N$ but fixed, satisfying $D^j q(t_\ell) = D^j x(t_\ell)$; $0 \le j \le 2, \ \ell = i, i+1$. Then, it can be shown by integration by parts that

$$\int_{t_i}^{t_{i+1}} [D^3 q(t)]^2 \, dt + \int_{t_i}^{t_{i+1}} [D^3 q(t) - D^3 x(t)]^2 \, dt = \int_{t_i}^{t_{i+1}} [D^3 x(t)]^2 \, dt. \tag{5.3.61}$$

Next, since $q(t)-x(t)\in PC^{3,2}[t_i,t_{i+1}]$, $D^j(q(t_i)-x(t_i)) = D^j(q(t_{i+1})-x(t_{i+1}))=0$, $j=0,1,2$ from Theorem 5.2.4 we have

$$\int_{t_i}^{t_{i+1}}[D^k(q(t)-x(t))]^2\,dt \le \left(\frac{t_{i+1}-t_i}{\lambda_{3,k}}\right)^{6-2k}\int_{t_i}^{t_{i+1}}[D^3(q(t)-x(t))]^2\,dt$$

$$(5.3.62)\qquad \le \left(\frac{h}{\lambda_{3,k}}\right)^{6-2k}\int_{t_i}^{t_{i+1}}[D^3x(t)]^2\,dt,$$

where in the last inequality we have used (5.3.61).

For all $t\in[t_i,t_{i+1}]$ we note that $H_2^{\Delta}x(t)=H_2^{\Delta}q(t)$, and hence we can write

$$x(t)-H_2^{\Delta}x(t)=x(t)-q(t)+q(t)-H_2^{\Delta}q(t).$$

Thus, by the triangle inequality, (5.3.62), (5.3.56) for $n=4$, $0\le k\le 2$, Theorem 5.2.6 for $n=2$ and (5.3.61) it follows that

$$\left\{\int_{t_i}^{t_{i+1}}[D^k(x(t)-H_2^{\Delta}x(t))]^2\,dt\right\}^{\frac12}$$

$$\le \left\{\int_{t_i}^{t_{i+1}}[D^k(x(t)-q(t))]^2\,dt\right\}^{\frac12}+\left\{\int_{t_i}^{t_{i+1}}[D^k(q(t)-H_2^{\Delta}q(t))]^2\,dt\right\}^{\frac12}$$

$$\le \frac{h^{3-k}}{\lambda_{3,k}^{3-k}}\left\{\int_{t_i}^{t_{i+1}}[D^3x(t)]^2\,dt\right\}^{\frac12}+\bar c_{2,4,k}\,h^{4-k}\left\{\int_{t_i}^{t_{i+1}}[D^4q(t)]^2\,dt\right\}^{\frac12}$$

$$\le \frac{h^{3-k}}{\lambda_{3,k}^{3-k}}\left\{\int_{t_i}^{t_{i+1}}[D^3x(t)]^2\,dt\right\}^{\frac12}+\bar c_{2,4,k}\,2\sqrt{15}\,h^{3-k}\left\{\int_{t_i}^{t_{i+1}}[D^3q(t)]^2\,dt\right\}^{\frac12}$$

$$\le \left(\lambda_{3,k}^{-3+k}+\bar c_{2,4,k}\,2\sqrt{15}\right)h^{3-k}\left\{\int_{t_i}^{t_{i+1}}[D^3x(t)]^2\,dt\right\}^{\frac12},\quad 0\le k\le 2.$$

The required inequalities now follow by squaring both sides of the above inequality, summing i from 0 to N, and then taking the square root of both sides. ∎

Theorem 5.3.15. Let $x(t)\in PC^{n,2}[a,b]$, $3\le n\le 6$. Then,

$$(5.3.63)\qquad \|\,D^k(x-H_3^{\Delta}x)\,\|_2\le \bar c_{3,n,k}\,h^{n-k}\,\|\,D^nx\,\|_2\,,\ 0\le k\le 3$$

where the constants $\bar c_{3,n,k}$ in terms of $\lambda_{n,k}$ (see Table 5.2.1) and k_n, $\tilde k_{n,\ell}$ (see Remarks 5.2.4 - 5.2.6) are given in the following table.

Table 5.3.12.

$k \backslash n$	3	4	5	6
0	$\lambda_{3,0}^{-3}$ $= 0.403145-2$	$\lambda_{4,0}^{-4} + 4\sqrt{k_{3,2}}\lambda_{3,0}^{-6}$ $= 0.175716-2$	$\lambda_{5,0}^{-5} + 2\sqrt{k_4}\lambda_{3,0}^{-6}$ $= 0.330962-3$	$\lambda_{3,0}^{-6}$ $= 0.162526-4$
1	$\lambda_{3,1}^{-2}$ $= 0.162146-1$	$\lambda_{4,1}^{-3} + 4\sqrt{k_{3,2}}\lambda_{3,0}^{-3}\lambda_{3,1}^{-2}$ $= 0.718473-2$	$\lambda_{5,1}^{-4} + 2\sqrt{k_4}\lambda_{3,0}^{-3}\lambda_{3,1}^{-2}$ $= 0.134306-2$	$\lambda_{3,0}^{-3}\lambda_{3,1}^{-2}$ $= 0.653682-4$
2	$\lambda_{3,2}^{-1}$ $= 0.111274+0$	$\lambda_{4,2}^{-2} + 4\sqrt{k_{3,2}}\lambda_{3,0}^{-3}\lambda_{3,2}^{-1}$ $= 0.501240-1$	$\lambda_{5,2}^{-3} + 2\sqrt{k_4}\lambda_{3,0}^{-3}\lambda_{3,2}^{-1}$ $= 0.931026-2$	$\lambda_{3,0}^{-3}\lambda_{3,2}^{-1}$ $= 0.448595-3$
3	1	$\lambda_{4,3}^{-1} + 4\sqrt{k_{3,2}}\lambda_{3,0}^{-3}$ $= 0.456241+0$	$\lambda_{5,3}^{-2} + 2\sqrt{k_4}\lambda_{3,0}^{-3}$ $= 0.844747-1$	$\lambda_{3,0}^{-3}$ $= 0.403145-2$

Proof. For $n = 3,\ k = 3$ the inequality (5.3.63) follows directly from the first integral relation (5.3.47). For $n = 3,\ 0 \le k \le 2$ we note that $D^j(x(t_i) - H_3^{\Delta}x(t_i)) = 0;\ 0 \le i \le N+1,\ 0 \le j \le 2$ and hence from Theorem 5.2.4, we have

$$\int_{t_i}^{t_{i+1}} [D^k(x(t) - H_3^{\Delta}x(t))]^2\, dt \le \frac{(t_{i+1}-t_i)^{6-2k}}{\lambda_{3,k}^{6-2k}} \int_{t_i}^{t_{i+1}} [D^3(x(t) - H_3^{\Delta}x(t))]^2\, dt;$$

$$0 \le i \le N,\ 0 \le k \le 2.$$

Summing both sides of the above inequality with respect to i from 0 to N, taking the square root of both sides of the resulting inequality, and bounding the right side of the resulting inequality by (5.3.63) for $n = 3,\ k = 3$ we obtain

$$\text{(5.3.64)} \qquad \| D^k(x - H_3^{\Delta}x) \|_2 \le \frac{h^{3-k}}{\lambda_{3,k}^{3-k}} \| D^3 x \|_2\ ,\ 0 \le k \le 2.$$

For the case $n = 6,\ k = 3$ we apply the Cauchy - Schwarz inequality to the second integral relation (5.3.49) for $m = 3$, to get

$$\| D^3(x - H_3^{\Delta}x) \|_2^2 \le \| D^6 x \|_2 \| x - H_3^{\Delta}x \|_2\ .$$

From this inequality it immediately follows that

$$\text{(5.3.65)} \qquad \| D^3(x - H_3^{\Delta}x) \|_2 \le \frac{h^3}{\lambda_{3,0}^3} \| D^6 x \|_2\ .$$

For $n = 6$, $0 \le k \le 2$ we use Corollary 5.2.5, to obtain

$$\int_{t_i}^{t_{i+1}} [D^k(x(t) - H_3^\Delta x(t))]^2\, dt \le \frac{(t_{i+1} - t_i)^{12-2k}}{\lambda_{3,k}^{6-2k} \lambda_{3,0}^6} \int_{t_i}^{t_{i+1}} [D^6(x(t) - H_3^\Delta x(t))]^2\, dt.$$

However, since $H_3^\Delta x(t)$ is a quintic polynomial in $[t_i, t_{i+1}]$, $0 \le i \le N$ we find that

$$\| D^k(x - H_3^\Delta x) \|_2 \le \frac{h^{6-k}}{\lambda_{3,k}^{3-k} \lambda_{3,0}^3} \| D^6 x \|_2\ ,\ 0 \le k \le 2. \tag{5.3.66}$$

For $n = 4$, $0 \le k \le 3$ let $q(t)$ be the unique septic polynomial in $[t_i, t_{i+1}]$, $0 \le i \le N$ but fixed, satisfying $D^j q(t_\ell) = D^j x(t_\ell)$; $0 \le j \le 3$, $\ell = i, i+1$. We shall show that

$$\int_{t_i}^{t_{i+1}} [D^4 q(t)]^2\, dt + \int_{t_i}^{t_{i+1}} [D^4 q(t) - D^4 x(t)]^2\, dt = \int_{t_i}^{t_{i+1}} [D^4 x(t)]^2\, dt. \tag{5.3.67}$$

Since

$$\begin{aligned}\int_{t_i}^{t_{i+1}} [D^4 x(t)]^2\, dt &= \left(D^4(x - q + q), D^4(x - q + q)\right)\\ &= (D^4(x-q), D^4(x-q)) + 2\,(D^4(x-q), D^4 q) + (D^4 q, D^4 q)\end{aligned}$$

(5.3.67) holds if and only if

$$\int_{t_i}^{t_{i+1}} [D^4 x(t) - D^4 q(t)] D^4 q(t)\, dt = 0.$$

For this, integration by parts and the fact that $q(t)$ is septic gives that

$$\begin{aligned}&\int_{t_i}^{t_{i+1}} [D^4 x(t) - D^4 q(t)] D^4 q(t)\, dt\\ &\quad = [D^3 x(t) - D^3 q(t)] D^4 q(t) \Big|_{t_i}^{t_{i+1}} - \int_{t_i}^{t_{i+1}} [D^3 x(t) - D^3 q(t)] D^5 q(t)\, dt\\ &\quad = -\int_{t_i}^{t_{i+1}} [D^3 x(t) - D^3 q(t)] D^5 q(t)\, dt\\ &\quad = (-1)^2 \int_{t_i}^{t_{i+1}} [D^2 x(t) - D^2 q(t)] D^6 q(t)\, dt\\ &\quad = (-1)^3 \int_{t_i}^{t_{i+1}} [D x(t) - D q(t)] D^7 q(t)\, dt\\ &\quad = (-1)^4 \int_{t_i}^{t_{i+1}} [x(t) - q(t)] D^8 q(t)\, dt\ =\ 0.\end{aligned}$$

Next, since $q(t) - x(t) \in PC^{4,2}[t_i, t_{i+1}]$, $D^j(q(t_i) - x(t_i)) = D^j(q(t_{i+1}) - x(t_{i+1})) = 0$, $0 \le j \le 3$ from Theorem 5.2.4, we have

$$\int_{t_i}^{t_{i+1}} [D^k(q(t) - x(t))]^2 \, dt \le \left(\frac{t_{i+1} - t_i}{\lambda_{4,k}}\right)^{8-2k} \int_{t_i}^{t_{i+1}} [D^4(q(t) - x(t))]^2 \, dt$$

$$(5.3.68) \qquad \le \left(\frac{h}{\lambda_{4,k}}\right)^{8-2k} \int_{t_i}^{t_{i+1}} [D^4 x(t)]^2 \, dt,$$

where in the last inequality we have used (5.3.67).

For all $t \in [t_i, t_{i+1}]$ we note that $H_3^{\Delta} x(t) = H_3^{\Delta} q(t)$, and hence we can write

$$x(t) - H_3^{\Delta} x(t) = x(t) - q(t) + q(t) - H_3^{\Delta} q(t).$$

Thus, by the triangle inequality, (5.3.68), (5.3.63) for $n = 6$, $0 \le k \le 3$, Remark 5.2.6 for $n = 3$, $\ell = 2$ and (5.3.67) it follows that

$$\left\{\int_{t_i}^{t_{i+1}} [D^k(x(t) - H_3^{\Delta} x(t))]^2 \, dt\right\}^{\frac{1}{2}}$$

$$\le \left\{\int_{t_i}^{t_{i+1}} [D^k(x(t) - q(t))]^2 \, dt\right\}^{\frac{1}{2}} + \left\{\int_{t_i}^{t_{i+1}} [D^k(q(t) - H_3^{\Delta} q(t))]^2 \, dt\right\}^{\frac{1}{2}}$$

$$\le \frac{h^{4-k}}{\lambda_{4,k}^{4-k}} \left\{\int_{t_i}^{t_{i+1}} [D^4 x(t)]^2 \, dt\right\}^{\frac{1}{2}} + \bar{c}_{3,6,k} \, h^{6-k} \left\{\int_{t_i}^{t_{i+1}} [D^6 q(t)]^2 \, dt\right\}^{\frac{1}{2}}$$

$$\le \frac{h^{4-k}}{\lambda_{4,k}^{4-k}} \left\{\int_{t_i}^{t_{i+1}} [D^4 x(t)]^2 \, dt\right\}^{\frac{1}{2}} + \bar{c}_{3,6,k} \, 4\sqrt{\tilde{k}_{3,2}} \, h^{4-k} \left\{\int_{t_i}^{t_{i+1}} [D^4 q(t)]^2 \, dt\right\}^{\frac{1}{2}}$$

$$\le \left(\lambda_{4,k}^{-4+k} + 4\bar{c}_{3,6,k} \sqrt{\tilde{k}_{3,2}}\right) h^{4-k} \left\{\int_{t_i}^{t_{i+1}} [D^4 x(t)]^2 \, dt\right\}^{\frac{1}{2}}, \quad 0 \le k \le 3.$$

The required inequalities now follow by squaring both sides of the above inequality, summing i from 0 to N, and then taking the square root of both sides.

For $n = 5$, $0 \le k \le 3$ let $q(t)$ be the unique nonic polynomial in $[t_i, t_{i+1}]$, $0 \le i \le N$ but fixed, satisfying $D^j q(t_\ell) = D^j x(t_\ell)$; $0 \le j \le 4$, $\ell = i, i+1$. Then, as earlier by integration by parts it can be shown that

$$(5.3.69) \quad \int_{t_i}^{t_{i+1}} [D^5 q(t)]^2 \, dt + \int_{t_i}^{t_{i+1}} [D^5 q(t) - D^5 x(t)]^2 \, dt = \int_{t_i}^{t_{i+1}} [D^5 x(t)]^2 \, dt.$$

Rest of the proof is similar to that for the case $n = 4$. ∎

A result which includes the previous Theorems 5.3.13 - 5.3.15 is the following :

Theorem 5.3.16. Let $x(t) \in PC^{m+j,2}[a,b]$, $0 \le j \le m$. Then,

$$\text{(5.3.70)} \qquad \| D^k(x - H_m^{\Delta} x) \|_2 \le \bar{c}_{m,m+j,k}\ h^{m+j-k} \| D^{m+j} x \|_2 \ , \ 0 \le k \le m$$

where

$$\bar{c}_{m,m,k} = \lambda_{m,k}^{-m+k}$$

$$\bar{c}_{m,2m,k} = \lambda_{m,k}^{-m+k} \lambda_{m,0}^{-m}$$

$$\bar{c}_{m,m+j,k} = \lambda_{m+j,k}^{-m-j+k} + 2^{m-j} \left(\tilde{k}_{m+j-1,m-j} \right)^{\frac{1}{2}} \bar{c}_{m,2m,k}, \quad 1 \le j \le m-1$$

and the constants $\lambda_{n,k}$ and $\tilde{k}_{n,k}$ are defined in Theorem 5.2.4 and Remark 5.2.6, respectively.

Proof. The proof is similar to that of Theorem 5.3.15. ∎

It is clear that for $m \ge 4$ an upper bound for $\bar{c}_{m,m+j,k}$ can be obtained by using the right sides of (5.2.9), (5.2.12) and Remarks 5.2.4 - 5.2.6.

Our final result in this section is the following :

Theorem 5.3.17. Let $p, q \ge 1$, $r = \max\{p, q, 2\}$, and $x(t) \in PC^{m+j,r}[a,b]$, $0 \le j \le m$. Then, for $1 \le p \le 2 \le q$

$$\text{(5.3.71)} \quad \| D^k(x - H_m^{\Delta} x) \|_p \le \bar{c}_{m,m+j,k}\ h^{m+j-k} (b-a)^{p^{-1}-q^{-1}} \| D^{m+j} x \|_q \ , \qquad 0 \le k \le m$$

and for $p \ge 2$

$$\text{(5.3.72)} \quad \| D^k(x - H_m^{\Delta} x) \|_p \le \frac{1}{2}\ \bar{c}_{m,m+j,k+1}\ h^{m+j-k-2^{-1}+p^{-1}} \| D^{m+j} x \|_2 \ , \qquad 0 \le k \le m-1.$$

Proof. For any function $\omega(t) \in PC^{0,q}[a,b]$ and $q \ge p \ge 1$, Hölder's inequality gives

$$\| \omega \|_p \le (b-a)^{p^{-1}-q^{-1}} \| \omega \|_q \ .$$

Therefore, for $1 \le p \le 2 \le q$, from the inequality (5.3.70) it follows that

$$\left[(b-a)^{p^{-1}-2^{-1}}\right]^{-1} \| D^k(x - H_m^{\Delta}x) \|_p \le \bar{c}_{m,m+j,k}\, h^{m+j-k} \times$$

$$(b-a)^{2^{-1}-q^{-1}} \| D^{m+j}x \|_q ,$$

which is the same as (5.3.71).

To prove (5.3.72) we let $\omega(t) \in PC^{1,q}[a,b]$ satisfying $\omega(t_i) = 0,\ 0 \le i \le N+1$. For $t \in (t_i, t_{i+1})$, we have

$$\omega(t) = \frac{1}{2}\left[\int_{t_i}^{t} D\omega(s)\, ds - \int_{t}^{t_{i+1}} D\omega(s)\, ds\right].$$

Thus, by Hölder's inequality we find that

$$| \omega(t) | \le \frac{1}{2}\, h^{1-q^{-1}}\, z_i$$

where $z_i = \left(\int_{t_i}^{t_{i+1}} | D\omega(t) |^q\, dt\right)^{1/q}$, and hence for any $p \ge 1$ it follows that

$$\left(\sum_{i=0}^{N} \int_{t_i}^{t_{i+1}} | \omega(t) |^p\, dt\right)^{\frac{1}{p}} \le \frac{1}{2}\, h^{1-q^{-1}+p^{-1}} \left(\sum_{i=0}^{N} z_i^p\right)^{\frac{1}{p}}.$$

Therefore, as an application of Jenson's inequality [6, p.18]

$$\left(\sum_{i=0}^{N} z_i^p\right)^{\frac{1}{p}} \le \left(\sum_{i=0}^{N} z_i^q\right)^{\frac{1}{q}},\quad q \le p$$

we obtain

$$\| \omega \|_p \le \frac{1}{2}\, h^{1-q^{-1}+p^{-1}} \| D\omega \|_q ,\quad 1 \le q \le p.$$

Now combining the above inequality with (5.3.70), for $0 \le k \le m-1$ and $p \ge 2$, we get

$$\begin{aligned} \| D^k(x - H_m^{\Delta}x) \|_p &\le \frac{1}{2}\, h^{1-2^{-1}+p^{-1}} \| D^{k+1}(x - H_m^{\Delta}x) \|_2 \\ &\le \frac{1}{2}\, h^{2^{-1}+p^{-1}}\, \bar{c}_{m,m+j,k+1}\, h^{m+j-k-1} \| D^{m+j}x \|_2 , \end{aligned}$$

which is the same as (5.3.72). ∎

Remark 5.3.5. In (5.3.71) and (5.3.72) the exponents of h are the best possible for the set $PC^{m+j,r}[a,b]$. ∎

5.4 PIECEWISE LIDSTONE INTERPOLATION

For a fixed Δ, we define the set $L_m(\Delta) = \{h(t) \in C[a,b] : h(t)$ is a polynomial of degree at most $(2m-1)$ in each subinterval $[t_i, t_{i+1}]$, $0 \le i \le N\}$. It is clear that $L_m(\Delta)$ is of dimension $[2m(N+1) - N]$.

Definition 5.4.1. For a given $x(t) \in C^{(2m-2)}[a,b]$ we say $L_m^{\Delta}x(t)$ is the $L_m(\Delta)$ - *interpolate of* $x(t)$, also known as *Lidstone interpolate of* $x(t)$ if $L_m^{\Delta}x(t) \in L_m(\Delta)$ with $D^{2k}L_m^{\Delta}x(t_i) = x^{(2k)}(t_i) = x_i^{(2k)}$; $0 \le k \le m-1$, $0 \le i \le N+1$.

From Chapter 1 it is clear that for $x(t) \in C^{(2m-2)}[a,b]$, $L_m^{\Delta}x(t)$ uniquely exists and in the subinterval $[t_i, t_{i+1}]$ can be explicitly expressed as

$$L_m^{\Delta}x(t) = \sum_{k=0}^{m-1}\left[x_i^{(2k)}\,\Lambda_k\left(\frac{t_{i+1}-t}{h}\right) + x_{i+1}^{(2k)}\,\Lambda_k\left(\frac{t-t_i}{h}\right)\right]h^{2k}. \tag{5.4.1}$$

Therefore, it follows that

$$L_m^{\Delta}x(t) = \sum_{i=0}^{N+1}\sum_{j=0}^{m-1} r_{m,i,j}(t)\, x_i^{(2j)}, \tag{5.4.2}$$

where $r_{m,i,j}(t)$, $0 \le i \le N+1$ are the basic elements of $L_m(\Delta)$ satisfying

$$D^{2\nu}r_{m,i,j}(t_\mu) = \delta_{i\mu}\,\delta_{2\nu,j}\ ;\ 0 \le \nu \le m-1,\ 0 \le \mu \le N+1 \tag{5.4.3}$$

and appear as

$$\begin{aligned} r_{m,i,j}(t) &= \Lambda_j\left(\frac{t_{i+1}-t}{h}\right)h^{2j}, \quad t_i \le t \le t_{i+1},\ 0 \le i \le N \\ &= \Lambda_j\left(\frac{t-t_{i-1}}{h}\right)h^{2j}, \quad t_{i-1} \le t \le t_i,\ 1 \le i \le N+1 \\ &= 0, \qquad \text{otherwise.} \end{aligned} \tag{5.4.4}$$

In what follows $d_{2m,k}$ represents the numbers

$$d_{2m,k} = \begin{cases} \dfrac{(-1)^{m-i}E_{2m-2i}}{2^{2m-2i}(2m-2i)!}, & k = 2i,\ 0 \le i \le m \\[2ex] (-1)^{m-i+1}\dfrac{2(2^{2m-2i}-1)}{(2m-2i)!}\,B_{2m-2i}, & k = 2i+1, \\ & 0 \le i \le m-1 \\[2ex] 2, \quad k = 2m+1. \end{cases} \tag{5.4.5}$$

It is clear that $d_{2m,k+2} = d_{2m-2,k}$, $0 \le k \le 2m-1$.

The following result for the basic elements of $L_m(\Delta)$ will be used in the next chapter.

Lemma 5.4.1. For $0 \le i \le N$ and $0 \le k \le 2j+1$, $0 \le j \le m-1$ the following equalities hold

$$\max_{t_i \le t \le t_{i+1}} \left[\mid D^k r_{m,i,j}(t) \mid + \mid D^k r_{m,i+1,j}(t) \mid \right] = d_{2j,k}\, h^{2j-k}. \tag{5.4.6}$$

Proof. For $t_i \le t \le t_{i+1}$, from (5.4.4), (1.2.22) and (1.2.2), we find that

$$\begin{aligned} \mid D^k r_{m,i,j}(t) \mid &= h^{2j} \left| D^k \Lambda_j \left(1 - \frac{t-t_i}{h} \right) \right| \\ &= h^{2j-k} \left| \frac{d^k}{d\theta^k} \int_0^1 g_j(\theta,\phi)(1-\phi)d\phi \right|, \quad \theta = \frac{t-t_i}{h} \\ &\le h^{2j-k} \int_0^1 \left| \frac{\partial^k g_j(\theta,\phi)}{\partial \theta^k} \right| (1-\phi)d\phi, \quad 0 \le k \le 2j-1 \end{aligned}$$

and

$$\mid D^k r_{m,i+1,j}(t) \mid \le h^{2j-k} \int_0^1 \left| \frac{\partial^k g_j(\theta,\phi)}{\partial \theta^k} \right| \phi\, d\phi, \quad 0 \le k \le 2j-1.$$

Therefore, it follows that

$$\begin{aligned} &\max_{t_i \le t \le t_{i+1}} \left[\mid D^k r_{m,i,j}(t) \mid + \mid D^k r_{m,i+1,j}(t) \mid \right] \\ &\qquad \le h^{2j-k} \max_{0 \le \theta \le 1} \int_0^1 \left| \frac{\partial^k g_j(\theta,\phi)}{\partial \theta^k} \right| d\phi, \quad 0 \le k \le 2j-1. \end{aligned}$$

Now, (5.4.6) for $0 \le k \le 2j-1$ follows from Lemmas 1.2.7 and 1.2.8.

For $k = 2j$, from (5.4.4) and (1.2.1), we have

$$\begin{aligned} &\max_{t_i \le t \le t_{i+1}} \left[\mid D^{2j} r_{m,i,j}(t) \mid + \mid D^{2j} r_{m,i+1,j}(t) \mid \right] \\ &\qquad = \max_{t_i \le t \le t_{i+1}} \left[\left| \frac{t_{i+1}-t}{h} \right| + \left| \frac{t-t_i}{h} \right| \right] = 1 = d_{2j,2j}. \end{aligned}$$

Finally, for $k = 2j+1$ it suffices to note that $D^{2j+1} r_{m,i,j}(t) = -\frac{1}{h}$ and $D^{2j+1} r_{m,i+1,j}(t) = \frac{1}{h}$. ∎

The following result is immediate from Theorem 1.5.1.

Theorem 5.4.2. Let $x(t) \in PC^{2m,\infty}[a,b]$. Then,

(5.4.7) $\| D^k(x - L_m^{\Delta}x) \|_\infty \le d_{2m,k}\, h^{2m-k} \| D^{2m}x \|_\infty\,,\ 0 \le k \le 2m-1.$ ∎

In view of Remark 1.5.1 it is clear that the constants $d_{2m,k}$ defined in (5.4.5) are the best possible in the inequalities (5.4.7).

Theorem 5.4.3. Let $x(t) \in PC^{2m-2,\infty}[a,b]$. Then,

$$\| D^k(x - L_m^{\Delta}x) \|_\infty$$

$$\le d_{2m-2,k}\, h^{2m-2-k} \times \tag{5.4.8}$$

$$\max_{0\le i\le N} \sup_{t\in(t_i,t_{i+1})} \left| D^{2m-2}x(t) - \left(\frac{t_{i+1}-t}{h}\right) x_i^{(2m-2)} - \left(\frac{t-t_i}{h}\right) x_{i+1}^{(2m-2)} \right|$$

$$\le 2d_{2m-2,k}\, h^{2m-2-k} \| D^{2m-2}x \|_\infty\,,\ 0 \le k \le 2m-2. \tag{5.4.9}$$

Proof. Without loss of generality we assume that $a = 0,\ b = 1$ and $h = 1$, so that

$$L_m^{\Delta}x(t) = \sum_{k=0}^{m-1} [x^{(2k)}(0)\Lambda_k(1-t) + x^{(2k)}(1)\Lambda_k(t)].$$

Therefore, from Theorem 5.2.1 for any $x(t) \in PC^{2m,1}[0,1]$ it follows that

$$x(t) - L_m^{\Delta}x(t) = \int_0^1 g_m(t,s) D^{2m}x(s)\, ds, \tag{5.4.10}$$

where

$$g_m(t,s) = -\begin{cases} \displaystyle\sum_{k=0}^{m-1} \Lambda_k(t)\, \frac{(1-s)^{2m-2k-1}}{(2m-2k-1)!}, & t \le s \\ \displaystyle\sum_{k=0}^{m-1} \Lambda_k(1-t)\, \frac{s^{2m-2k-1}}{(2m-2k-1)!}, & s \le t. \end{cases} \tag{5.4.11}$$

From the uniqueness of $g_m(t,s)$ it is clear that (1.2.4) for $n = m$ and (5.4.11) are two different representations of the same Peano's kernel.

An integration by parts of the right side of (5.4.10) and the fact that $g_m(t,0) = g_m(t,1) = 0$ lead to

$$x(t) - L_m^{\Delta}x(t) = \int_0^1 \bar{g}_m(t,s) D^{2m-1}x(s)\, ds, \tag{5.4.12}$$

where

$$(5.4.13)\quad \bar{g}_m(t,s) = -\frac{\partial g_m(t,s)}{\partial s} = \begin{cases} -\sum_{k=0}^{m-1} \Lambda_k(t)\, \frac{(1-s)^{2m-2k-2}}{(2m-2k-2)!}, & t \le s \\ \sum_{k=0}^{m-1} \Lambda_k(1-t)\, \frac{s^{2m-2k-2}}{(2m-2k-2)!}, & s \le t. \end{cases}$$

It is clear that

$$(5.4.14)\qquad \bar{g}_m(t,0) = \Lambda_{m-1}(1-t), \qquad \bar{g}_m(t,1) = -\Lambda_{m-1}(t).$$

Now an integration by parts of the right side of (5.4.12) gives that

$$x(t) - L_m^{\Delta}x(t) = -\bar{g}_m(t,0)x^{(2m-2)}(0) + \bar{g}_m(t,1)x^{(2m-2)}(1) - \int_0^1 \frac{\partial \bar{g}_m(t,s)}{\partial s} D^{2m-2}x(s)\, ds,$$

which in view of (5.4.14) and $\frac{\partial \bar{g}_m(t,s)}{\partial s} = -g_{m-1}(t,s)$ is the same as

$$x(t) - L_m^{\Delta}x(t) = \int_0^1 g_{m-1}(t,s) D^{2m-2}x(s)\, ds - \Lambda_{m-1}(1-t)x^{(2m-2)}(0) - \Lambda_{m-1}(t)x^{(2m-2)}(1).$$

Using (1.2.2) and (1.2.22) for $n = m-1$ in the above equality, we get

$$(5.4.15)\qquad x(t) - L_m^{\Delta}x(t) = \int_0^1 g_{m-1}(t,s) \left[D^{2m-2}x(s) - (1-s)x^{(2m-2)}(0) - s\, x^{(2m-2)}(1)\right] ds.$$

Now from the recurrence relation (1.2.4) and Lemmas 1.2.7 and 1.2.8 it immediately follows that

$$(5.4.16)\quad \| D^k(x - L_m^{\Delta}x) \|_\infty \le d_{2m-2,k} \sup_{t\in(0,1)} \mid D^{2m-2}x(t) - (1-t)x^{(2m-2)}(0) - t\, x^{(2m-2)}(1) \mid,\ 0 \le k \le 2m-3. \qquad \blacksquare$$

Remark 5.4.1. For each k the inequality (5.4.8) is the best possible, as equality holds for the function $x(t) = E_{2m-2}(t)$ for $a = 0$, $b = 1$ and $h = 1$, whose Lidstone interpolate is $L_m^{\Delta}E_{2m-2}(t) = E_{2m-2}(t)$, and only for this function up to a constant factor. We also note that the inequality connecting the right sides of (5.4.8) and (5.4.9) is also the best possible. For this, it suffices to note that for the continuous function

$$x(t) = \begin{cases} -1+4t, & 0 \le t \le \frac{1}{2} \\ 3-4t, & \frac{1}{2} \le t \le 1 \end{cases}$$

the equality $\max_{0\le t\le 1} | x(t) - (1-t)x(0) - t\,x(1) | = 2 \max_{0\le t\le 1} | x(t) |$ holds. However, as such the sharpness of (5.4.9) remains undecided. ∎

Theorem 5.4.4. Let $x(t) \in PC^{2m-1,\infty}[a,b]$, $1 \le m \le 3$. Then,

(5.4.17) $\| D^k(x - L_m^{\Delta} x) \|_\infty \le e_{2m-1,k}\, h^{2m-1-k} \| D^{2m-1} x \|_\infty$, $0 \le k \le 2m-2$

where the constants $e_{2m-1,k}$ are given in the following table.

Table 5.4.1.

k \ m	1	2	3
0	$\frac{1}{2}$	$\frac{1}{24}$	$\frac{1}{240}$
1		$\frac{2}{9}\sqrt{\frac{1}{3}}$	$\frac{1}{225}\left(1-\sqrt{\frac{8}{15}}\right)^{\frac{1}{2}}\left(2+5\sqrt{\frac{8}{15}}\right)$
2		$\frac{1}{2}$	$\frac{1}{24}$
3			$\frac{2}{9}\sqrt{\frac{1}{3}}$
4			$\frac{1}{2}$

Proof. Without loss of generality we assume that $a = 0$, $b = 1$ and $h = 1$. Thus, in view of (5.4.12) it follows that

$$| D^k(x - L_m^{\Delta} x)(t) | \le \left(\int_0^1 \left| \frac{\partial^k \bar{g}_m(t,s)}{\partial t^k} \right| ds \right) \| D^{2m-1} x \|_\infty , \tag{5.4.18}$$

$$0 \le k \le 2m-2$$

where $\bar{g}_m(t,s)$ is defined in (5.4.13). Hence, for $m = 3$ it suffices to show that

$$\max_{0\le t\le 1} \int_0^1 \left|\frac{\partial^k \bar{g}_3(t,s)}{\partial t^k}\right| ds = e_{5,k},\ 0 \le k \le 4. \tag{5.4.19}$$

Further, since $\frac{\partial^2 \bar{g}_m(t,s)}{\partial t^2} = \bar{g}_{m-1}(t,s)$ the proof of (5.4.19) will automatically include the proof for $m = 2$ and 1. For this, from a direct computation we find that

$$\int_0^1 |\ \bar{g}_3(t,s)\ |\ ds$$

$$= \frac{1}{225}\ t\ p_1(t)\left[7 - 10t^2 + 3t^4 - 5p_1^2(t)(1-t^2)\right] = \phi_1(t),$$

$$p_1^2(t) = (1-t^2) - \sqrt{\frac{4}{15}(1-t^2)(2-3t^2)},\ \ 0 \le t \le \frac{1}{2}$$

$$= \phi_1(1-t),\ \ \frac{1}{2} \le t \le 1$$

$$\le \phi_1\left(\frac{1}{2}\right) = \frac{1}{240};$$

$$\int_0^1 \left|\frac{\partial \bar{g}_3(t,s)}{\partial t}\right| ds$$

$$= \frac{1}{225}\ p_2(t)\left[7 - 30t^2 + 15t^4 - 5p_2^2(t)(1-3t^2)\right] = \phi_2(t),$$

$$p_2^2(t) = (1-3t^2) - \sqrt{\frac{4}{15}(30t^4 - 15t^2 + 2)},\ \ 0 \le t \le 1 - \left(1 - \sqrt{\frac{8}{15}}\right)^{\frac{1}{2}}$$

$$= \phi_2(t) + \frac{1}{225}\ p_3(t)\left[-4 + 60t^2 - 60t^3 + 15t^4\right.$$

$$\left. + 5p_3^2(t)(2 - 6t + 3t^2)\right] = \phi_3(t),$$

$$p_3^2(t) = (-2 + 6t - 3t^2) - \sqrt{\frac{4}{15}(30t^4 - 120t^3 + 165t^2 - 90t + 17)},$$

$$1 - \left(1 - \sqrt{\frac{8}{15}}\right)^{\frac{1}{2}} \le t \le \left(1 - \sqrt{\frac{8}{15}}\right)^{\frac{1}{2}}$$

$$= \phi_2(1-t), \quad \left(1-\sqrt{\frac{8}{15}}\right)^{\frac{1}{2}} \le t \le 1$$

$$\le \phi_2(0) = \frac{1}{225}\left(1-\sqrt{\frac{8}{15}}\right)^{\frac{1}{2}}\left(2+5\sqrt{\frac{8}{15}}\right);$$

$$\int_0^1 \left|\frac{\partial^2 \bar{g}_3(t,s)}{\partial t^2}\right| ds = \frac{2\sqrt{3}}{27}\, t(1-t^2)^{\frac{3}{2}}, \quad 0 \le t \le \frac{1}{2}$$

$$= \frac{2\sqrt{3}}{27}(1-t)(2t-t^2)^{\frac{3}{2}}, \quad \frac{1}{2} \le t \le 1$$

$$\le \frac{1}{24};$$

$$\int_0^1 \left|\frac{\partial^3 \bar{g}_3(t,s)}{\partial t^3}\right| ds = \frac{2}{9}\sqrt{\frac{1}{3}}(1-3t^2)^{\frac{3}{2}}, \quad 0 \le t \le 1-\frac{1}{\sqrt{3}}$$

$$= \frac{2}{9}\sqrt{\frac{1}{3}}\left\{(1-3t^2)^{\frac{3}{2}} + \left[1-3(1-t)^2\right]^{\frac{3}{2}}\right\},$$

$$1-\frac{1}{\sqrt{3}} \le t \le \frac{1}{\sqrt{3}}$$

$$= \frac{2}{9}\sqrt{\frac{1}{3}}\left[1-3(1-t)^2\right]^{\frac{3}{2}}, \quad \frac{1}{\sqrt{3}} \le t \le 1$$

$$\le \frac{2}{9}\sqrt{\frac{1}{3}};$$

$$\int_0^1 \left|\frac{\partial^4 \bar{g}_3(t,s)}{\partial t^4}\right| ds = 2t(1-t) \le \frac{1}{2}. \qquad \blacksquare$$

Now we shall obtain error bounds for the derivatives of the interpolation error $x(t) - L_m^{\Delta}x(t)$ in terms of the derivatives of $x(t)$ in L_2 - norm.

Theorem 5.4.5. Let $x(t) \in PC^{2m-1,2}[a,b]$. Then,

$$(5.4.20) \quad \| D^k(x - L_m^{\Delta}x) \|_2 \le \frac{1}{\pi^{2m-1-k}}\, h^{2m-1-k} \| D^{2m-1}x \|_2\,, \; 0 \le k \le 2m-1.$$

Proof. By an integration by parts it can be shown that for $0 \le i \le N$

$$\int_{t_i}^{t_{i+1}} [D^{2m-1} L_m^{\Delta} x(t)]^2 \, dt + \int_{t_i}^{t_{i+1}} [D^{2m-1} x(t) - D^{2m-1} L_m^{\Delta} x(t)]^2 \, dt = \int_{t_i}^{t_{i+1}} [D^{2m-1} x(t)]^2 \, dt. \tag{5.4.21}$$

Indeed, the above equality holds from the fact that

$$\begin{aligned}
&\int_{t_i}^{t_{i+1}} D^{2m-1}(x(t) - L_m^{\Delta} x(t)) \, D^{2m-1} L_m^{\Delta} x(t) \, dt \\
&= D^{2m-2}(x(t) - L_m^{\Delta} x(t)) \, D^{2m-1} L_m^{\Delta} x(t) \Big|_{t_i}^{t_{i+1}} \\
&\quad - \int_{t_i}^{t_{i+1}} D^{2m-2}(x(t) - L_m^{\Delta} x(t)) \, D^{2m} L_m^{\Delta} x(t) \, dt \\
&= 0.
\end{aligned}$$

Now since for a fixed i such that $0 \le i \le N$, $D^{2j}(x(t_\ell) - L_m^{\Delta}(t_\ell)) = 0$; $0 \le j \le m-1$, $\ell = i, i+1$ from Corollary 5.2.3 and Theorem 5.2.2 successively it follows that

$$\begin{aligned}
&\int_{t_i}^{t_{i+1}} [D^{2j}(x(t) - L_m^{\Delta} x(t))]^2 \, dt \\
&\le \left(\frac{t_{i+1} - t_i}{\pi}\right)^4 \int_{t_i}^{t_{i+1}} [D^{2j+2}(x(t) - L_m^{\Delta} x(t))]^2 \, dt \\
&\le \left(\frac{h}{\pi}\right)^{4\times 2} \int_{t_i}^{t_{i+1}} [D^{2j+4}(x(t) - L_m^{\Delta} x(t))]^2 \, dt \\
&\quad \cdots \\
&\le \left(\frac{h}{\pi}\right)^{4(m-j-1)} \int_{t_i}^{t_{i+1}} [D^{2m-2}(x(t) - L_m^{\Delta} x(t))]^2 \, dt \\
&\le \left(\frac{h}{\pi}\right)^{2(2m-2j-1)} \int_{t_i}^{t_{i+1}} [D^{2m-1}(x(t) - L_m^{\Delta} x(t))]^2 \, dt
\end{aligned}$$

$$\le \left(\frac{h}{\pi}\right)^{2(2m-2j-1)} \int_{t_i}^{t_{i+1}} [D^{2m-1} x(t)]^2 \, dt, \tag{5.4.22}$$

where in the last inequality we have used (5.4.21).

The required inequalities (5.4.20) for $k = 2j$, $0 \le j \le m-1$ now follow by summing the above inequality from $i = 0$ to N, and then taking the square root of both sides.

Next, we shall consider the case $k = 2j+1,\ 0 \le j \le m-2$. For this, we have

$$\int_{t_i}^{t_{i+1}} [D^{2j+1}(x(t) - L_m^{\Delta}x(t))]^2\, dt$$

$$= -\int_{t_i}^{t_{i+1}} D^{2j}(x(t) - L_m^{\Delta}x(t))\ D^{2j+2}(x(t) - L_m^{\Delta}x(t))\ dt.$$

Thus, by the Cauchy - Schwarz inequality and (5.4.22) it follows that

$$\int_{t_i}^{t_{i+1}} [D^{2j+1}(x(t) - L_m^{\Delta}x(t))]^2\, dt$$

$$(5.4.23) \qquad \le \left\{\int_{t_i}^{t_{i+1}} [D^{2j}(x(t) - L_m^{\Delta}x(t))]^2\, dt\right\}^{\frac{1}{2}} \times$$

$$\left\{\int_{t_i}^{t_{i+1}} [D^{2j+2}(x(t) - L_m^{\Delta}x(t))]^2\, dt\right\}^{\frac{1}{2}}$$

$$\le \left(\frac{h}{\pi}\right)^{2m-2j-1} \left(\frac{h}{\pi}\right)^{2m-2j-3} \int_{t_i}^{t_{i+1}} [D^{2m-1}x(t)]^2\, dt$$

$$= \left(\frac{h}{\pi}\right)^{2[2m-(2j+1)-1]} \int_{t_i}^{t_{i+1}} [D^{2m-1}x(t)]^2\, dt.$$

From the above inequality, (5.4.20) for $k = 2j+1,\ 0 \le j \le m-2$ follows as earlier.

Finally, we note that (5.4.20) for $k = 2m-1$ is immediate from (5.4.21). ∎

Theorem 5.4.6. Let $x(t) \in PC^{2m,2}[a,b]$. Then,

$$(5.4.24) \qquad \| D^k(x - L_m^{\Delta}x) \|_2 \le \frac{1}{\pi^{2m-k}}\, h^{2m-k}\, \| D^{2m}x \|_2\ ,\ 0 \le k \le 2m.$$

Proof. The inequality (5.4.24) for $k = 2m$ follows from the fact that in each subinterval $[t_i, t_{i+1}]$, $0 \le i \le N$, $L_m^{\Delta}x(t)$ is a polynomial of degree $(2m-1)$. Using this fact further, as in Theorem 5.4.5 we find that

$$\int_{t_i}^{t_{i+1}} [D^{2j}(x(t) - L_m^{\Delta}x(t))]^2\, dt \le \left(\frac{h}{\pi}\right)^{2(2m-2j)} \int_{t_i}^{t_{i+1}} [D^{2m}(x(t) - L_m^{\Delta}x(t))]^2\, dt$$

$$(5.4.25) \qquad = \left(\frac{h}{\pi}\right)^{2(2m-2j)} \int_{t_i}^{t_{i+1}} [D^{2m}x(t)]^2\, dt$$

from which (5.4.24) is immediate for $k = 2j,\ 0 \le j \le m-1$.

Finally, we note that since $x(t) \in PC^{2m,2}[a,b]$ inequality (5.4.23) now holds for $0 \le j \le m-1$. Thus, on combining (5.4.23) and (5.4.25), we obtain

$$\int_{t_i}^{t_{i+1}} [D^{2j+1}(x(t) - L_m^{\Delta} x(t))]^2\, dt \le \left(\frac{h}{\pi}\right)^{2[2m-(2j+1)]} \int_{t_i}^{t_{i+1}} [D^{2m} x(t)]^2\, dt,$$

which as earlier gives (5.4.24) for $k = 2j+1,\ 0 \le j \le m-1$. ∎

Our final result in this section is the following :

Theorem 5.4.7. Let $p, q \ge 1,\ r = \max\{p, q, 2\}$, and $x(t) \in PC^{2m-1+j,r}[a,b]$; $j = 0, 1$. Then, for $1 \le p \le 2 \le q$

$$\text{(5.4.26)} \qquad \| D^k(x - L_m^{\Delta} x) \|_p \le \frac{1}{\pi^{2m-1+j-k}}\, h^{2m-1+j-k}(b-a)^{p^{-1}-q^{-1}} \times \\ \| D^{2m-1+j} x \|_q\,,\ 0 \le k \le 2m-1+j$$

and for $p \ge 2$

$$\text{(5.4.27)} \qquad \| x - L_m^{\Delta} x \|_p \le \frac{1}{2}\, \frac{1}{\pi^{2m-2+j}}\, h^{2m-\frac{3}{2}+j+p^{-1}} \| D^{2m-1+j} x \|_2\ .$$

Proof. The proof is similar to that of Theorem 5.3.17. ∎

5.5 TWO VARIABLE PIECEWISE HERMITE INTERPOLATION

For a fixed ρ, we define the set $H_m(\rho)$ as follows :

$$\begin{aligned} H_m(\rho) &= H_m(\Delta) \oplus H_m(\Delta') \quad \text{(the tensor product)} \\ &= \text{Span}\,\{h_{m,i,\mu}(t) h_{m,j,\nu}(u)\}_{i=0}^{N+1}\ {}_{\mu=0}^{m-1}\ {}_{j=0}^{M+1}\ {}_{\nu=0}^{m-1} \\ &= \Big\{ h(t,u) \in C^{(m-1,m-1)}([a,b] \times [c,d]) : h(t,u) \text{ is a two - dimensional polynomial of degree at most } (2m-1) \text{ in each variable and in each subrectangle } [t_i, t_{i+1}] \times [u_j, u_{j+1}]; \\ &\qquad 0 \le i \le N,\ 0 \le j \le M \Big\}. \end{aligned}$$

Since $H_m(\rho)$ is the tensor product of $H_m(\Delta)$ and $H_m(\Delta')$ which are of dimensions $m(N+2)$ and $m(M+2)$ respectively, $H_m(\rho)$ is of dimension $m^2(N+2)(M+2)$.

Definition 5.5.1. For a given $f(t,u) \in C^{(m-1,m-1)}([a,b] \times [c,d])$, we say $H_m^\rho f(t,u)$ is the $H_m(\rho)$ - *interpolate of* $f(t,u)$ if $H_m^\rho f(t,u) \in H_m(\rho)$ with $D_t^\mu D_u^\nu H_m^\rho f(t_i,u_j) = f_{i,j}^{(\mu,\nu)}$; $0 \le i \le N+1$, $0 \le j \le M+1$, $0 \le \mu,\nu \le m-1$.

For $f(t,u) \in C^{(m-1,m-1)}([a,b] \times [c,d])$ it is clear that $H_m^\rho f(t,u)$ uniquely exists and can be explicitly expressed as

$$H_m^\rho f(t,u) = \sum_{i=0}^{N+1} \sum_{\mu=0}^{m-1} \sum_{j=0}^{M+1} \sum_{\nu=0}^{m-1} h_{m,i,\mu}(t) h_{m,j,\nu}(u) f_{i,j}^{(\mu,\nu)}. \tag{5.5.1}$$

The following result provides a characterization of $H_m^\rho f(t,u)$ in terms of one - dimensional interpolation schemes.

Lemma 5.5.1. If $f(t,u) \in C^{(m-1,m-1)}([a,b] \times [c,d])$, then

$$H_m^\rho f(t,u) = H_m^{\Delta'} H_m^{\Delta} f(t,u) = H_m^{\Delta} H_m^{\Delta'} f(t,u). \tag{5.5.2}$$

Proof. By definition

$$\begin{aligned} H_m^{\Delta'} H_m^{\Delta} f(t,u) &= H_m^{\Delta'} \left(\sum_{i=0}^{N+1} \sum_{\mu=0}^{m-1} h_{m,i,\mu}(t)\ D^\mu f(t_i,u) \right) \\ &= \sum_{j=0}^{M+1} \sum_{\nu=0}^{m-1} \left(\sum_{i=0}^{N+1} \sum_{\mu=0}^{m-1} h_{m,i,\mu}(t)\ D^\mu D^\nu f(t_i,u_j) \right) h_{m,j,\nu}(u) \\ &= \sum_{i=0}^{N+1} \sum_{\mu=0}^{m-1} \sum_{j=0}^{M+1} \sum_{\nu=0}^{m-1} h_{m,i,\mu}(t) h_{m,j,\nu}(u) f_{i,j}^{(\mu,\nu)} \\ &= H_m^\rho f(t,u). \end{aligned}$$

The proof of the second equality in (5.5.2) is similar. ∎

Now let $f(t,u) \in C^{(m-1,m-1)}([a,b] \times [c,d])$ be an arbitrary function. From Lemma 5.5.1 we have

$$f - H_m^\rho f$$

$$(5.5.3) \qquad = (f - H_m^\Delta f) + H_m^\Delta (f - H_m^{\Delta'} f)$$

$$(5.5.4) \qquad = (f - H_m^\Delta f) + \left[H_m^\Delta (f - H_m^{\Delta'} f) - (f - H_m^{\Delta'} f)\right] + (f - H_m^{\Delta'} f)$$

$$(5.5.5) \qquad = (f - H_m^\Delta f) + \left[H_m^{\Delta'} (f - H_m^{\Delta} f) - (f - H_m^{\Delta} f)\right] + (f - H_m^{\Delta'} f).$$

From the above relations we shall deduce the following error estimates.

Theorem 5.5.2. Let $f(t,u) \in PC^{2,\infty}([a,b] \times [c,d])$. Then,

$$(5.5.6) \qquad \| f - H_1^\rho f \|_\infty \le c_{1,2,0} \left(h^2 \| D_t^2 f \|_\infty + \ell^2 \| D_u^2 f \|_\infty\right),$$

where the constant $c_{1,2,0}$ is defined in Theorem 5.3.6.

Proof. Since for $t_i \le t \le t_{i+1}$,

$$H_1^\Delta x(t) = \frac{1}{h}\left[(t_{i+1} - t)x_i + (t - t_i)x_{i+1}\right]$$

it follows that

$$| H_1^\Delta x(t) | \le \| x \|_\infty .$$

Thus, from (5.5.3) and Theorem 5.3.6 we find that

$$\begin{aligned} \| f - H_1^\rho f \|_\infty &\le \| f - H_1^\Delta f \|_\infty + \| H_1^\Delta (f - H_1^{\Delta'} f) \|_\infty \\ &\le \| f - H_1^\Delta f \|_\infty + \| f - H_1^{\Delta'} f \|_\infty \\ &\le c_{1,2,0}\, h^2 \| D_t^2 f \|_\infty + c_{1,2,0}\, \ell^2 \| D_u^2 f \|_\infty . \qquad \blacksquare \end{aligned}$$

Theorem 5.5.3. Let $f(t,u) \in PC^{4,\infty}([a,b] \times [c,d])$. Then,

$$(5.5.7) \qquad \| f - H_2^\rho f \|_\infty \le c_{2,4,0}\, h^4 \| D_t^4 f \|_\infty + c_{2,2,0}^2\, h^2 \ell^2 \| D_u^2 D_t^2 f \|_\infty + c_{2,4,0}\, \ell^4 \| D_u^4 f \|_\infty$$

and

$$(5.5.8) \quad \| D_t(f - H_2^\rho f) \|_\infty \le c_{2,4,1}\, h^3 \| D_t^4 f \|_\infty + c_{2,2,0} c_{2,2,1}\, h\ell^2 \| D_u^2 D_t^2 f \|_\infty + c_{2,3,0}\, \ell^3 \| D_u^3 D_t f \|_\infty ,$$

where the constants $c_{2,n,k}$ are defined in Table 5.3.4.

Proof. Since as a function of t, $(f - H_2^{\Delta'} f) \in PC^{2,\infty}[a,b]$, from (5.5.4) and Theorem 5.3.7 it follows that

$$\| f - H_2^{\rho} f \|_\infty \le \| f - H_2^{\Delta} f \|_\infty + \| H_2^{\Delta}(f - H_2^{\Delta'} f) - (f - H_2^{\Delta'} f) \|_\infty$$

$$+ \ \| f - H_2^{\Delta'} f \|_\infty$$

$$\le c_{2,4,0}\, h^4 \, \| D_t^4 f \|_\infty + c_{2,2,0}\, h^2 \, \| D_t^2 (f - H_2^{\Delta'} f) \|_\infty \tag{5.5.9}$$

$$+ \ c_{2,4,0}\, \ell^4 \, \| D_u^4 f \|_\infty \ .$$

Now since as a function of u, $D_t^2 f \in PC^{2,\infty}[c,d]$, and $D_t^2 H_2^{\Delta'} f = H_2^{\Delta'} D_t^2 f$, Theorem 5.3.7 can be used to obtain

$$\| D_t^2 (f - H_2^{\Delta'} f) \|_\infty \le c_{2,2,0}\, \ell^2 \, \| D_u^2 D_t^2 f \|_\infty \ . \tag{5.5.10}$$

Using (5.5.10) in (5.5.9) we find (5.5.7). Finally, since

$$\| D_t (f - H_2^{\rho} f) \|_\infty \le \| D_t (f - H_2^{\Delta} f) \|_\infty + \| D_t [H_2^{\Delta}(f - H_2^{\Delta'} f) - (f - H_2^{\Delta'} f)] \|_\infty$$

$$+ \ \| D_t (f - H_2^{\Delta'} f) \|_\infty$$

the proof of (5.5.8) is similar to that of (5.5.7). ∎

Theorem 5.5.4. Let $f(t,u) \in PC^{4,\infty}([a,b] \times [c,d])$. Then,

$$\| f - H_3^{\rho} f \|_\infty \le c_{3,4,0}\, h^4 \, \| D_t^4 f \|_\infty + c_{3,2,0}^2\, h^2 \ell^2 \, \| D_u^2 D_t^2 f \|_\infty \tag{5.5.11}$$

$$+ \ c_{3,4,0}\, \ell^4 \, \| D_u^4 f \|_\infty$$

and

$$\| D_t (f - H_3^{\rho} f) \|_\infty \le c_{3,4,1}\, h^3 \, \| D_t^4 f \|_\infty + c_{3,2,0} c_{3,2,1}\, h \ell^2 \, \| D_u^2 D_t^2 f \|_\infty \tag{5.5.12}$$

$$+ \ c_{3,3,0}\, \ell^3 \, \| D_u^3 D_t f \|_\infty \ ,$$

where the constants $c_{3,n,k}$ are defined in Table 5.3.5.

Proof. The proof is similar to that of Theorem 5.5.3. ∎

Theorem 5.5.5. Let $f(t,u) \in PC^{5,\infty}([a,b] \times [c,d])$. Then,

$$\| f - H_3^{\rho} f \|_\infty \le c_{3,5,0}\, h^5 \, \| D_t^5 f \|_\infty + c_{3,2,0} c_{3,3,0}\, h^3 \ell^2 \, \| D_u^2 D_t^3 f \|_\infty \tag{5.5.13}$$

$$+ c_{3,5,0}\ \ell^5 \parallel D_u^5 f \parallel_\infty$$

(5.5.14) $$\parallel f - H_3^\rho f \parallel_\infty \leq c_{3,5,0}\ h^5 \parallel D_t^5 f \parallel_\infty + c_{3,2,0}c_{3,3,0}\ h^2\ell^3 \parallel D_u^3 D_t^2 f \parallel_\infty + c_{3,5,0}\ \ell^5 \parallel D_u^5 f \parallel_\infty$$

(5.5.15) $$\parallel D_t(f - H_3^\rho f) \parallel_\infty \leq c_{3,5,1}\ h^4 \parallel D_t^5 f \parallel_\infty + c_{3,2,0}c_{3,3,1}\ h^2\ell^2 \parallel D_u^2 D_t^3 f \parallel_\infty + c_{3,4,0}\ \ell^4 \parallel D_u^4 D_t f \parallel_\infty$$

(5.5.16) $$\parallel D_t(f - H_3^\rho f) \parallel_\infty \leq c_{3,5,1}\ h^4 \parallel D_t^5 f \parallel_\infty + c_{3,2,1}c_{3,3,0}\ h\ell^3 \parallel D_u^3 D_t^2 f \parallel_\infty + c_{3,4,0}\ \ell^4 \parallel D_u^4 D_t f \parallel_\infty$$

and

(5.5.17) $$\parallel D_t^2(f - H_3^\rho f) \parallel_\infty \leq c_{3,5,2}\ h^3 \parallel D_t^5 f \parallel_\infty + c_{3,2,0}c_{3,3,2}\ h\ell^2 \parallel D_u^2 D_t^3 f \parallel_\infty + c_{3,3,0}\ \ell^3 \parallel D_u^3 D_t^2 f \parallel_\infty ,$$

where the constants $c_{3,n,k}$ are defined in Table 5.3.5.

Proof. The proof is similar to that of Theorem 5.5.3. ∎

Theorem 5.5.6. Let $f(t,u) \in PC^{6,\infty}([a,b] \times [c,d])$. Then,

(5.5.18) $$\parallel f - H_3^\rho f \parallel_\infty \leq c_{3,6,0}\ h^6 \parallel D_t^6 f \parallel_\infty + c_{3,2,0}c_{3,4,0}\ h^4\ell^2 \parallel D_u^2 D_t^4 f \parallel_\infty + c_{3,6,0}\ \ell^6 \parallel D_u^6 f \parallel_\infty$$

(5.5.19) $$\parallel f - H_3^\rho f \parallel_\infty \leq c_{3,6,0}\ h^6 \parallel D_t^6 f \parallel_\infty + c_{3,3,0}^2\ h^3\ell^3 \parallel D_u^3 D_t^3 f \parallel_\infty + c_{3,6,0}\ \ell^6 \parallel D_u^6 f \parallel_\infty$$

(5.5.20) $$\parallel f - H_3^\rho f \parallel_\infty \leq c_{3,6,0}\ h^6 \parallel D_t^6 f \parallel_\infty + c_{3,2,0}c_{3,4,0}\ h^2\ell^4 \parallel D_u^4 D_t^2 f \parallel_\infty + c_{3,6,0}\ \ell^6 \parallel D_u^6 f \parallel_\infty$$

(5.5.21) $$\parallel D_t(f - H_3^\rho f) \parallel_\infty \leq c_{3,6,1}\ h^5 \parallel D_t^6 f \parallel_\infty + c_{3,2,0}c_{3,4,1}\ h^3\ell^2 \parallel D_u^2 D_t^4 f \parallel_\infty + c_{3,5,0}\ \ell^5 \parallel D_u^5 D_t f \parallel_\infty$$

(5.5.22) $$\parallel D_t(f - H_3^\rho f) \parallel_\infty \leq c_{3,6,1}\ h^5 \parallel D_t^6 f \parallel_\infty + c_{3,3,0}c_{3,3,1}\ h^2\ell^3 \parallel D_u^3 D_t^3 f \parallel_\infty$$

$$+ c_{3,5,0}\ \ell^5 \parallel D_u^5 D_t f \parallel_\infty$$

(5.5.23) $\parallel D_t(f - H_3^\rho f) \parallel_\infty \leq c_{3,6,1}\ h^5 \parallel D_t^6 f \parallel_\infty + c_{3,2,1} c_{3,4,0}\ h\ell^4 \parallel D_u^4 D_t^2 f \parallel_\infty$

$$+ c_{3,5,0}\ \ell^5 \parallel D_u^5 D_t f \parallel_\infty$$

(5.5.24) $\parallel D_t^2(f - H_3^\rho f) \parallel_\infty \leq c_{3,6,2}\ h^4 \parallel D_t^6 f \parallel_\infty + c_{3,2,0} c_{3,4,2}\ h^2\ell^2 \parallel D_u^2 D_t^4 f \parallel_\infty$

$$+ c_{3,4,0}\ \ell^4 \parallel D_u^4 D_t^2 f \parallel_\infty$$

(5.5.25) $\parallel D_t^2(f - H_3^\rho f) \parallel_\infty \leq c_{3,6,2}\ h^4 \parallel D_t^6 f \parallel_\infty + c_{3,3,0} c_{3,3,2}\ h\ell^3 \parallel D_u^3 D_t^3 f \parallel_\infty$

$$+ c_{3,4,0}\ \ell^4 \parallel D_u^4 D_t^2 f \parallel_\infty$$

and

(5.5.26) $\parallel D_t^3(f - H_3^\rho f) \parallel_\infty \leq c_{3,6,3}\ h^3 \parallel D_t^6 f \parallel_\infty + c_{3,2,0} c_{3,4,3}\ h\ell^2 \parallel D_u^2 D_t^4 f \parallel_\infty$

$$+ c_{3,3,0}\ \ell^3 \parallel D_u^3 D_t^3 f \parallel_\infty ,$$

where the constants $c_{3,n,k}$ are defined in Table 5.3.5.

Proof. The proof is similar to that of Theorem 5.5.3. ∎

Now we shall provide error bounds in L_2 - norm.

Theorem 5.5.7. Let $f(t,u) \in PC^{2,2}([a,b] \times [c,d])$. Then,

(5.5.27) $$\parallel f - H_1^\rho f \parallel_2 \leq \bar{c}_{1,2,0}\ h^2 \parallel D_t^2 f \parallel_2 + \bar{c}_{1,1,0}^2\ h\ell \parallel D_u D_t f \parallel_2$$

$$+ \bar{c}_{1,2,0}\ \ell^2 \parallel D_u^2 f \parallel_2$$

and

(5.5.28) $$\parallel D_t(f - H_1^\rho f) \parallel_2 \leq \bar{c}_{1,2,1}\ h \parallel D_t^2 f \parallel_2 + 2\ \bar{c}_{1,1,0}\ \ell \parallel D_u D_t f \parallel_2 ,$$

where the constants $\bar{c}_{1,n,k}$ are defined in Table 5.3.10.

Proof. From (5.5.4), the triangle inequality and Theorem 5.3.13, we have

$$\parallel f - H_1^\rho f \parallel_2 \leq \parallel f - H_1^\Delta f \parallel_2 + \parallel H_1^\Delta(f - H_1^{\Delta'} f) - (f - H_1^{\Delta'} f) \parallel_2$$

$$+ \parallel f - H_1^{\Delta'} f \parallel_2$$

$$(5.5.29)\quad \le \bar{c}_{1,2,0}\, h^2 \parallel D_t^2 f \parallel_2 + \bar{c}_{1,1,0}\, h \parallel D_t(f - H_1^{\Delta'} f) \parallel_2 + \bar{c}_{1,2,0}\, \ell^2 \parallel D_u^2 f \parallel_2 .$$

Further, since $D_t H_1^{\Delta'} f = H_1^{\Delta'} D_t f$, we find that

$$\parallel D_t(f - H_1^{\Delta'} f) \parallel_2 \le \bar{c}_{1,1,0}\, \ell \parallel D_u D_t f \parallel_2 .$$

Using the above inequality in (5.5.29) we obtain (5.5.27).

Once again from (5.5.4), the triangle inequality and Theorem 5.3.13, we have

$$\begin{aligned}
&\parallel D_t(f - H_1^{\rho} f) \parallel_2 \\
&\quad \le \parallel D_t(f - H_1^{\Delta} f) \parallel_2 + \parallel D_t[H_1^{\Delta}(f - H_1^{\Delta'} f) - (f - H_1^{\Delta'} f)] \parallel_2 \\
&\qquad + \parallel D_t(f - H_1^{\Delta'} f) \parallel_2 \\
&\quad \le \bar{c}_{1,2,1}\, h \parallel D_t^2 f \parallel_2 + 2 \parallel D_t(f - H_1^{\Delta'} f) \parallel_2 \\
&\quad \le \bar{c}_{1,2,1}\, h \parallel D_t^2 f \parallel_2 + 2\, \bar{c}_{1,1,0}\, \ell \parallel D_u D_t f \parallel_2 . \quad \blacksquare
\end{aligned}$$

Theorem 5.5.8. Let $f(t,u) \in PC^{4,2}([a,b] \times [c,d])$. Then,

$$(5.5.30)\qquad \parallel f - H_2^{\rho} f \parallel_2 \le \bar{c}_{2,4,0}\, h^4 \parallel D_t^4 f \parallel_2 + \bar{c}_{2,2,0}^2\, h^2 \ell^2 \parallel D_u^2 D_t^2 f \parallel_2 + \bar{c}_{2,4,0}\, \ell^4 \parallel D_u^4 f \parallel_2$$

$$(5.5.31)\quad \parallel D_t(f - H_2^{\rho} f) \parallel_2 \le \bar{c}_{2,4,1}\, h^3 \parallel D_t^4 f \parallel_2 + \bar{c}_{2,2,0} \bar{c}_{2,2,1}\, h \ell^2 \parallel D_u^2 D_t^2 f \parallel_2 + \bar{c}_{2,3,0}\, \ell^3 \parallel D_u^3 D_t f \parallel_2$$

and

$$(5.5.32)\quad \parallel D_t^2(f - H_2^{\rho} f) \parallel_2 \le \bar{c}_{2,4,2}\, h^2 \parallel D_t^4 f \parallel_2 + 2\, \bar{c}_{2,2,0}\, \ell^2 \parallel D_u^2 D_t^2 f \parallel_2 ,$$

where the constants $\bar{c}_{2,n,k}$ are defined in Table 5.3.11.

Proof. The proof is similar to that of Theorems 5.5.3 and 5.5.7. $\blacksquare$

Theorem 5.5.9. Let $f(t,u) \in PC^{6,2}([a,b] \times [c,d])$. Then,

$$(5.5.33) \qquad \| f - H_3^\rho f \|_2 \le \bar{c}_{3,6,0}\; h^6 \| D_t^6 f \|_2 + \bar{c}_{3,3,0}^2\; h^3 \ell^3 \| D_u^3 D_t^3 f \|_2 + \bar{c}_{3,6,0}\; \ell^6 \| D_u^6 f \|_2$$

$$(5.5.34) \qquad \| D_t(f - H_3^\rho f) \|_2 \le \bar{c}_{3,6,1}\; h^5 \| D_t^6 f \|_2 + \bar{c}_{3,3,0}\bar{c}_{3,3,1}\; h^2 \ell^3 \| D_u^3 D_t^3 f \|_2 + \bar{c}_{3,5,0}\; \ell^5 \| D_u^5 D_t f \|_2$$

$$(5.5.35) \qquad \| D_t^2(f - H_3^\rho f) \|_2 \le \bar{c}_{3,6,2}\; h^4 \| D_t^6 f \|_2 + \bar{c}_{3,3,0}\bar{c}_{3,3,2}\; h\ell^3 \| D_u^3 D_t^3 f \|_2 + \bar{c}_{3,4,0}\; \ell^4 \| D_u^4 D_t^2 f \|_2$$

and

$$(5.5.36) \qquad \| D_t^3(f - H_3^\rho f) \|_2 \le \bar{c}_{3,6,3}\; h^3 \| D_t^6 f \|_2 + 2\, \bar{c}_{3,3,0}\; \ell^3 \| D_u^3 D_t^3 f \|_2 ,$$

where the constants $\bar{c}_{3,n,k}$ are defined in Table 5.3.12.

Proof. The proof is similar to that of Theorems 5.5.3 and 5.5.7. ∎

A result which includes the above Theorems 5.5.7 - 5.5.9 is the following :

Theorem 5.5.10. Let $f(t,u) \in PC^{2m,2}([a,b] \times [c,d])$. Then,

$$(5.5.37) \qquad \| D_t^k(f - H_m^\rho f) \|_2 \le \bar{c}_{m,2m,k}\; h^{2m-k} \| D_t^{2m} f \|_2 + \bar{c}_{m,m,0}\bar{c}_{m,m,k}\; h^{m-k}\ell^m \| D_u^m D_t^m f \|_2 + \bar{c}_{m,2m-k,0}\; \ell^{2m-k} \| D_u^{2m-k} D_t^k f \|_2 , \quad 0 \le k \le m-1$$

and

$$(5.5.38) \qquad \| D_t^m(f - H_m^\rho f) \|_2 \le \bar{c}_{m,2m,m}\; h^m \| D_t^{2m} f \|_2 + 2\, \bar{c}_{m,m,0}\; \ell^m \| D_u^m D_t^m f \|_2 ,$$

where the constants $\bar{c}_{m,n,k}$ are defined in Theorem 5.3.16.

Proof. The proof is similar to that of Theorem 5.5.7. ∎

Finally, we note that similar bounds for $\| D_u^k(f - H_m^\rho f) \|_\infty$ and $\| D_u^k(f - H_m^\rho f) \|_2$ can be obtained by interchanging t and u, and h and ℓ in the previous theorems.

5.6 TWO VARIABLE PIECEWISE LIDSTONE INTERPOLATION

For a fixed ρ, we define the set $L_m(\rho)$ as follows :

$$\begin{aligned} L_m(\rho) &= L_m(\Delta) \oplus L_m(\Delta') \quad \text{(the tensor product)} \\ &= \operatorname{Span}\{r_{m,i,\mu}(t) r_{m,j,\nu}(u)\}_{i=0}^{N+1}\ {}_{\mu=0}^{m-1}\ {}_{j=0}^{M+1}\ {}_{\nu=0}^{m-1} \\ &= \Big\{ h(t,u) \in C([a,b]\times[c,d]) : h(t,u) \text{ is a two - dimensional} \\ &\quad \text{polynomial of degree at most } (2m-1) \text{ in each variable} \\ &\quad \text{and in each subrectangle } [t_i, t_{i+1}] \times [u_j, u_{j+1}]; \\ &\qquad\qquad 0 \le i \le N,\ 0 \le j \le M \Big\}. \end{aligned}$$

Since $L_m(\rho)$ is the tensor product of $L_m(\Delta)$ and $L_m(\Delta')$ which are of dimensions $[2m(N+1)-N]$ and $[2m(M+1)-M]$ respectively, $L_m(\rho)$ is of dimension $[2m(N+1)-N] \times [2m(M+1)-M]$.

Definition 5.6.1. For a given $f(t,u) \in C^{(2m-2,2m-2)}([a,b]\times[c,d])$, we say $L_m^\rho f(t,u)$ is the $L_m(\rho)$ - *interpolate of* $f(t,u)$, also known as the *two - dimensional Lidstone interpolate of* $f(t,u)$ if $L_m^\rho f(t,u) \in L_m(\rho)$ with $D_t^{2\mu} D_u^{2\nu} L_m^\rho f(t_i, u_j) = f_{i,j}^{(2\mu,2\nu)};\ 0 \le i \le N+1,\ 0 \le j \le M+1,\ 0 \le \mu,\nu \le m-1$.

For $f(t,u) \in C^{(2m-2,2m-2)}([a,b]\times[c,d])$ it is clear that $L_m^\rho f(t,u)$ uniquely exists and can be explicitly expressed as

$$L_m^\rho f(t,u) = \sum_{i=0}^{N+1} \sum_{\mu=0}^{m-1} \sum_{j=0}^{M+1} \sum_{\nu=0}^{m-1} r_{m,i,\mu}(t) r_{m,j,\nu}(u) f_{i,j}^{(2\mu,2\nu)}. \tag{5.6.1}$$

The following result provides a characterization of $L_m^\rho f(t,u)$ in terms of one - dimensional interpolation schemes.

Lemma 5.6.1. If $f(t,u) \in C^{(2m-2,2m-2)}([a,b]\times[c,d])$, then

$$L_m^\rho f(t,u) = L_m^{\Delta'} L_m^{\Delta} f(t,u) = L_m^{\Delta} L_m^{\Delta'} f(t,u). \tag{5.6.2}$$

Proof. The proof is similar to that of Lemma 5.5.1. ∎

Now let $f(t,u) \in C^{(2m-2,2m-2)}([a,b]\times[c,d])$ be an arbitrary function. From Lemma 5.6.1 we have

$$f - L^{\rho}_m f = (f - L^{\Delta}_m f) + L^{\Delta}_m (f - L^{\Delta'}_m f) \tag{5.6.3}$$

$$= (f - L^{\Delta}_m f) + \left[L^{\Delta}_m (f - L^{\Delta'}_m f) - (f - L^{\Delta'}_m f)\right] + (f - L^{\Delta'}_m f) \tag{5.6.4}$$

$$= (f - L^{\Delta}_m f) + \left[L^{\Delta'}_m (f - L^{\Delta}_m f) - (f - L^{\Delta}_m f)\right] + (f - L^{\Delta'}_m f). \tag{5.6.5}$$

We shall use the above relations to acquire the following error estimates.

Theorem 5.6.2. Let $f(t,u) \in PC^{2m-2,2m-2,\infty}([a,b] \times [c,d])$. Then,

$$\begin{aligned} \| D^k_t (f - L^{\rho}_m f) \|_\infty & \\ \le 2\ d_{2m-2,k}\ h^{2m-2-k} \| D^{2m-2}_t f \|_\infty & \\ + 4\ d_{2m-2,k} d_{2m-2,0}\ h^{2m-2-k} \ell^{2m-2} \| D^{2m-2}_u D^{2m-2}_t f \|_\infty & \\ + 2\ d_{2m-2,0}\ \ell^{2m-2} \| D^{2m-2}_u D^k_t f \|_\infty\ , \quad 0 \le k \le 2m-2 & \end{aligned} \tag{5.6.6}$$

where the constants $d_{2m-2,k}$ are defined in (5.4.5).

Proof. Since as a function of t, $(f - L^{\Delta'}_m f) \in PC^{2m-2,\infty}[a,b]$, from (5.6.4) and Theorem 5.4.3 it follows that

$$\begin{aligned} \| D^k_t (f - L^{\rho}_m f) \|_\infty &\le \| D^k_t (f - L^{\Delta}_m f) \|_\infty + \| D^k_t [L^{\Delta}_m (f - L^{\Delta'}_m f) - (f - L^{\Delta'}_m f)] \|_\infty \\ &\quad + \| D^k_t (f - L^{\Delta'}_m f) \|_\infty \end{aligned}$$

$$\begin{aligned} &\le 2\ d_{2m-2,k}\ h^{2m-2-k} \| D^{2m-2}_t f \|_\infty \\ &\quad + 2\ d_{2m-2,k}\ h^{2m-2-k} \| D^{2m-2}_t (f - L^{\Delta'}_m f) \|_\infty \\ &\quad + \| D^k_t (f - L^{\Delta'}_m f) \|_\infty\ . \end{aligned} \tag{5.6.7}$$

Now since as a function of u, for each $0 \le k \le 2m-2$, $D^k_t f \in PC^{2m-2,\infty}[c,d]$, and $D^k_t L^{\Delta'}_m f = L^{\Delta'}_m D^k_t f$, Theorem 5.4.3 can be used again to obtain

$$\| D^k_t (f - L^{\Delta'}_m f) \|_\infty \le 2\ d_{2m-2,0}\ \ell^{2m-2} \| D^{2m-2}_u D^k_t f \|_\infty\ , \quad 0 \le k \le 2m-2. \tag{5.6.8}$$

Now using (5.6.8) in (5.6.7), we obtain (5.6.6). ∎

Theorem 5.6.3. Let $f(t,u) \in PC^{1,\infty}([a,b] \times [c,d])$. Then,

$$\| f - L^{\rho}_1 f \|_\infty \le \frac{1}{2} (h \| D_t f \|_\infty + \ell \| D_u f \|_\infty). \tag{5.6.9}$$

Proof. Since $L_1^\rho f = H_1^\rho f$, following as in Theorem 5.5.2 and using Theorem 5.4.4, we find that

$$\begin{aligned} \| f - L_1^\rho f \|_\infty &\le \| f - L_1^\Delta f \|_\infty + \| f - L_1^{\Delta'} f \|_\infty \\ &\le \frac{1}{2}\, h \| D_t f \|_\infty + \frac{1}{2}\, \ell \| D_u f \|_\infty . \quad \blacksquare \end{aligned}$$

Theorem 5.6.4. Let $f(t,u) \in PC^{2m-1,2m-1,\infty}([a,b] \times [c,d]),\ m = 2,3$. Then,

(5.6.10) $\| D_t^k (f - L_m^\rho f) \|_\infty$

$$\begin{aligned} &\le e_{2m-1,k}\, h^{2m-1-k} \| D_t^{2m-1} f \|_\infty \\ &\quad + e_{2m-1,k} e_{2m-1,0}\, h^{2m-1-k} \ell^{2m-1} \| D_u^{2m-1} D_t^{2m-1} f \|_\infty \\ &\quad + e_{2m-1,0}\, \ell^{2m-1} \| D_u^{2m-1} D_t^k f \|_\infty ; \quad 0 \le k \le 2m-2,\ m = 2,3 \end{aligned}$$

where the constants $e_{2m-1,k}$ are defined in Table 5.4.1.

Proof. The proof is similar to that of Theorem 5.6.2. ∎

Theorem 5.6.5. Let $f(t,u) \in PC^{2m,2m,\infty}([a,b] \times [c,d])$. Then,

(5.6.11) $\| D_t^k (f - L_m^\rho f) \|_\infty$

$$\begin{aligned} &\le d_{2m,k}\, h^{2m-k} \| D_t^{2m} f \|_\infty \\ &\quad + d_{2m,k} d_{2m,0}\, h^{2m-k} \ell^{2m} \| D_u^{2m} D_t^{2m} f \|_\infty \\ &\quad + d_{2m,0}\, \ell^{2m} \| D_u^{2m} D_t^k f \|_\infty , \quad 0 \le k \le 2m-1 \end{aligned}$$

where the constants $d_{2m,k}$ are defined in (5.4.5).

Proof. The proof is similar to that of Theorem 5.6.2. ∎

Theorem 5.6.6. Let $f(t,u) \in PC^{2m-1,2m-1,2}([a,b] \times [c,d])$. Then,

$$\begin{aligned} (5.6.12) \qquad \| D_t^k (f - L_m^\rho f) \|_2 &\le \frac{1}{\pi^{2m-1-k}}\, h^{2m-1-k} \| D_t^{2m-1} f \|_2 \\ &\quad + \frac{1}{\pi^{2(2m-1)-k}}\, h^{2m-1-k} \ell^{2m-1} \| D_u^{2m-1} D_t^{2m-1} f \|_2 \\ &\quad + \frac{1}{\pi^{2m-1}}\, \ell^{2m-1} \| D_u^{2m-1} D_t^k f \|_2 , \quad 0 \le k \le 2m-1. \end{aligned}$$

Proof. Following the proof of Theorem 5.6.2 and using Theorem 5.4.5, we find that

$$\| D_t^k(f - L_m^\rho f) \|_2 \le \| D_t^k(f - L_m^\Delta f) \|_2 + \| D_t^k[L_m^\Delta(f - L_m^{\Delta'} f) - (f - L_m^{\Delta'} f)] \|_2$$

$$+ \ \| D_t^k(f - L_m^{\Delta'} f) \|_2$$

$$(5.6.13) \qquad \le \frac{1}{\pi^{2m-1-k}} \, h^{2m-1-k} \, \| D_t^{2m-1} f \|_2$$

$$+ \ \frac{1}{\pi^{2m-1-k}} \, h^{2m-1-k} \, \| D_t^{2m-1}(f - L_m^{\Delta'} f) \|_2$$

$$+ \| D_t^k(f - L_m^{\Delta'} f) \|_2 \ .$$

Further, since

$$(5.6.14) \qquad \| D_t^k(f - L_m^{\Delta'} f) \|_2 \le \frac{1}{\pi^{2m-1}} \, \ell^{2m-1} \, \| D_u^{2m-1} D_t^k f \|_2 \, ,$$

$$0 \le k \le 2m - 1$$

(5.6.12) follows on combining (5.6.13) and (5.6.14). ∎

Theorem 5.6.7. Let $f(t,u) \in PC^{2m,2m,2}([a,b] \times [c,d])$. Then,

$$(5.6.15) \ \| D_t^k(f - L_m^\rho f) \|_2 \le \frac{1}{\pi^{2m-k}} \, h^{2m-k} \, \| D_t^{2m} f \|_2$$

$$+ \ \frac{1}{\pi^{4m-k}} \, h^{2m-k} \ell^{2m} \, \| D_u^{2m} D_t^{2m} f \|_2$$

$$+ \ \frac{1}{\pi^{2m}} \, \ell^{2m} \, \| D_u^{2m} D_t^k f \|_2 \, , \quad 0 \le k \le 2m.$$

Proof. The proof is similar to that of Theorem 5.6.6. ∎

Finally, we remark that similar bounds for $\| D_u^k(f - L_m^\rho f) \|_\infty$ and $\| D_u^k(f - L_m^\rho f) \|_2$ can be obtained by interchanging t and u, and h and ℓ in the previous results.

REFERENCES

1. **R.P.Agarwal** and **R.C.Gupta,** Essentials of Ordinary Differential Equations, McGraw-Hill, Singapore, 1991.

2. **R.P.Agarwal** and **P.J.Y.Wong,** Explicit error bounds for the derivatives of piecewise - Hermite interpolation in L_2 - norm, to appear.

3. **R.P.Agarwal** and **P.J.Y.Wong,** Explicit error bounds for the derivatives of piecewise - Lidstone interpolation, to appear.

4. **J.H.Ahlberg, E.N.Nilson** and **J.L.Walsh,** Convergence properties of generalized splines, Proc. Nat. Acad. Sci., U.S.A. 54(1965), 344-350.

5. **J.H.Ahlberg, E.N.Nilson** and **J.L.Walsh,** The Theory of Splines and their Applications, Academic Press, New York, 1967.

6. **E.Beckenbach** and **R.Bellman,** Inequalities, Springer-Verlag, Berlin, 1965.

7. **R.Bellman,** A note on an inequality of E. Schmidt, Bull. Amer. Math. Soc. 50(1944), 734-736.

8. **G.Birkhoff, M.H.Schultz** and **R.S.Varga,** Piecewise Hermite interpolation in one and two variables with applications to partial differential equations, Numerische Mathematik 11(1968), 232-256.

9. **J.Brink,** Inequalities involving $\| f \|_p$ and $\| f^{(n)} \|_q$ for f with n zeros, Pacific J. Math. 42(1972), 289-311.

10. **P.G.Ciarlet, M.H.Schultz** and **R.S.Varga,** Numerical methods of high order accuracy for nonlinear boundary value problems, I. one dimensional problem, Numerische Mathematik 9(1967), 394-430.

11. **P.G.Ciarlet, M.H.Schultz** and **R.S.Varga,** Numerical methods of high order accuracy for nonlinear boundary value problems, II. nonlinear boundary conditions, Numerische Mathematik 11(1968), 331-345.

12. **G.Cimmino,** Su una questione di minimo, Boll. Un. Mat. Ital 9(1930), 1-6.

13. **K.M.Das** and **A.S.Vatsala,** Green's function for n - n boundary value problem and an analogue of Hartman's result, J. Math. Anal. Appl. 51(1975), 670-677.

14. **P.Dörfler,** New inequalities of Markov type, SIAM J. Math. Anal. 18(1987), 490-494.

15. **P.Forster,** Existenzaussagen und Fehlerabschätzungen bei gewissen nichtlinearen Randwertaufgaben mit gewöhnlichen Differentialgleichungen, Numerishe Mathematik 10(1967), 410-422.

16. **P.Goetgheluck,** On the Markov inequality in L^p - spaces, J. Approximation Theory 62(1990), 197-205.

17. **G.H.Hardy, J.E.Littlewood** and **G.Polya,** Inequalities, Cambridge, 1934.

18. **E.Hille, G.Szegö** and **J.D.Tamarkin,** On some generalization of a theorem of A. Markoff, Duke Math. J. 3(1937), 729-739.

19. **D.S.Mitrinović, J.E.Pečarić** and **A.M.Fink,** Inequalities Involving Functions and their Integrals and Derivatives, Kluwer Academic Publishers, Dordrecht, 1991.

20. **E.Schmidt,** Über die nebst ihren Ableitungen orthogonalen Polynomensysteme und das zugehörige Extremum, Math. Ann. 119(1944), 165-204.

21. **M.H.Schultz,** Spline Analysis, Prentice-Hall, Englewood Cliffs, N.J., 1973.

22. **A.K.Varma** and **K.L.Katsifarakis,** Optimal error bounds for Hermite interpolation, J. Approximation Theory 51(1987), 350-359.

23. **P.J.Y.Wong** and **R.P.Agarwal,** Explicit error estimates for quintic and biquintic spline interpolation, Computers Math. Applic. 18(1989), 701-722.

CHAPTER 6

SPLINE INTERPOLATION

6.1 INTRODUCTION

Spline interpolation is an improvement over piecewise - polynomial interpolation. It uses less information of the given function, yet furnishes smoother interpolates. The plan of this chapter is as follows : In Section 6.2 we shall define the spline space $S_{m,\tau}(\Delta)$, and for a given function $x(t)$ the spline and Lidstone - spline interpolates $S^{\Delta}_{m,\tau}x(t)$ and $LS^{\Delta}_{m,2m-2}x(t)$, respectively. Here, we shall also show that to acquire the bounds for $\| D^k(x - S^{\Delta}_{m,\tau}x) \|_\infty$ and $\| D^k(x - LS^{\Delta}_{m,2m-2}x) \|_\infty$ in terms of the derivatives of $x(t)$ it is necessary to estimate several terms. While some of these terms can be estimated by the results of Chapter 5, other terms which require a different analysis for each m and τ, need to be bounded. In Sections 6.3, 6.4 and 6.6 respectively, we shall consider the cases $m = 2,\ \tau = 2;\ m = 3,\ \tau = 4$ and $m = 3,\ \tau = 3$. These cases correspond to cubic in the class $C^{(2)}[a,b]$, and quintic in the classes $C^{(4)}[a,b]$ and $C^{(3)}[a,b]$ spline interpolates. For the cases $m = 3,\ \tau = 4$ and

$m = 3$, $\tau = 3$ in Sections 6.5 and 6.7 respectively, we shall discuss the construction of approximated quintic splines, and for these interpolates we will provide explicit error bounds in L_∞ - norm. For the cubic and quintic Lidstone - spline interpolates error bounds in L_∞ - norm are obtained in Sections 6.8 and 6.9, respectively. In Section 6.10 we shall extend Theorems 5.3.16 and 5.3.17 for the spline interpolate $S^{\Delta}_{m,\tau}x(t)$. Results of Sections 6.3, 6.4, 6.6, 6.10 and of Sections 6.8, 6.9 are generalized to two variables in Sections 6.11 and 6.12, respectively. Finally, in Section 6.13 we shall present some applications of the results established in the previous sections.

We remark that the asymptotic error bounds for more general spline interpolates compare to what we consider in this chapter are known, e.g., [2,3,9,11, 13,15-17]; however, the bounds obtained here are explicitly calculated and these improve on, or supplement, those given by Hall [7] and Schultz [12,14].

6.2 PRELIMINARIES

For a fixed Δ and an integer τ such that $m-1 \le \tau \le 2m-2$, we define the spline space $S_{m,\tau}(\Delta) = \{s(t) \in C^{(\tau)}[a,b] : s(t)$ is a polynomial of degree at most $(2m-1)$ in each subinterval $[t_i, t_{i+1}]$, $0 \le i \le N\}$. It is clear that $S_{m,\tau}(\Delta)$ is of dimension $[(2m-1-\tau)N + 2m]$.

Definition 6.2.1. For a given $x(t) \in C^{(m-1)}[a,b]$ we say $S^{\Delta}_{m,\tau}x(t)$ is the $S_{m,\tau}(\Delta)$ - *interpolate of* $x(t)$, also known as *spline interpolate of* $x(t)$ if $S^{\Delta}_{m,\tau}x(t) \in S_{m,\tau}(\Delta)$ with

$$(6.2.1) \quad D^k S^{\Delta}_{m,\tau}x(t_i) = x^{(k)}(t_i) = x_i^{(k)}; \; 1 \le i \le N, \; 0 \le k \le 2m-2-\tau$$
$$i = 0, N+1, \; 0 \le k \le m-1.$$

The above definitions of the spline space and the spline interpolate are due to Schoenberg [10]. Further, the mapping $S^{\Delta}_{m,\tau} : C^{(m-1)}[a,b] \to S_{m,\tau}(\Delta)$ corresponds to the Type 1 interpolation of Ahlberg, Nilson and Walsh [2].

It is clear that for $\tau = m-1$, $S_{m,m-1}(\Delta) = H_m(\Delta)$ and $S^{\Delta}_{m,m-1}x(t) \equiv$

$H^{\Delta}_{m}x(t)$, which has been the subject matter of our previous chapter. For $m-1<\tau\le 2m-2$ we have the inclusion relation $S_{m,\tau}(\Delta)\subset H_m(\Delta)$. Thus, we can represent every $S^{\Delta}_{m,\tau}x(t)$ in terms of the basic elements $h_{m,i,j}(t)$; $0\le i\le N+1$, $0\le j\le m-1$ of $H_m(\Delta)$ even though these functions may not belong to $S_{m,\tau}(\Delta)$. In fact, we have

$$(6.2.2)\qquad S^{\Delta}_{m,\tau}x(t)=\sum_{i=0}^{N+1}\sum_{j=0}^{m-1}h_{m,i,j}(t)\ D^jS^{\Delta}_{m,\tau}x(t_i).$$

It is known [2,15] that in (6.2.2), $D^jS^{\Delta}_{m,\tau}x(t_i)$, $1\le i\le N$, $2m-1-\tau\le j\le m-1$ exist uniquely. In fact, for $m=2$ and 3 we shall show that these unknown constants are the solutions of diagonally dominant systems of linear algebraic equations, and hence can be obtained explicitly in terms of known quantities.

As in Chapter 5 here also we note that in the above Definition 6.2.1 as well as in the representation (6.2.2) for the function $x(t)$ it is sufficient to assume that $x_i^{(k)}$; $1\le i\le N$, $0\le k\le 2m-2-\tau$, $i=0,N+1$, $0\le k\le m-1$ exist.

Let $x(t)\in PC^{n,\infty}[a,b]$, $m-1\le n\le 2m$. To acquire upper bounds for $\| D^k(x-S^{\Delta}_{m,\tau}x)\|_{\infty}$, $0\le k\le n-1$ in terms of $\| D^nx\|_{\infty}$ we begin with the equality

$$(6.2.3)\qquad x(t)-S^{\Delta}_{m,\tau}x(t)=(x(t)-H^{\Delta}_{m}x(t))+(H^{\Delta}_{m}x(t)-S^{\Delta}_{m,\tau}x(t)).$$

In (6.2.3) the term $(H^{\Delta}_{m}x(t)-S^{\Delta}_{m,\tau}x(t))$ belongs to $H_m(\Delta)$, and

$$D^k(H^{\Delta}_{m}x(t_i)-S^{\Delta}_{m,\tau}x(t_i))=0;\ 1\le i\le N,\ 0\le k\le 2m-2-\tau$$
$$i=0,N+1,\ 0\le k\le m-1.$$

Hence, from (5.3.1) and (6.2.2) it follows that

$$(6.2.4)\qquad H^{\Delta}_{m}x(t)-S^{\Delta}_{m,\tau}x(t)=\sum_{i=1}^{N}\sum_{j=2m-1-\tau}^{m-1}h_{m,i,j}(t)\ e_i^j,$$

where $e_i^j=x_i^{(j)}-D^jS^{\Delta}_{m,\tau}x(t_i)$. Thus, on substituting (6.2.4) into (6.2.3) and differentiating the resulting relation k times, $0\le k\le n-1$ at $t\in[t_i,t_{i+1}]$, $0\le i\le N$ we obtain

$$(6.2.5)\ D^k(x(t)-S^{\Delta}_{m,\tau}x(t))=D^k(x(t)-H^{\Delta}_{m}x(t))+\sum_{i=1}^{N}\sum_{j=2m-1-\tau}^{m-1}h^{(k)}_{m,i,j}(t)\ e_i^j.$$

Let the vector $[e_i^j]$ be denoted as e^j. Then, from the triangle inequality, we find that

$$(6.2.6)\quad \| D^k(x - S^{\Delta}_{m,\tau}x) \|_\infty \le \| D^k(x - H^{\Delta}_m x) \|_\infty + \sum_{j=2m-1-\tau}^{m-1} \| e^j \|_\infty \times$$
$$\max_{0\le i\le N}\ \max_{t_i\le t\le t_{i+1}} \left[|h^{(k)}_{m,i,j}(t)| + |h^{(k)}_{m,i+1,j}(t)| \right],\ 0 \le k \le n-1$$

where we have used the fact that $h^{(k)}_{m,i,j}(t)$ is nonzero only in the interval $[t_{i-1}, t_i] \bigcup [t_i, t_{i+1}]$.

In the right side of (6.2.6) the equalities and the inequalities obtained in Chapter 5 can be used, and hence what remains is to compute only a priori estimates for $\| e^j \|_\infty$, $2m-1-\tau \le j \le m-1$.

Definition 6.2.2. For a given $x(t) \in C^{(2m-2)}[a,b]$ we say $LS^{\Delta}_{m,2m-2}x(t)$ is the *Lidstone* $S_{m,2m-2}(\Delta)$ *- interpolate of* $x(t)$, also known as *Lidstone - spline interpolate of* $x(t)$ if $LS^{\Delta}_{m,2m-2}x(t) \in S_{m,2m-2}(\Delta)$ with

$$(6.2.7)\quad \begin{aligned} &LS^{\Delta}_{m,2m-2}x(t_i) = x(t_i) = x_i,\quad 1 \le i \le N \\ &D^{2k}LS^{\Delta}_{m,2m-2}x(t_i) = x^{(2k)}(t_i) = x_i^{(2k)};\quad i = 0, N+1,\quad 0 \le k \le m-1. \end{aligned}$$

We can represent $LS^{\Delta}_{m,2m-2}x(t)$ in terms of the basic elements $r_{m,i,j}(t);\ 0 \le i \le N+1,\ 0 \le j \le m-1$ of $L_m(\Delta)$ as follows

$$(6.2.8)\qquad LS^{\Delta}_{m,2m-2}x(t) = \sum_{i=0}^{N+1} \sum_{j=0}^{m-1} r_{m,i,j}(t)\ D^{2j}LS^{\Delta}_{m,2m-2}x(t_i).$$

Let $x(t) \in PC^{n,\infty}[a,b]$, $2m-2 \le n \le 2m$. To obtain upper bounds for $\| D^k(x - LS^{\Delta}_{m,2m-2}x) \|_\infty$, $0 \le k \le n-1$ in terms of $\| D^n x \|_\infty$ we begin with the equality

$$(6.2.9)\quad x(t) - LS^{\Delta}_{m,2m-2}x(t) = (x(t) - L^{\Delta}_m x(t)) + (L^{\Delta}_m x(t) - LS^{\Delta}_{m,2m-2}x(t)).$$

In (6.2.9) the term $(L^{\Delta}_m x(t) - LS^{\Delta}_{m,2m-2}x(t))$ belongs to $L_m(\Delta)$, and

$$\begin{aligned} D^{2k}(L^{\Delta}_m x(t_i) - LS^{\Delta}_{m,2m-2}x(t_i)) = 0;\ &1 \le i \le N,\ k = 0 \\ &i = 0, N+1,\ 0 \le k \le m-1. \end{aligned}$$

Hence, from (5.4.2) and (6.2.8) it follows that

$$L_m^{\Delta}x(t) - LS_{m,2m-2}^{\Delta}x(t) = \sum_{i=1}^{N}\sum_{j=1}^{m-1} r_{m,i,j}(t)\, d_i^{2j}, \tag{6.2.10}$$

where $d_i^{2j} = x_i^{(2j)} - D^{2j}LS_{m,2m-2}^{\Delta}x(t_i)$. Thus, as earlier we find that

$$\| D^k(x - LS_{m,2m-2}^{\Delta}x) \|_\infty \le \| D^k(x - L_m^{\Delta}x) \|_\infty + \sum_{j=1}^{m-1} \| d^{2j} \|_\infty \times$$
$$\max_{0\le i\le N}\ \max_{t_i\le t\le t_{i+1}} \left[|r_{m,i,j}^{(k)}(t)| + |r_{m,i+1,j}^{(k)}(t)|\right],\ 0\le k\le n-1 \tag{6.2.11}$$

where $d^{2j} = [d_i^{2j}]$.

In the right side of (6.2.11) the equalities and the inequalities obtained in Chapter 5 can be used, and hence we only need to compute upper estimates for $\| d^{2j} \|_\infty$, $1 \le j \le m-1$.

6.3 CUBIC SPLINE INTERPOLATION

We need the following results.

Lemma 6.3.1. Let $1 \le i \le N$ but fixed, and $p(t)$, $q(t)$ be two cubic polynomials in $[t_{i-1}, t_i]$ and $[t_i, t_{i+1}]$, respectively. Suppose $D^kp(t_i) = D^kq(t_i) = x_i^{(k)}$; $k = 0, 1$ then $D^2p(t_i) = D^2q(t_i)$ if and only if

$$h[Dp(t_{i-1}) + 4\ x_i' + Dq(t_{i+1})] = -3[p(t_{i-1}) - q(t_{i+1})]. \tag{6.3.1}$$

Proof. By construction

$$p(t) = \alpha_1 + \beta_1(t-t_i) + \gamma_1(t-t_i)^2 + \delta_1(t-t_i)^3,$$

where

$$\alpha_1 = p(t_i), \qquad \beta_1 = Dp(t_i),$$

$$\gamma_1 = \frac{3}{h^2}[p(t_{i-1}) - p(t_i)] + \frac{1}{h}[Dp(t_{i-1}) + 2Dp(t_i)],$$

$$\delta_1 = \frac{2}{h^3}[p(t_{i-1}) - p(t_i)] + \frac{1}{h^2}[Dp(t_{i-1}) + Dp(t_i)]$$

and hence $D^2p(t_i) = 2\,\gamma_1$.

Similarly,

$$q(t) = \alpha_2 + \beta_2(t - t_i) + \gamma_2(t - t_i)^2 + \delta_2(t - t_i)^3,$$

where

$$\alpha_2 = q(t_i), \qquad \beta_2 = Dq(t_i),$$

$$\gamma_2 = -\frac{3}{h^2}[q(t_i) - q(t_{i+1})] - \frac{1}{h}[2Dq(t_i) + Dq(t_{i+1})],$$

$$\delta_2 = \frac{2}{h^3}[q(t_i) - q(t_{i+1})] + \frac{1}{h^2}[Dq(t_i) + Dq(t_{i+1})]$$

and hence $D^2q(t_i) = 2\,\gamma_2$.

Therefore, $D^2p(t_i) = D^2q(t_i)$ if and only if $\gamma_1 = \gamma_2$, which is the same as (6.3.1). ∎

Lemma 6.3.2. For a given $h(t) \in H_2(\Delta)$, we define $c_i = h(t_i)$, $c_i' = Dh(t_i)$, $0 \le i \le N+1$. The function $h(t) \in S_{2,2}(\Delta)$ if and only if c_i', $1 \le i \le N$ satisfy the following relations

$$c_{i-1}' + 4\,c_i' + c_{i+1}' = -\frac{3}{h}(c_{i-1} - c_{i+1}),\ 1 \le i \le N. \tag{6.3.2}$$

Moreover, from the system (6.3.2) the unknowns c_i', $1 \le i \le N$ can be obtained uniquely in terms of c_i, $0 \le i \le N+1$, c_0' and c_{N+1}'.

Proof. By Lemma 6.3.1 the continuity of $D^2h(t)$ is equivalent to the equations (6.3.1), which are the same as (6.3.2). The system (6.3.2) in matrix form can be written as

$$B\,c^1 = k, \tag{6.3.3}$$

where $c^1 = [c_i']$,

$$B = h\begin{bmatrix} 4 & 1 & & & \\ 1 & 4 & 1 & & \\ & 1 & 4 & 1 & \\ & \cdots & \cdots & \cdots & \\ & & 1 & 4 & 1 \\ & & & 1 & 4 \end{bmatrix} \tag{6.3.4}$$

and $k = [k_i]$,

$$k_i = -h\, c_0' - 3(c_0 - c_2), \quad i = 1 \tag{6.3.5}$$

$$= -3(c_{i-1} - c_{i+1}), \quad 2 \le i \le N-1$$

$$= -h\, c_{N+1}' - 3(c_{N-1} - c_{N+1}), \quad i = N.$$

Since the matrix B is strictly diagonally dominant, the system (6.3.3) has a unique solution. ∎

Lemma 6.3.3. For a given $x(t) \in C^{(1)}[a,b]$, $S_{2,2}^{\Delta}x(t)$ exists and is unique.

Proof. For a given $g(t) \in C^{(1)}[a,b]$, $H_2^{\Delta}g(t)$ exists and is unique. Further, by Lemma 6.3.2 for the given set of numbers $c_i = x_i$, $0 \le i \le N+1$, $c_i' = x_i'$; $i = 0, N+1$ there exist unique c_i', $1 \le i \le N$ satisfying (6.3.2). Now, let $g(t) \in C^{(1)}[a,b]$ be such that $g(t_i) = c_i$, $Dg(t_i) = c_i'$, $0 \le i \le N+1$. Then, again by Lemma 6.3.2, $H_2^{\Delta}g(t) \in S_{2,2}(\Delta)$. However, from the definition this $H_2^{\Delta}g(t)$ is the same as $S_{2,2}^{\Delta}x(t)$. ∎

Remark 6.3.1. From the proof of Lemma 6.3.3 and (6.2.2) it is clear that $S_{2,2}^{\Delta}x(t)$ can be expressed as

$$S_{2,2}^{\Delta}x(t) = \sum_{i=0}^{N+1} h_{2,i,0}(t)x_i + h_{2,0,1}(t)x_0' + h_{2,N+1,1}(t)x_{N+1}' + \sum_{i=1}^{N} h_{2,i,1}(t)c_i', \tag{6.3.6}$$

where c_i', $1 \le i \le N$ satisfy (6.3.2). ∎

Remark 6.3.2. It is possible to describe a basis for $S_{2,2}(\Delta)$, namely the 'cardinal splines', $\{s_i(t)\}_{i=0}^{N+3}$, defined by the following interpolation conditions :

$$s_j(t_i) = \delta_{ij}, \quad Ds_j(a) = Ds_j(b) = 0, \quad 0 \le i,j \le N+1$$

$$s_{N+2}(t_i) = s_{N+3}(t_i) = 0, \quad 0 \le i \le N+1$$

$$Ds_{N+2}(a) = Ds_{N+3}(b) = 1, \quad \text{and} \quad Ds_{N+2}(b) = Ds_{N+3}(a) = 0.$$

Obviously, $S_{2,2}^{\Delta}x(t)$ can be explicitly expressed as

$$S_{2,2}^{\Delta}x(t) = \sum_{i=0}^{N+1} s_i(t)x_i + s_{N+2}(t)x_0' + s_{N+3}(t)x_{N+1}'. \tag{6.3.7}$$ ∎

Lemma 6.3.4. [20] Let A be a square matrix such that $\| A \|_\infty < 1$. Then, $(I \pm A)$ is nonsingular and

$$\| (I \pm A)^{-1} \|_\infty \leq (1 - \| A \|_\infty)^{-1},$$

where I is the identity matrix. ∎

Lemma 6.3.5. If $x(t) \in PC^{n,\infty}[a,b]$, $2 \leq n \leq 4$ then

$$\| e^1 \|_\infty \leq a_n \, h^{n-1} \, \| D^n x \|_\infty \,, \tag{6.3.8}$$

where $a_2 = \frac{5}{3}$, $a_3 = \frac{4}{27}$, and $a_4 = \frac{1}{24}$.

Proof. Since the proofs for the cases $n = 2$ and $n = 4$ are available in Schultz [14] and Hall [7] respectively, we shall consider the case $n = 3$ only. Let $r = [r_i(x)]$ be an $N \times 1$ vector defined by

$$r = B \, e^1, \tag{6.3.9}$$

where the matrix B is given in (6.3.4). Then, it follows that

$$B \, x^1 = k + r, \tag{6.3.10}$$

where k is defined in (6.3.5) and $x^1 = [x_i']$ is an $N \times 1$ vector.

For $1 \leq i \leq N$, from (6.3.4) and (6.3.5), we have

$$r_i(x) = h(x_{i-1}' + 4 \, x_i' + x_{i+1}') + 3(x_{i-1} - x_{i+1}), \tag{6.3.11}$$

which in view of Theorem 5.2.1 can be written as

$$r_i(x) = \frac{1}{2} \int_{t_{i-1}}^{t_{i+1}} (r_i)_t (t-s)_+^2 \; x'''(s) \; ds, \tag{6.3.12}$$

where

$$
\begin{aligned}
&(r_i)_t(t-s)_+^2 \\
&= 2h \left[(t_{i-1} - s)_+ + 4(t_i - s)_+ + (t_{i+1} - s)_+ \right] + 3 \left[(t_{i-1} - s)_+^2 - (t_{i+1} - s)_+^2 \right] \\
&= \begin{cases} S(-3S + 2h); & S = s - t_{i-1}, \; s \in [t_{i-1}, t_i], \; S \in [0, h] \\ (h - \bar{S})(3\bar{S} - h); & \bar{S} = s - t_i, \; s \in [t_i, t_{i+1}], \; \bar{S} \in [0, h]. \end{cases}
\end{aligned}
$$

Therefore, on using the proper sgn $(r_i)_t(t-s)_+^2$ in (6.3.12), we find that

$$\begin{aligned} | r_i(x) | &\le \frac{1}{2} \| D^3 x \|_\infty \int_{t_{i-1}}^{t_{i+1}} | (r_i)_t(t-s)_+^2 | ds \\ &= \frac{1}{2} \| D^3 x \|_\infty \left\{ \int_0^{\frac{2h}{3}} (-3S^2 + 2hS) dS + \int_{\frac{2h}{3}}^{h} (3S^2 - 2hS) dS \right. \\ &\quad \left. + \int_0^{\frac{h}{3}} (3\bar{S}^2 - 4h\bar{S} + h^2) d\bar{S} + \int_{\frac{h}{3}}^{h} (-3\bar{S}^2 + 4h\bar{S} - h^2) d\bar{S} \right\} \end{aligned}$$

$$(6.3.13) \qquad = \frac{8}{27} h^3 \| D^3 x \|_\infty , \quad 1 \le i \le N.$$

Now multiplying both sides of (6.3.9) by the diagonal matrix $\square = [d_{ij}]$, where $d_{ii} = 1/(4h)$, $1 \le i \le N$, we obtain

$$(6.3.14) \qquad \| e^1 \|_\infty \le \| (\square\, B)^{-1} \|_\infty \| \square\, r \|_\infty .$$

Writing $\square\, B = I + A$, we find that $\| A \|_\infty = \frac{1}{2}$, and thus in view of Lemma 6.3.4, we have

$$(6.3.15) \qquad \| (\square\, B)^{-1} \|_\infty = \| (I + A)^{-1} \|_\infty \le 2.$$

Using (6.3.13) and (6.3.15) in (6.3.14), we get

$$\| e^1 \|_\infty \le \frac{4}{27} h^2 \| D^3 x \|_\infty . \qquad \blacksquare$$

Theorem 6.3.6. Let $x(t) \in PC^{n,\infty}[a,b]$, $2 \le n \le 4$. Then,

$$(6.3.16) \qquad \| D^k (x - S_{2,2}^{\Delta} x) \|_\infty \le \alpha_{2,n,k}\, h^{n-k} \| D^n x \|_\infty , \quad 0 \le k \le n-1$$

where the constants $\alpha_{2,n,k}$ are given in the following table.

Table 6.3.1.

k \ n	2	3	4
0	$\frac{23}{48}$	$\frac{41}{864}$	$\frac{5}{384}$
1	$0.191816432 + 1$	$\frac{13\sqrt{13} - 42}{27}$	$\frac{9 + \sqrt{3}}{216}$
2		$\frac{32}{27}$	$\frac{1}{3}$
3			1

Proof. Using Lemma 5.3.2, Theorem 5.3.7 and Lemma 6.3.5 in (6.2.6) the inequalities (6.3.16) are immediate. ∎

6.4 QUINTIC SPLINE INTERPOLATION : $\tau = 4$

We need the following results.

Lemma 6.4.1. Let $1 \le i \le N$ but fixed, and $p(t)$, $q(t)$ be two quintic polynomials in $[t_{i-1}, t_i]$ and $[t_i, t_{i+1}]$, respectively. Suppose $D^k p(t_i) = D^k q(t_i) = x_i^{(k)}$, $0 \le k \le 2$ then $D^3 p(t_i) = D^3 q(t_i)$ if and only if

$$\text{(6.4.1)} \qquad \frac{20}{h^2}[p(t_{i-1}) - 2\, x_i + q(t_{i+1})]$$
$$= -\frac{8}{h}[Dp(t_{i-1}) - Dq(t_{i+1})] + 6\, x_i'' - [D^2 p(t_{i-1}) + D^2 q(t_{i+1})],$$

and $D^4 p(t_i) = D^4 q(t_i)$ if and only if

$$\text{(6.4.2)} \qquad -\frac{15}{h^2}[p(t_{i-1}) - q(t_{i+1})]$$
$$= \frac{7}{h}[Dp(t_{i-1}) + Dq(t_{i+1})] + \frac{16}{h}\, x_i' + [D^2 p(t_{i-1}) - D^2 q(t_{i+1})].$$

Proof. By construction

$$p(t) = \alpha_1 + \beta_1(t - t_i) + \gamma_1(t - t_i)^2 + \delta_1(t - t_i)^3 + \eta_1(t - t_i)^4 + \nu_1(t - t_i)^5,$$

where

$$\alpha_1 = p(t_i), \qquad \beta_1 = Dp(t_i), \qquad \gamma_1 = \frac{1}{2}\, D^2 p(t_i),$$

$$\delta_1 = -\frac{10}{h^3}[p(t_{i-1}) - p(t_i)] - \frac{1}{h^2}[4Dp(t_{i-1}) + 6Dp(t_i)]$$
$$-\frac{1}{2h}[D^2 p(t_{i-1}) - 3D^2 p(t_i)],$$

$$\eta_1 = -\frac{15}{h^4}[p(t_{i-1}) - p(t_i)] - \frac{1}{h^3}[7Dp(t_{i-1}) + 8Dp(t_i)]$$
$$-\frac{1}{2h^2}[2D^2 p(t_{i-1}) - 3D^2 p(t_i)],$$

$$\nu_1 = -\frac{6}{h^5}[p(t_{i-1}) - p(t_i)] - \frac{3}{h^4}[Dp(t_{i-1}) + Dp(t_i)] - \frac{1}{2h^3}[D^2 p(t_{i-1}) - D^2 p(t_i)]$$

and hence $D^3p(t_i) = 6\ \delta_1$ and $D^4p(t_i) = 24\ \eta_1$.

Similarly,

$$q(t) = \alpha_2 + \beta_2(t-t_i) + \gamma_2(t-t_i)^2 + \delta_2(t-t_i)^3 + \eta_2(t-t_i)^4 + \nu_2(t-t_i)^5,$$

where

$$\alpha_2 = q(t_i), \qquad \beta_2 = Dq(t_i), \qquad \gamma_2 = \frac{1}{2}\ D^2q(t_i),$$

$$\delta_2 = -\frac{10}{h^3}[q(t_i) - q(t_{i+1})] - \frac{1}{h^2}[6Dq(t_i) + 4Dq(t_{i+1})] - \frac{1}{2h}[3D^2q(t_i) - D^2q(t_{i+1})],$$

$$\eta_2 = \frac{15}{h^4}[q(t_i) - q(t_{i+1})] + \frac{1}{h^3}[8Dq(t_i) + 7Dq(t_{i+1})] + \frac{1}{2h^2}[3D^2q(t_i) - 2D^2q(t_{i+1})],$$

$$\nu_2 = -\frac{6}{h^5}[q(t_i) - q(t_{i+1})] - \frac{3}{h^4}[Dq(t_i) + Dq(t_{i+1})] - \frac{1}{2h^3}[D^2q(t_i) - D^2q(t_{i+1})]$$

and hence $D^3q(t_i) = 6\ \delta_2$ and $D^4q(t_i) = 24\ \eta_2$.

Hence, $D^3p(t_i) = D^3q(t_i)$ if and only if $\delta_1 = \delta_2$ which is the same as (6.4.1), and $D^4p(t_i) = D^4q(t_i)$ if and only if $\eta_1 = \eta_2$ which is the same as (6.4.2). ∎

Lemma 6.4.2. For a given $h(t) \in H_3(\Delta)$, we define $c_i = h(t_i)$, $c_i' = Dh(t_i)$ and $c_i'' = D^2h(t_i)$, $0 \le i \le N+1$. The function $h(t) \in S_{3,4}(\Delta)$ if and only if c_i' and c_i'', $1 \le i \le N$ satisfy the following relations

(6.4.3) $$h(227\ c_1' + 79\ c_2' + 3\ c_3') = -111h\ c_0' - 16h^2\ c_0'' - 235\ c_0 + 65\ c_1 + 155\ c_2 + 15\ c_3$$

(6.4.4) $$h(c_{i-2}' + 26\ c_{i-1}' + 66\ c_i' + 26\ c_{i+1}' + c_{i+2}') = -5\ c_{i-2} - 50\ c_{i-1} + 50\ c_{i+1} + 5\ c_{i+2}, \quad 2 \le i \le N-1$$

(6.4.5) $$h(3\ c_{N-2}' + 79\ c_{N-1}' + 227\ c_N') = -111h\ c_{N+1}' + 16h^2\ c_{N+1}'' - 15\ c_{N-2} - 155\ c_{N-1} - 65\ c_N + 235\ c_{N+1}$$

$$(6.4.6)\quad \frac{h^2}{20}(354\ c_1'' + 201\ c_2'' + 8\ c_3'')$$
$$= 12h\ c_0' + \frac{23}{20}\ h^2\ c_0'' + 37\ c_0 - 54\ c_1 + 9\ c_2 + 8\ c_3$$

$$(6.4.7)\quad \frac{h^2}{20}(c_{i-2}'' + 26\ c_{i-1}'' + 66\ c_i'' + 26\ c_{i+1}'' + c_{i+2}'')$$
$$= c_{i-2} + 2\ c_{i-1} - 6\ c_i + 2\ c_{i+1} + c_{i+2},\quad 2 \le i \le N-1$$

$$(6.4.8)\quad \frac{h^2}{20}(8\ c_{N-2}'' + 201\ c_{N-1}'' + 354\ c_N'')$$
$$= -12h\ c_{N+1}' + \frac{23}{20}\ h^2\ c_{N+1}'' + 8\ c_{N-2} + 9\ c_{N-1} - 54\ c_N + 37\ c_{N+1}.$$

Moreover, from the systems (6.4.3) - (6.4.5) and (6.4.6) - (6.4.8) the unknowns c_i' and c_i'', $1 \le i \le N$ can be obtained uniquely in terms of c_i, $0 \le i \le N+1$, c_0', c_{N+1}', c_0'' and c_{N+1}''.

Proof. By Lemma 6.4.1 the continuity of $D^3h(t)$ and $D^4h(t)$ is equivalent to the equations (6.4.1) and (6.4.2), respectively, which are better written as

$$(6.4.9)\quad P_i:\ -c_{i-1}'' + 6\ c_i'' - c_{i+1}'' = \frac{20}{h^2}(c_{i-1} - 2\ c_i + c_{i+1}) + \frac{8}{h}\ c_{i-1}' - \frac{8}{h}\ c_{i+1}'$$

and

$$(6.4.10)\quad Q_i:\ c_{i-1}'' - c_{i+1}'' = -\frac{15}{h^2}(c_{i-1} - c_{i+1}) - \frac{7}{h}\ c_{i-1}' - \frac{16}{h}\ c_i' - \frac{7}{h}\ c_{i+1}'.$$

Now the operations $P_1 + 3P_2 + 17Q_1 - 3Q_2$, $-P_{i-1} + P_{i+1} - Q_{i-1} + 6Q_i - Q_{i+1}$ and $3P_{N-1} + P_N + 3Q_{N-1} - 17Q_N$ give (6.4.3) - (6.4.5), respectively.

Next, to obtain the system (6.4.6) - (6.4.8) we rewrite (6.4.9) and (6.4.10) as follows

$$(6.4.11)\quad R_i:\ -c_{i-1}' + c_{i+1}' = \frac{h}{8}\left[\frac{20}{h^2}(c_{i-1} - 2\ c_i + c_{i+1}) + c_{i-1}'' - 6\ c_i'' + c_{i+1}''\right]$$

and

$$(6.4.12)\quad S_i:\ 7\ c_{i-1}' + 16\ c_i' + 7\ c_{i+1}' = h\left[-\frac{15}{h^2}(c_{i-1} - c_{i+1}) - c_{i-1}'' + c_{i+1}''\right].$$

The operations $-\frac{79}{8}\ R_1 - 7R_2 - \frac{7}{8}\ S_1 + S_2$, $-7R_{i-1} - 16R_i - 7R_{i+1} - S_{i-1} + S_{i+1}$ and $7R_{N-1} - \frac{79}{8}\ R_N + S_{N-1} - \frac{7}{8}\ S_N$ give (6.4.6) - (6.4.8), respectively.

The system (6.4.3) - (6.4.5) in matrix form can be written as

$$B_1\, c^1 = k^1, \tag{6.4.13}$$

where $c^1 = [c_i']$,

$$B_1 = h \begin{bmatrix} 227 & 79 & 3 & & & & & \\ 26 & 66 & 26 & 1 & & & & \\ 1 & 26 & 66 & 26 & 1 & & & \\ & 1 & 26 & 66 & 26 & 1 & & \\ & & \cdots & \cdots & \cdots & \cdots & & \\ & & & 1 & 26 & 66 & 26 & 1 \\ & & & & 1 & 26 & 66 & 26 \\ & & & & & 3 & 79 & 227 \end{bmatrix} \tag{6.4.14}$$

and $k^1 = [k_i^1]$,

$$\begin{aligned} k_i^1 &= -111h\, c_0' - 16h^2\, c_0'' - 235\, c_0 + 65\, c_1 + 155\, c_2 + 15\, c_3, \quad i = 1 \\ &= -h\, c_0' - 5\, c_0 - 50\, c_1 + 50\, c_3 + 5\, c_4, \quad i = 2 \\ &= -5\, c_{i-2} - 50\, c_{i-1} + 50\, c_{i+1} + 5\, c_{i+2}, \quad 3 \le i \le N-2 \\ &= -h\, c_{N+1}' - 5\, c_{N-3} - 50\, c_{N-2} + 50\, c_N + 5\, c_{N+1}, \quad i = N-1 \\ &= -111h\, c_{N+1}' + 16h^2\, c_{N+1}'' - 15\, c_{N-2} - 155\, c_{N-1} - 65\, c_N \\ &\qquad + 235\, c_{N+1}, \quad i = N. \end{aligned} \tag{6.4.15}$$

Since the matrix B_1 is strictly diagonally dominant, the system (6.4.13) has a unique solution.

Similarly, the system (6.4.6) - (6.4.8) in matrix form can be written as

$$B_2\, c^2 = k^2, \tag{6.4.16}$$

where $c^2 = [c_i'']$,

$$B_2 = \frac{h^2}{20} \begin{bmatrix} 354 & 201 & 8 & & & & & \\ 26 & 66 & 26 & 1 & & & & \\ 1 & 26 & 66 & 26 & 1 & & & \\ & 1 & 26 & 66 & 26 & 1 & & \\ & & \cdots & \cdots & \cdots & \cdots & & \\ & & & 1 & 26 & 66 & 26 & 1 \\ & & & & 1 & 26 & 66 & 26 \\ & & & & & 8 & 201 & 354 \end{bmatrix} \tag{6.4.17}$$

and $k^2 = [k_i^2]$,

(6.4.18) $$k_i^2 = 12h\, c_0' + \frac{23}{20}\, h^2\, c_0'' + 37\, c_0 - 54\, c_1 + 9\, c_2 + 8\, c_3, \quad i = 1$$

$$= -\frac{h^2}{20}\, c_0'' + c_0 + 2\, c_1 - 6\, c_2 + 2\, c_3 + c_4, \quad i = 2$$

$$= c_{i-2} + 2\, c_{i-1} - 6\, c_i + 2\, c_{i+1} + c_{i+2}, \quad 3 \le i \le N-2$$

$$= -\frac{h^2}{20}\, c_{N+1}'' + c_{N-3} + 2\, c_{N-2} - 6\, c_{N-1} + 2\, c_N + c_{N+1}, \quad i = N-1$$

$$= -12h\, c_{N+1}' + \frac{23}{20}\, h^2\, c_{N+1}'' + 8\, c_{N-2} + 9\, c_{N-1} - 54\, c_N + 37\, c_{N+1}, \quad i = N.$$

Once again the matrix B_2 is strictly diagonally dominant, the system (6.4.16) therefore has a unique solution. ∎

Lemma 6.4.3. For a given $x(t) \in C^{(2)}[a, b]$, $S_{3,4}^{\Delta} x(t)$ exists and is unique.

Proof. The proof is similar to that of Lemma 6.3.3. ∎

Remark 6.4.1. From Lemma 6.4.3 and (6.2.2) it is clear that $S_{3,4}^{\Delta} x(t)$ can be expressed as

(6.4.19) $$S_{3,4}^{\Delta} x(t) = \sum_{i=0}^{N+1} h_{3,i,0}(t) x_i + h_{3,0,1}(t) x_0' + h_{3,N+1,1}(t) x_{N+1}' + \sum_{i=1}^{N} h_{3,i,1}(t) c_i'$$
$$+ h_{3,0,2}(t) x_0'' + h_{3,N+1,2}(t) x_{N+1}'' + \sum_{i=1}^{N} h_{3,i,2}(t) c_i'',$$

where c_i' and c_i'', $1 \le i \le N$ satisfy (6.4.3) - (6.4.8). ∎

Remark 6.4.2. It is possible to describe a basis for $S_{3,4}(\Delta)$, namely the 'cardinal splines', $\{s_i(t)\}_{i=0}^{N+5}$, defined by the following interpolation conditions :

$$s_j(t_i) = \delta_{ij}, \quad D^k s_j(a) = D^k s_j(b) = 0; \quad k = 1, 2, \quad 0 \le i, j \le N+1$$

$$s_j(t_i) = 0; \quad N+2 \le j \le N+5, \quad 0 \le i \le N+1$$

$$D s_{N+2}(a) = 1, \quad D s_{N+2}(b) = 0, \quad D^2 s_{N+2}(a) = D^2 s_{N+2}(b) = 0$$

$$Ds_{N+3}(a) = 0, \quad Ds_{N+3}(b) = 1, \quad D^2 s_{N+3}(a) = D^2 s_{N+3}(b) = 0$$

$$D^2 s_{N+4}(a) = 1, \quad D^2 s_{N+4}(b) = 0, \quad Ds_{N+4}(a) = Ds_{N+4}(b) = 0$$

$$D^2 s_{N+5}(a) = 0, \quad D^2 s_{N+5}(b) = 1, \quad Ds_{N+5}(a) = Ds_{N+5}(b) = 0.$$

Obviously, $S^{\Delta}_{3,4}x(t)$ can be explicitly expressed as

(6.4.20)
$$S^{\Delta}_{3,4}x(t) = \sum_{i=0}^{N+1} s_i(t)x_i + s_{N+2}(t)x'_0 + s_{N+3}(t)x'_{N+1} + s_{N+4}(t)x''_0 + s_{N+5}(t)x''_{N+1}. \quad \blacksquare$$

Lemma 6.4.4. If $x(t) \in PC^{n,\infty}[a,b]$, $2 \le n \le 6$ then

(6.4.21)
$$\| e^1 \|_\infty \le \bar{a}_n \; h^{n-1} \; \| D^n x \|_\infty ,$$

where $\bar{a}_2 = 0.10369640 + 2$, $\bar{a}_3 = 0.13014691 + 1$, $\bar{a}_4 = 0.24476058 + 0$, $\bar{a}_5 = 0.51735859 - 1$ and $\bar{a}_6 = \frac{1}{72}$.

Proof. We shall provide the proof only for the case $n = 2$. Let $r = [r_i(x)]$ be an $N \times 1$ vector defined by

(6.4.22)
$$r = B_1 \; e^1,$$

where the matrix B_1 is given in (6.4.14). Then, it follows that

(6.4.23)
$$B_1 \; x^1 = k^1 + r,$$

where k^1 is defined in (6.4.15) and $x^1 = [x'_i]$ is an $N \times 1$ vector.

For $i = 1$, from (6.4.23), (6.4.14) and (6.4.15), we have

(6.4.24)
$$r_1(x) = h(111 \; x'_0 + 227 \; x'_1 + 79 \; x'_2 + 3 \; x'_3) + 16h^2 \; x''_0 + 235 \; x_0 - 65 \; x_1 - 155 \; x_2 - 15 \; x_3.$$

By (6.4.24), $r_1(p) = 0$ for all quintic polynomials $p(t)$, and hence by Theorem 5.2.1 it follows that

(6.4.25)
$$r_1(x) = \int_{t_0}^{t_3} (r_1)_t(t-s)_+ \; x''(s) \; ds + 16h^2 \; x''_0,$$

where

$$(6.4.26)\quad (r_1)_t(t-s)_+$$

$$= h\left[111(t_0-s)_+^0 + 227(t_1-s)_+^0 + 79(t_2-s)_+^0 + 3(t_3-s)_+^0\right]$$

$$+ 235(t_0-s)_+ - 65(t_1-s)_+ - 155(t_2-s)_+ - 15(t_3-s)_+$$

$$= \begin{cases} 3h - 15(t_3-s) = \alpha(s), \quad s \in [t_2, t_3] \\ 82h - 155(t_2-s) - 15(t_3-s) = \beta(s), \quad s \in [t_1, t_2] \\ 309h - 65(t_1-s) - 155(t_2-s) - 15(t_3-s) = \gamma(s), \\ \qquad\qquad s \in [t_0, t_1]. \end{cases}$$

Clearly, $\alpha(s) \geq 0$ if $s \in [t_3 - \frac{1}{5}h, t_3]$ and $\alpha(s) \leq 0$ if $s \in [t_2, t_3 - \frac{1}{5}h]$; $\beta(s) \geq 0$ if $s \in [t_2 - \frac{67}{170}h, t_2]$ and $\beta(s) \leq 0$ if $s \in [t_1, t_2 - \frac{67}{170}h]$; $\gamma(s) \geq 0$ if $s \in [t_1 - \frac{124}{235}h, t_1]$ and $\gamma(s) \leq 0$ if $s \in [t_0, t_1 - \frac{124}{235}h]$. Thus, it follows that

$$| r_1(x) | \leq \| D^2x \|_\infty \left\{ \int_{t_0}^{t_3} | (r_1)_t(t-s)_+ | \, ds + 16h^2 \right\}$$

$$= \| D^2x \|_\infty \left\{ \int_{t_0}^{t_1 - \frac{124}{235}h} -\gamma(s)\, ds + \int_{t_1 - \frac{124}{235}h}^{t_1} \gamma(s)\, ds \right.$$

$$+ \int_{t_1}^{t_2 - \frac{67}{170}h} -\beta(s)\, ds + \int_{t_2 - \frac{67}{170}h}^{t_2} \beta(s)\, ds$$

$$\left. + \int_{t_2}^{t_3 - \frac{1}{5}h} -\alpha(s)\, ds + \int_{t_3 - \frac{1}{5}h}^{t_3} \alpha(s)\, ds + 16h^2 \right\}$$

$$(6.4.27)\qquad \leq (0.12443567 + 3)h^2 \| D^2x \|_\infty \; .$$

For $2 \leq i \leq N-1$, again from (6.4.23), (6.4.14) and (6.4.15), we have

$$(6.4.28)\qquad r_i(x) = h(x'_{i-2} + 26\, x'_{i-1} + 66\, x'_i + 26\, x'_{i+1} + x'_{i+2})$$

$$+ 5\, x_{i-2} + 50\, x_{i-1} - 50\, x_{i+1} - 5\, x_{i+2}$$

and since for each $2 \leq i \leq N-1$, $r_i(p) = 0$ for all quintic polynomials $p(t)$, from Theorem 5.2.1 it follows that

$$(6.4.29)\qquad r_i(x) = \int_{t_{i-2}}^{t_{i+2}} (r_i)_t(t-s)_+ \, x''(s)\, ds,$$

where

$$(r_i)_t(t-s)_+$$

$$= h\left[(t_{i-2}-s)_+^0 + 26(t_{i-1}-s)_+^0 + 66(t_i-s)_+^0 + 26(t_{i+1}-s)_+^0 + (t_{i+2}-s)_+^0\right]$$

$$+ 5(t_{i-2}-s)_+ + 50(t_{i-1}-s)_+ - 50(t_{i+1}-s)_+ - 5(t_{i+2}-s)_+$$

$$= \begin{cases} h - 5(t_{i+2}-s) = \alpha(s), & s \in [t_{i+1}, t_{i+2}] \\ 27h - 50(t_{i+1}-s) - 5(t_{i+2}-s) = \beta(s), & s \in [t_i, t_{i+1}] \\ 93h - 50(t_{i+1}-s) - 5(t_{i+2}-s) = \gamma(s), & s \in [t_{i-1}, t_i] \\ 119h + 50(t_{i-1}-s) - 50(t_{i+1}-s) - 5(t_{i+2}-s) = \delta(s), & s \in [t_{i-2}, t_{i-1}]. \end{cases}$$

Clearly, $\alpha(s) \geq 0$ if $s \in [t_{i+2} - \frac{1}{5}h, t_{i+2}]$ and $\alpha(s) \leq 0$ if $s \in [t_{i+1}, t_{i+2} - \frac{1}{5}h]$; $\beta(s) \geq 0$ if $s \in [t_{i+1} - \frac{2}{5}h, t_{i+1}]$ and $\beta(s) \leq 0$ if $s \in [t_i, t_{i+1} - \frac{2}{5}h]$; $\gamma(s) \geq 0$ if $s \in [t_i - \frac{3}{5}h, t_i]$ and $\gamma(s) \leq 0$ if $s \in [t_{i-1}, t_i - \frac{3}{5}h]$; $\delta(s) \geq 0$ if $s \in [t_{i-1} - \frac{4}{5}h, t_{i-1}]$ and $\delta(s) \leq 0$ if $s \in [t_{i-2}, t_{i-1} - \frac{4}{5}h]$. Using the obtained sgn $(r_i)_t(t-s)_+$ and following as in the case $i = 1$, we find

$$\text{(6.4.30)} \qquad |\, r_i(x) \,| \leq 32\, h^2 \,\|\, D^2x \,\|_\infty\,, \quad 2 \leq i \leq N-1.$$

Finally, for $i = N$ we have

$$r_N(x) = h(3\, x'_{N-2} + 79\, x'_{N-1} + 227\, x'_N + 111\, x'_{N+1}) - 16h^2\, x''_{N+1} + 15\, x_{N-2}$$

$$+ 155\, x_{N-1} + 65\, x_N - 235\, x_{N+1}$$

and hence as earlier in the case $i = 1$, we obtain

$$\text{(6.4.31)} \qquad |\, r_N(x) \,| \leq (0.12443567 + 3)h^2 \,\|\, D^2x \,\|_\infty\,.$$

Now multiplying both sides of (6.4.22) by the diagonal matrix $\square = [d_{ij}]$, where $d_{ii} = 1/(ah)$, $a \in \Re^+$, $1 \leq i \leq N$ to obtain $\square\, B_1\, e^1 = \square\, r$, which gives that

$$\text{(6.4.32)} \qquad \|\, e^1 \,\|_\infty \leq \|\, (\square\, B_1)^{-1} \,\|_\infty \,\|\, \square\, r \,\|_\infty\,.$$

Writing $\square\, B_1 = I + A$, where A is an $N \times N$ matrix with the property that $\| A \|_\infty < 1$, it follows from (6.4.32) and Lemma 6.3.4 that

$$\| e^1 \|_\infty \leq \frac{1}{1- \| A \|_\infty} \| \square\, r \|_\infty$$

$$\leq \frac{1}{1- \| A \|_\infty} \frac{1}{ah} \max_{1\leq i \leq N} | r_i(x) | . \tag{6.4.33}$$

To obtain the smallest bound in (6.4.33), we need to maximize $(1- \| A \|_\infty)\, a$ over $a \in \Re^+$. For this, from (6.4.14) we have

$$\| A \|_\infty = \max\left\{ \left|\frac{227}{a} - 1\right| + \frac{79+3}{a}, \left|\frac{66}{a} - 1\right| + \frac{2(26+1)}{a}.\right\}$$

$$= \begin{cases} \dfrac{309}{a} - 1, & 0 < a \leq 66 \\[2ex] \max\left\{ \dfrac{309}{a} - 1,\ 1 - \dfrac{12}{a}\right\}, & 66 \leq a \leq 227 \\[2ex] 1 - \dfrac{12}{a}, & a \geq 227 \end{cases}$$

$$= \begin{cases} \dfrac{309}{a} - 1, & 0 < a \leq 160.5 \\[2ex] 1 - \dfrac{12}{a}, & a \geq 160.5. \end{cases} \tag{6.4.34}$$

Thus, the condition $\| A \|_\infty < 1$ is equivalent to $\frac{309}{a} - 1 < 1$ $(1 - \frac{12}{a} < 1$ for all $a \geq 160.5)$ which gives that $a > 154.5$. Hence, on using (6.4.34) we find

$$\max_{a>154.5} (1- \| A \|_\infty)\, a$$

$$= \max\left\{ \max_{154.5<a\leq 160.5} (1- \| A \|_\infty)\, a,\ \max_{a\geq 160.5} (1- \| A \|_\infty)\, a \right\}$$

$$= \max\left\{ \max_{154.5<a\leq 160.5} (2a - 309),\ \max_{a\geq 160.5} (12) \right\}$$

$$= 12. \tag{6.4.35}$$

Using (6.4.27), (6.4.30), (6.4.31) and (6.4.35) in (6.4.33), we obtain

$$\| e^1 \|_\infty \leq \frac{1}{12}(0.12443567 + 3)h \| D^2 x \|_\infty$$

$$= (0.10369640 + 2)h \| D^2 x \|_\infty .$$

This completes the proof of Lemma 6.4.4. ∎

Lemma 6.4.5. If $x(t) \in PC^{n,\infty}[a,b]$, $2 \le n \le 6$ then

$$(6.4.36) \qquad \| e^2 \|_\infty \le \bar{b}_n \, h^{n-2} \, \| D^n x \|_\infty ,$$

where $\bar{b}_2 = 0.10031982 + 3$, $\bar{b}_3 = 0.11649771 + 2$, $\bar{b}_4 = 0.16544732 + 1$, $\bar{b}_5 = 0.31953197 + 0$ and $\bar{b}_6 = 0.86126455 - 1$.

Proof. The proof is similar to that of Lemma 6.4.4. ∎

Theorem 6.4.6. Let $x(t) \in PC^{n,\infty}[a,b]$, $2 \le n \le 6$. Then,

$$(6.4.37) \qquad \| D^k(x - S^{\Delta}_{3,4}x) \|_\infty \le \alpha_{3,n,k} \, h^{n-k} \, \| D^n x \|_\infty , \quad 0 \le k \le n-1$$

where the constants $\alpha_{3,n,k}$ are given in the following table.

Table 6.4.1.

k \ n	2	3	4	5	6
0	0.64731632 + 1	0.77694933 + 0	0.12886698 + 0	0.26250486 − 1	0.70534309 − 2
1	0.20381558 + 2	0.24445242 + 1	0.40642248 + 0	0.82823292 − 1	0.22250946 − 1
2		0.19372140 + 2	0.30861176 + 1	0.62083326 + 0	0.16683483 + 0
3			0.34774464 + 2	0.69730952 + 1	0.18751841 + 1
4				0.38191889 + 2	0.10267587 + 2
5					0.20835175 + 2

Proof. Using Lemma 5.3.4, Theorem 5.3.8, Lemmas 6.4.4 and 6.4.5 in (6.2.6) the inequalities (6.4.37) are immediate. ∎

Remark 6.4.3. The sharpness of the inequalities (6.4.37) remains undecided. However, in Tables 6.4.2 and 6.4.3 we compute the actual values of $\| D^k(x - S^{\Delta}_{3,4}x) \|_\infty$ for some simple functions and compare these with the corresponding right side bounds in (6.4.37).

Table 6.4.2.

$$x(t) = t^5 \sin \frac{1}{t}, \quad t \in \left[-\frac{1}{4}, \frac{1}{4}\right], \quad n = 2$$

N	7	15	24
$\Vert x - S^{\Delta}_{3,4}x \Vert_\infty$	$0.24180548 - 4$	$0.74092632 - 5$	$0.25445696 - 5$
Bound	$0.70679268 - 2$	$0.17669816 - 2$	$0.72375570 - 3$
$\Vert D(x - S^{\Delta}_{3,4}x) \Vert_\infty$	$0.14584234 - 2$	$0.70241544 - 3$	$0.38727821 - 3$
Bound	$0.35606791 + 0$	$0.17803395 + 0$	$0.11394174 + 0$

Table 6.4.3.

$$x(t) = t^6 \mid t \mid, \quad t \in [-1, 1], \quad n = 6$$

N	9	49	99
$\Vert x - S^{\Delta}_{3,4}x \Vert_\infty$	$0.21828938 - 4$	$0.17843074 - 8$	$0.28675645 - 10$
Bound	$0.22751512 - 2$	$0.14560967 - 6$	$0.22751512 - 8$
$\Vert D(x - S^{\Delta}_{3,4}x) \Vert_\infty$	$0.31529258 - 3$	$0.12975679 - 6$	$0.41767449 - 8$
Bound	$0.35886252 - 1$	$0.11483600 - 4$	$0.35886252 - 6$
$\Vert D^2(x - S^{\Delta}_{3,4}x) \Vert_\infty$	$0.85510914 - 2$	$0.17823579 - 4$	$0.11502702 - 5$
Bound	$0.13453574 + 1$	$0.21525719 - 2$	$0.13453574 - 3$
$\Vert D^3(x - S^{\Delta}_{3,4}x) \Vert_\infty$	$0.37555457 + 0$	$0.29940004 - 2$	$0.37399607 - 3$
Bound	$0.75607258 + 2$	$0.60485806 + 0$	$0.75607258 - 1$
$\Vert D^4(x - S^{\Delta}_{3,4}x) \Vert_\infty$	$0.21626967 + 2$	$0.88471955 + 0$	$0.22176471 + 0$
Bound	$0.20699482 + 4$	$0.82797926 + 2$	$0.20699482 + 2$
$\Vert D^5(x - S^{\Delta}_{3,4}x) \Vert_\infty$	$0.51609688 + 3$	$0.10657294 + 3$	$0.53493416 + 2$
Bound	$0.21001882 + 5$	$0.42003763 + 4$	$0.21001882 + 4$

∎

6.5 APPROXIMATED QUINTIC SPLINES : $\tau = 4$

For a given function $x(t) \in PC^{6,\infty}[a,b]$ and a fixed partition Δ we shall construct approximates for the spline interpolate $S^{\Delta}_{3,4}x(t)$ when

1. the values of x'_i, $i = 0, N+1$ are unknown;
2. the values of x''_i, $i = 0, N+1$ are unknown;
3. the values of x'_i and x''_i, $i = 0, N+1$ are unknown; and
4. the values of x_i, $0 \le i \le N+1$, x'_i, and x''_i, $i = 0, N+1$ are unknown.

Case 1 Since $x(t) \in PC^{6,\infty}[a,b]$ we can determine the constants α_i, $0 \le i \le 5$ in the representation

$$x'_0 = \sum_{i=0}^{5} \alpha_i \ x_i \tag{6.5.1}$$

so that the truncation error is $O(h^6)$. Indeed, in (6.5.1) for each $1 \le i \le 5$ we use Taylor's formula

$$x_i = \sum_{j=0}^{5} \frac{(ih)^j}{j!} \ x_0^{(j)} + O(h^6) \tag{6.5.2}$$

and require the resulting equation to be true for all $x(t) \in PC^{6,\infty}[a,b]$. This leads to a system of six equations

$$\sum_{i=0}^{5} \alpha_i = 0, \quad \sum_{i=1}^{5} i \ \alpha_i = \frac{1}{h}, \quad \sum_{i=1}^{5} i^j \ \alpha_i = 0, \quad 2 \le j \le 5.$$

The above system can be solved to obtain uniquely $\alpha_0 = -\frac{137}{60h}$, $\alpha_1 = \frac{5}{h}$, $\alpha_2 = -\frac{5}{h}$, $\alpha_3 = \frac{10}{3h}$, $\alpha_4 = -\frac{5}{4h}$, $\alpha_5 = \frac{1}{5h}$. Substituting these values in (6.5.1), we find that

$$x'_0 = -\frac{137}{60h} \ x_0 + \frac{5}{h} \ x_1 - \frac{5}{h} \ x_2 + \frac{10}{3h} \ x_3 - \frac{5}{4h} \ x_4 + \frac{1}{5h} \ x_5. \tag{6.5.3}$$

Similarly, we approximate x'_{N+1} by $\sum_{i=0}^{5} \beta_i \ x_{N-4+i}$, and obtain

$$x'_{N+1} = -\frac{1}{5h} \ x_{N-4} + \frac{5}{4h} \ x_{N-3} - \frac{10}{3h} \ x_{N-2} + \frac{5}{h} \ x_{N-1} - \frac{5}{h} \ x_N + \frac{137}{60h} \ x_{N+1}, \tag{6.5.4}$$

which also has $O(h^6)$ truncation error.

Definition 6.5.1. We say $S_1x(t)$ is an *approximate* for $S^{\Delta}_{3,4}x(t)$ if $S_1x(t) \in S_{3,4}(\Delta)$ with $S_1x(t_i) = x_i$, $0 \le i \le N+1$ and $DS_1x(t_i) = \tilde{x}'_i$, $D^2S_1x(t_i) = x''_i$, $i = 0, N+1$ where $\tilde{x}'_0$ and $\tilde{x}'_{N+1}$ are the right sides of (6.5.3) and (6.5.4), respectively.

We use (6.5.3) and (6.5.4) to replace x'_0 and x'_{N+1} in the systems (6.4.3) - (6.4.5) and (6.4.6) - (6.4.8) and observe that the resulting unknowns c'_{1i} and c''_{1i}, say, can be obtained uniquely in terms of x_i, $0 \le i \le N+1$, and x''_0 and x''_{N+1}. Further, by Remark 6.4.1, $S_1x(t)$ can be explicitly expressed as

$$(6.5.5)\quad S_1x(t) = \sum_{i=0}^{N+1} h_{3,i,0}(t)x_i + h_{3,0,1}(t)\sum_{i=0}^{5} \alpha_i\, x_i + h_{3,N+1,1}(t)\sum_{i=0}^{5} \beta_i\, x_{N-4+i}$$
$$+ \sum_{i=1}^{N} h_{3,i,1}(t)c'_{1i} + h_{3,0,2}(t)x''_0 + h_{3,N+1,2}(t)x''_{N+1} + \sum_{i=1}^{N} h_{3,i,2}(t)c''_{1i}\ .$$

To obtain a priori bound for $\| D^k(x - S_1x) \|_\infty$, $0 \le k \le 5$ we use the inequality

$$(6.5.6)\quad \| D^k(x - S_1x) \|_\infty \le \| D^k(x - S^{\Delta}_{3,4}x) \|_\infty + \| D^k(S^{\Delta}_{3,4}x - S_1x) \|_\infty\ , \qquad 0 \le k \le 5$$

in which the first term of the right side can be estimated from Theorem 6.4.6, whereas for the second term we proceed as follows : from Remark 6.4.1 and (6.5.5), we have

$$(6.5.7)\qquad (S_1x - S^{\Delta}_{3,4}x)(t) = h_{3,0,1}(t)\ \theta'_0 + h_{3,N+1,1}(t)\ \theta'_{N+1}$$
$$+\sum_{i=1}^{N} h_{3,i,1}(t)\ \theta'_i + \sum_{i=1}^{N} h_{3,i,2}(t)\ \theta''_i,$$

where

$$\theta'_0 = \sum_{i=0}^{5} \alpha_i\, x_i - x'_0, \qquad \theta'_{N+1} = \sum_{i=0}^{5} \beta_i\, x_{N-4+i} - x'_{N+1},$$
$$\theta'_i = c'_{1i} - c'_i \qquad \text{and} \qquad \theta''_i = c''_{1i} - c''_i.$$

Denoting the vectors $[\theta'_i]$ and $[\theta''_i]$ by θ^1 and θ^2 respectively, from (6.5.7) we find that

(6.5.8) $\| D^k(S_1 x - S^{\Delta}_{3,4}x) \|_\infty$

$$\leq \left(| \theta'_0 | + | \theta'_{N+1} |\right) \max_{0\leq i\leq N} \max_{t_i\leq t\leq t_{i+1}} | h^{(k)}_{3,i,1}(t) |$$

$$+ \| \theta^1 \|_\infty \max_{0\leq i\leq N} \max_{t_i\leq t\leq t_{i+1}} \left[| h^{(k)}_{3,i,1}(t) | + | h^{(k)}_{3,i+1,1}(t) |\right]$$

$$+ \| \theta^2 \|_\infty \max_{0\leq i\leq N} \max_{t_i\leq t\leq t_{i+1}} \left[| h^{(k)}_{3,i,2}(t) | + | h^{(k)}_{3,i+1,2}(t) |\right], \quad 0 \leq k \leq 5$$

where we have used the facts that $| h^{(k)}_{3,i,1}(t) |$ is symmetrical about the line $t = t_i$, and $| h^{(k)}_{3,i,1}(t) |$, $| h^{(k)}_{3,i,2}(t) |$ are nonzero only in the interval $[t_{i-1}, t_i] \bigcup [t_i, t_{i+1}]$.

Lemma 6.5.1. If $x(t) \in PC^{6,\infty}[a, b]$, then

(6.5.9) $$| \theta'_0 | \leq \frac{1}{6} h^5 \| D^6 x \|_\infty$$

and

(6.5.10) $$| \theta'_{N+1} | \leq \frac{1}{6} h^5 \| D^6 x \|_\infty .$$

Proof. We shall prove only (6.5.9), whereas the proof of (6.5.10) is similar. Denoting θ'_0 by $r(x)$, we have

(6.5.11) $$r(x) = \frac{1}{h}\left(-\frac{137}{60} x_0 + 5 x_1 - 5 x_2 + \frac{10}{3} x_3 - \frac{5}{4} x_4 + \frac{1}{5} x_5\right) - x'_0$$

and hence by Theorem 5.2.1 it follows that

(6.5.12) $$r(x) = \frac{1}{5!}\int_{t_0}^{t_5} (r)_t(t-s)^5_+ \; D^6 x(s)\, ds,$$

where

$$(r)_t(t-s)^5_+ = \frac{1}{60h}\left[-137(t_0-s)^5_+ + 300(t_1-s)^5_+ - 300(t_2-s)^5_+ \right.$$
$$\left. + 200(t_3-s)^5_+ - 75(t_4-s)^5_+ + 12(t_5-s)^5_+ - 300h(t_0-s)^4_+\right],$$

which is nonnegative for all $s \in [t_0, t_5]$. Thus, from (6.5.12) we find that

$$| r(x) | \leq \frac{1}{5!} \| D^6 x \|_\infty \int_{t_0}^{t_5} (r)_t(t-s)^5_+ \, ds$$
$$= \frac{1}{6} h^5 \| D^6 x \|_\infty . \quad \blacksquare$$

Lemma 6.5.2. If $x(t) \in PC^{6,\infty}[a,b]$, then

$$\| \theta^1 \|_\infty \le \frac{37}{24} h^5 \| D^6 x \|_\infty . \tag{6.5.13}$$

Proof. Letting $c^1 = [c_i']$ and $c_1^1 = [c_{1i}']$, we have the systems $B_1\, c^1 = k^1$, and $B_1\, c_1^1 = k_1^1$, where the matrix B_1 is defined in (6.4.14), k^1 is given in (6.4.15), and k_1^1 is the vector obtained from k^1 by replacing x_0' and x_{N+1}' with (6.5.3) and (6.5.4), respectively. Thus, it follows that

$$B_1(c_1^1 - c^1) = k_1^1 - k^1. \tag{6.5.14}$$

Now, multiplying both sides of (6.5.14) by the diagonal matrix $\Box = [d_{ij}]$, where $d_{ii} = 1/(ah)$, $a \in \Re^+$, $1 \le i \le N$ we find that

$$\| \theta^1 \|_\infty = \| c_1^1 - c^1 \|_\infty \le \| (\Box\, B_1)^{-1} \|_\infty \, \| \Box(k_1^1 - k^1) \|_\infty . \tag{6.5.15}$$

Writing $\Box\, B_1 = I + A$, where A is an $N \times N$ matrix having the property that $\| A \|_\infty < 1$, from (6.5.15), Lemmas 6.3.4 and 6.5.1 we obtain

$$\begin{aligned} \| \theta^1 \|_\infty &\le \frac{1}{1 - \| A \|_\infty} \| \Box(k_1^1 - k^1) \|_\infty \\ &= \frac{1}{1 - \| A \|_\infty} \frac{1}{ah} \max\Big\{111h \mid \theta_0' \mid,\ \ h \mid \theta_0' \mid, \\ &\qquad\qquad h \mid \theta_{N+1}' \mid,\ \ 111h \mid \theta_{N+1}' \mid \Big\} \\ &\le \frac{1}{1 - \| A \|_\infty} \frac{1}{ah}\, 111h \left(\frac{1}{6}\, h^5 \| D^6 x \|_\infty\right) \\ &= \frac{1}{(1 - \| A \|_\infty) a} \frac{111}{6} h^5 \| D^6 x \|_\infty . \end{aligned} \tag{6.5.16}$$

To obtain the smallest bound in (6.5.16), we need to maximize $(1 - \| A \|_\infty)\, a$ over $a \in \Re^+$. For this, the proof of Lemma 6.4.4 provides that

$$\max_{a \in \Re^+} (1 - \| A \|_\infty)\, a = 12$$

(subject to the restriction that $\| A \|_\infty < 1$). Hence, from (6.5.16) we have

$$\| \theta^1 \|_\infty \le \frac{1}{12} \frac{111}{6} h^5 \| D^6 x \|_\infty = \frac{37}{24} h^5 \| D^6 x \|_\infty . \qquad \blacksquare \tag{6.5.17}$$

Lemma 6.5.3. If $x(t) \in PC^{6,\infty}[a,b]$, then

$$\| \theta^2 \|_\infty \leq \frac{10}{3} h^4 \| D^6 x \|_\infty . \tag{6.5.18}$$

Proof. The proof is similar to that of Lemma 6.5.2. ∎

Theorem 6.5.4. If $x(t) \in PC^{6,\infty}[a,b]$, then

$$\| D^k(x - S_1 x) \|_\infty \leq \alpha^1_{3,6,k}\, h^{6-k} \| D^6 x \|_\infty , \quad 0 \leq k \leq 5 \tag{6.5.19}$$

where $\alpha^1_{3,6,0} = 0.65883455+0$, $\alpha^1_{3,6,1} = 0.22180011+1$, $\alpha^1_{3,6,2} = 0.13714396+2$, $\alpha^1_{3,6,3} = 0.14637518+3$, $\alpha^1_{3,6,4} = 0.82926759+3$, and $\alpha^1_{3,6,5} = 0.16508352+4$.

Proof. We use Lemmas 5.3.4, 5.3.5, 6.5.1, 6.5.2 and 6.5.3 in (6.5.8) to obtain an upper estimate for $\| D^k(S_1 x - S^{\Delta}_{3,4} x) \|_\infty$, this estimate together with Theorem 6.4.6 in (6.5.6) then gives (6.5.19). ∎

Case 2 As in Case 1 we approximate x''_0 and x''_{N+1} by the formulae

$$\begin{aligned} x''_0 &= \frac{1}{h^2}\left(\frac{15}{4}\, x_0 - \frac{77}{6}\, x_1 + \frac{107}{6}\, x_2 - 13\, x_3 + \frac{61}{12}\, x_4 - \frac{5}{6}\, x_5\right) \\ &= \sum_{i=0}^{5} \gamma_i\, x_i \end{aligned} \tag{6.5.20}$$

and

$$\begin{aligned} x''_{N+1} &= \frac{1}{h^2}\left(-\frac{5}{6}\, x_{N-4} + \frac{61}{12}\, x_{N-3} - 13\, x_{N-2} + \frac{107}{6}\, x_{N-1}\right. \\ &\qquad \left. - \frac{77}{6}\, x_N + \frac{15}{4}\, x_{N+1}\right) \\ &= \sum_{i=0}^{5} \delta_i\, x_{N-4+i}, \end{aligned} \tag{6.5.21}$$

which have $O(h^6)$ truncation error.

Definition 6.5.2. We say $S_2 x(t)$ is an *approximate* for $S^{\Delta}_{3,4} x(t)$ if $S_2 x(t) \in S_{3,4}(\Delta)$ with $S_2 x(t_i) = x_i$, $0 \leq i \leq N+1$ and $DS_2 x(t_i) = x'_i$, $D^2 S_2 x(t_i) = \tilde{x}''_i$, $i = 0, N+1$ where $\tilde{x}''_0$ and $\tilde{x}''_{N+1}$ are the right sides of (6.5.20) and (6.5.21), respectively.

Theorem 6.5.5. If $x(t) \in PC^{6,\infty}[a,b]$, then

$$\| D^k(x - S_2x) \|_\infty \le \alpha^2_{3,6,k}\; h^{6-k} \| D^6 x \|_\infty\,, \quad 0 \le k \le 5 \tag{6.5.22}$$

where $\alpha^2_{3,6,0} = 0.39607444 + 0$, $\alpha^2_{3,6,1} = 0.12806265 + 1$, $\alpha^2_{3,6,2} = 0.90068894 + 1$, $\alpha^2_{3,6,3} = 0.93969629 + 2$, $\alpha^2_{3,6,4} = 0.51792870 + 3$, and $\alpha^2_{3,6,5} = 0.10178907 + 4$.

Proof. The proof is similar to that of Theorem 6.5.4. ∎

Case 3

Definition 6.5.3. We say $S_3x(t)$ is an *approximate* for $S^{\Delta}_{3,4}x(t)$ if $S_3x(t) \in S_{3,4}(\Delta)$ with $S_3x(t_i) = x_i$, $0 \le i \le N+1$ and $DS_3x(t_i) = \tilde{x}'_i$, $D^2S_2x(t_i) = \tilde{x}''_i$, $i = 0, N+1$ where $\tilde{x}'_0$, $\tilde{x}'_{N+1}$, $\tilde{x}''_0$ and $\tilde{x}''_{N+1}$ are the right sides of (6.5.3), (6.5.4), (6.5.20) and (6.5.21), respectively.

Theorem 6.5.6. If $x(t) \in PC^{6,\infty}[a,b]$, then

$$\| D^k(x - S_3x) \|_\infty \le \alpha^3_{3,6,k}\; h^{6-k} \| D^6 x \|_\infty\,, \quad 0 \le k \le 5 \tag{6.5.23}$$

where $\alpha^3_{3,6,0} = 0.32242154 + 0$, $\alpha^3_{3,6,1} = 0.11660016 + 1$, $\alpha^3_{3,6,2} = 0.79187860 + 1$, $\alpha^3_{3,6,3} = 0.81680740 + 2$, $\alpha^3_{3,6,4} = 0.43120648 + 3$, and $\alpha^3_{3,6,5} = 0.83644629 + 3$.

Proof. The proof is similar to that of Theorem 6.5.4. ∎

Remark 6.5.1. From Theorems 6.5.4 - 6.5.6 we find that

$$\begin{aligned} \text{bound for } \| D^k(x - S_3x) \|_\infty &< \text{bound for } \| D^k(x - S_2x) \|_\infty \\ &< \text{bound for } \| D^k(x - S_1x) \|_\infty\,, \quad 0 \le k \le 5. \end{aligned}$$

The first inequality comes as a surprise because it amounts to 'knowing more from less information'. ∎

Case 4

Definition 6.5.4. We say $S_4x(t)$ is an *approximate* for $S^{\Delta}_{3,4}x(t)$ if $S_4x(t) \in S_{3,4}(\Delta)$ with $S_4x(t_i) = g_i$, $0 \le i \le N+1$, where the given g_i, $0 \le i \le N+1$ are such that

$$\max_{0 \le i \le N+1} | x_i - g_i | = \xi,$$

and $DS_4x(t_i) = \tilde{x}'_i$, $D^2S_4x(t_i) = \tilde{x}''_i$, $i = 0, N+1$ where $\tilde{x}'_0$, $\tilde{x}'_{N+1}$, $\tilde{x}''_0$ and $\tilde{x}''_{N+1}$ are the right sides of (6.5.3), (6.5.4), (6.5.20) and (6.5.21) respectively, with x_i being replaced by g_i.

We use (6.5.3), (6.5.20), (6.5.4) and (6.5.21) to replace x'_0, x''_0, x'_{N+1} and x''_{N+1} in the systems (6.4.3) - (6.4.5) and (6.4.6) - (6.4.8), and then replace x_i by g_i everywhere. The resulting systems can be solved uniquely for the unknowns c'_{4i} and c''_{4i}, $1 \le i \le N$, say, in terms of the known g_i, $0 \le i \le N+1$. By Remark 6.4.1, $S_4x(t)$ can be explicitly expressed as

$$\text{(6.5.24)}\quad S_4x(t) = \sum_{i=0}^{N+1} h_{3,i,0}(t)g_i + h_{3,0,1}(t)\sum_{i=0}^{5}\alpha_i\, g_i + h_{3,N+1,1}(t)\sum_{i=0}^{5}\beta_i\, g_{N-4+i}$$
$$+\sum_{i=1}^{N} h_{3,i,1}(t)c'_{4i} + h_{3,0,2}(t)\sum_{i=0}^{5}\gamma_i\, g_i + h_{3,N+1,2}(t)\sum_{i=0}^{5}\delta_i\, g_{N-4+i}$$
$$+\sum_{i=1}^{N} h_{3,i,2}(t)c''_{4i}\ .$$

Theorem 6.5.7. If $x(t) \in PC^{6,\infty}[a,b]$, then

$$\text{(6.5.25)}\quad \| D^k(x - S_4x) \|_\infty \le \alpha^3_{3,6,k}\, h^{6-k} \| D^6x \|_\infty + \alpha^4_{3,6,k}\, h^{-k}\, \xi,\quad 0 \le k \le 5$$

where $\alpha^4_{3,6,0} = 0.39736194+2$, $\alpha^4_{3,6,1} = 0.14098689+3$, $\alpha^4_{3,6,2} = 0.87793911+3$, $\alpha^4_{3,6,3} = 0.90657999+4$, $\alpha^4_{3,6,4} = 0.49719433+5$, and $\alpha^4_{3,6,5} = 0.97339666+5$.

Proof. To obtain a bound for $\| D^k(x - S_4x) \|_\infty$, $0 \le k \le 5$ we use the inequality

$$\| D^k(x - S_4x) \|_\infty \le \| D^k(x - S_3x) \|_\infty + \| D^k(S_3x - S_4x) \|_\infty\ ,\quad 0 \le k \le 5.$$

In the above inequality for the first term we use Theorem 6.5.6, whereas for the second term we follow a similar technique as in the proof of Theorem 6.5.4. ∎

Remark 6.5.2. Like in our previous results the sharpness of the inequalities (6.5.19), (6.5.22), (6.5.23) and (6.5.25) remains undecided. However, in Tables 6.5.1 - 6.5.3 we compute the actual values of $\| D^k(x - S_ix) \|_\infty$; $1 \le i \le 3$, $0 \le k \le 5$ for the function $x(t) = t^6 \mid t \mid$ in the interval $[-1,1]$ and compare these with the corresponding right side bounds.

Table 6.5.1.

N	9	49	99
$\|\| x - S_1x \|\|_\infty$	0.81546801 − 2	0.83481420 − 6	0.13655007 − 7
Bound	0.21251367 + 0	0.13600875 − 4	0.21251367 − 6
$\|\| D(x - S_1x) \|\|_\infty$	0.15360010 + 0	0.78643200 − 4	0.25728012 − 5
Bound	0.35771921 + 1	0.11447014 − 2	0.35771921 − 4
$\|\| D^2(x - S_1x) \|\|_\infty$	0.23153921 + 1	0.59254415 − 2	0.38769432 − 3
Bound	0.11059288 + 3	0.17694862 + 0	0.11059288 − 1
$\|\| D^3(x - S_1x) \|\|_\infty$	0.74187803 + 2	0.95107505 + 0	0.12447521 + 0
Bound	0.59018472 + 4	0.47214778 + 2	0.59018472 + 1
$\|\| D^4(x - S_1x) \|\|_\infty$	0.13829211 + 4	0.88952221 + 2	0.23290201 + 2
Bound	0.16718034 + 6	0.66872138 + 4	0.16718034 + 4
$\|\| D^5(x - S_1x) \|\|_\infty$	0.97431430 + 4	0.30811523 + 4	0.16114632 + 4
Bound	0.16640418 + 7	0.33280837 + 6	0.16640418 + 6

Table 6.5.2.

N	9	49	99
$\|\| x - S_2x \|\|_\infty$	0.43575311 − 2	0.43834912 − 6	0.71605234 − 8
Bound	0.12775777 + 0	0.81764973 − 5	0.12775777 − 6
$\|\| D(x - S_2x) \|\|_\infty$	0.65026307 − 1	0.32698516 − 4	0.10682621 − 5
Bound	0.20653944 + 1	0.66092621 − 3	0.20653944 − 4
$\|\| D^2(x - S_2x) \|\|_\infty$	0.35840025 + 1	0.90030124 − 2	0.58822443 − 3
Bound	0.72631556 + 2	0.11621048 + 0	0.72631556 − 2
$\|\| D^3(x - S_2x) \|\|_\infty$	0.11999234 + 3	0.15051420 + 1	0.19666306 + 0
Bound	0.37888554 + 4	0.30310843 + 2	0.37888554 + 1
$\|\| D^4(x - S_2x) \|\|_\infty$	0.17614431 + 4	0.11010601 + 3	0.28766213 + 2
Bound	0.10441442 + 6	0.41765770 + 4	0.10441442 + 4
$\|\| D^5(x - S_2x) \|\|_\infty$	0.11082908 + 5	0.34227521 + 4	0.17868310 + 4
Bound	0.10260338 + 7	0.20520676 + 6	0.10260338 + 6

Table 6.5.3.

N	9	49	99
$\Vert x - S_3x \Vert_\infty$	0.42437823 − 2	0.44080505 − 6	0.72180631 − 8
Bound	0.10400029 + 0	0.66560186 − 5	0.10400029 − 6
$\Vert D(x - S_3x) \Vert_\infty$	0.15360031 + 0	0.78643219 − 4	0.25728005 − 5
Bound	0.18805273 + 1	0.60176876 − 3	0.18805273 − 4
$\Vert D^2(x - S_3x) \Vert_\infty$	0.35840027 + 1	0.90030128 − 2	0.58822432 − 3
Bound	0.63857090 + 2	0.10217134 + 0	0.63857090 − 2
$\Vert D^3(x - S_3x) \Vert_\infty$	0.45429108 + 2	0.55106635 + 0	0.71814127 − 1
Bound	0.32933674 + 4	0.26346939 + 2	0.32933674 + 1
$\Vert D^4(x - S_3x) \Vert_\infty$	0.35689011 + 3	0.20268804 + 2	0.52541733 + 1
Bound	0.86931226 + 5	0.34772490 + 4	0.86931226 + 3
$\Vert D^5(x - S_3x) \Vert_\infty$	0.17308821 + 4	0.43259441 + 3	0.22167000 + 3
Bound	0.84313786 + 6	0.16862757 + 6	0.84313786 + 5

■

6.6 QUINTIC SPLINE INTERPOLATION : $r = 3$

We need the following results.

Lemma 6.6.1. For a given $h(t) \in H_3(\Delta)$, we define $c_i = h(t_i)$, $c'_i = Dh(t_i)$ and $c''_i = D^2h(t_i)$, $0 \le i \le N+1$. Then, $h(t) \in S_{3,3}(\Delta)$ if and only if c''_i, $1 \le i \le N$ satisfy the following relations

$$(6.6.1) \quad h^2(c''_{i-1} - 6\, c''_i + c''_{i+1}) = -20(c_{i-1} - 2\, c_i + c_{i+1}) - 8h(c'_{i-1} - c'_{i+1}),$$

$$1 \le i \le N.$$

Moreover, from the system (6.6.1) the unknowns c''_i, $1 \le i \le N$ can be obtained uniquely in terms of c_i, c'_i, $0 \le i \le N+1$, c''_0 and c''_{N+1}.

Proof. By Lemma 6.4.1 the continuity of $D^3h(t)$ is equivalent to the system (6.6.1). This system in matrix form can be written as

$$B\,c^2 = k, \tag{6.6.2}$$

where $c^2 = [c_i'']$,

$$B = h^2 \begin{bmatrix} -6 & 1 & & & & \\ 1 & -6 & 1 & & & \\ & 1 & -6 & 1 & & \\ & \cdots & \cdots & \cdots & & \\ & & & 1 & -6 & 1 \\ & & & & 1 & -6 \end{bmatrix} \tag{6.6.3}$$

and $k = [k_i]$,

$$\begin{aligned} k_i &= -20(c_0 - 2c_1 + c_2) - 8h(c_0' - c_2') - h^2 c_0'', \; i = 1 \\ &= -20(c_{i-1} - 2c_i + c_{i+1}) - 8h(c_{i-1}' - c_{i+1}'), \; 2 \le i \le N-1 \\ &= -20(c_{N-1} - 2c_N + c_{N+1}) - 8h(c_{N-1}' - c_{N+1}') - h^2 c_{N+1}'', \; i = N. \end{aligned} \tag{6.6.4}$$

Since the matrix B is strictly diagonally dominant, the system (6.6.2) has a unique solution. ∎

Lemma 6.6.2. For a given $x(t) \in C^{(2)}[a,b]$, $S_{3,3}^{\Delta}x(t)$ exists and is unique.

Proof. The proof is similar to that of Lemma 6.3.3. ∎

Remark 6.6.1. From Lemma 6.6.2 and (6.2.2) it is clear that $S_{3,3}^{\Delta}x(t)$ can be expressed as

$$\begin{aligned} S_{3,3}^{\Delta}x(t) = \sum_{i=0}^{N+1} [h_{3,i,0}(t)x_i + h_{3,i,1}(t)x_i'] + h_{3,0,2}(t)x_0'' + h_{3,N+1,2}(t)x_{N+1}'' \\ + \sum_{i=1}^{N} h_{3,i,2}(t)c_i'', \end{aligned} \tag{6.6.5}$$

where c_i'', $1 \le i \le N$ satisfy (6.6.1). ∎

Remark 6.6.2. It is possible to describe a basis for $S_{3,3}(\Delta)$, namely the 'cardinal splines', $\{s_i(t)\}_{i=0}^{2N+5}$, defined by the following interpolation conditions :

$$s_j(t_i) = \delta_{ij},\ D^k s_j(a) = D^k s_j(b) = 0;\ k = 1,2,\ 0 \le i,j \le N+1$$

$$s_j(t_i) = 0,\ Ds_j(t_i) = \delta_{i,j-N-2},\ D^2 s_j(t_i) = 0;\ 0 \le i \le N+1,$$
$$N+2 \le j \le 2N+3$$

$$D^k s_{2N+4}(t_i) = 0;\ k = 0,1,\ 0 \le i \le N+1,\ D^2 s_{2N+4}(a) = 1,\ D^2 s_{2N+4}(b) = 0$$

$$D^k s_{2N+5}(t_i) = 0;\ k = 0,1,\ 0 \le i \le N+1,\ D^2 s_{2N+5}(a) = 0,\ D^2 s_{2N+5}(b) = 1.$$

Obviously, $S^{\Delta}_{3,3}x(t)$ can be explicitly expressed as

(6.6.6) $$S^{\Delta}_{3,3}x(t) = \sum_{i=0}^{N+1} [s_i(t)x_i + s_{i+N+2}(t)x'_i] + s_{2N+4}(t)x''_0 + s_{2N+5}(t)x''_{N+1}. \ \blacksquare$$

Lemma 6.6.3. If $x(t) \in PC^{n,\infty}[a,b]$, $2 \le n \le 6$ then

(6.6.7) $$\| e^2 \|_\infty \le \bar{b}_n\ h^{n-2} \| D^n x \|_\infty ,$$

where $\bar{b}_2 = \frac{23}{5}$, $\bar{b}_3 = \left(\frac{1}{6} + \frac{2\sqrt{6}}{25}\right)$, $\bar{b}_4 = \frac{2\sqrt{6}}{125}$, $\bar{b}_5 = \frac{18}{3125}$ and $\bar{b}_6 = \frac{1}{720}$.

Proof. The proof for the case $n = 6$ is indicated in Hall [7]. Here, we shall provide the proof only for the case $n = 3$. Let $r = [r_i(x)]$ be an $N \times 1$ vector defined by

(6.6.8) $$r = B\ e^2,$$

where the matrix B is given in (6.6.3). Then, it follows that

(6.6.9) $$B\ x^2 = k + r,$$

where k is defined in (6.6.4) and $x^2 = [x''_i]$ is an $N \times 1$ vector.

For $1 \le i \le N$, from (6.6.3) and (6.6.4), we have

(6.6.10) $$r_i(x) = h^2(x''_{i-1} - 6x''_i + x''_{i+1}) + 20(x_{i-1} - 2x_i + x_{i+1}) + 8h(x'_{i-1} - x'_{i+1}),$$

which in view of Theorem 5.2.1 can be written as (6.3.12), where

$(r_i)_t(t-s)_+^2$

$$= 2h^2\left[(t_{i-1}-s)_+^0 - 6(t_i-s)_+^0 + (t_{i+1}-s)_+^0\right]$$

$$+ 20\left[(t_{i-1}-s)_+^2 - 2(t_i-s)_+^2 + (t_{i+1}-s)_+^2\right] + 16h\left[(t_{i-1}-s)_+ - (t_{i+1}-s)_+\right]$$

$$= \begin{cases} -10h^2 - 40(t_i-s)^2 + 20(t_{i+1}-s)^2 - 16h(t_{i+1}-s) = \alpha(s), & s \in [t_{i-1}, t_i] \\ 2h^2 + 20(t_{i+1}-s)^2 - 16h(t_{i+1}-s) = \beta(s), & s \in [t_i, t_{i+1}]. \end{cases}$$

Since

$$|\,\alpha(s)\,| = \begin{cases} 20S^2 - 24hS + 6h^2, \; S = t_i - s, \; S \in \left[0, \frac{6-\sqrt{6}}{10}h\right] \cup \left[\frac{6+\sqrt{6}}{10}h, h\right] \\ -20S^2 + 24hS - 6h^2, \; S \in \left[\frac{6-\sqrt{6}}{10}h, \frac{6+\sqrt{6}}{10}h\right] \end{cases}$$

and

$$|\,\beta(s)\,| = \begin{cases} 20\bar{S}^2 - 16h\bar{S} + 2h^2, \; \bar{S} = t_{i+1} - s, \; \bar{S} \in \left[0, \frac{4-\sqrt{6}}{10}h\right] \cup \left[\frac{4+\sqrt{6}}{10}h, h\right] \\ -20\bar{S}^2 + 16h\bar{S} - 2h^2, \; \bar{S} \in \left[\frac{4-\sqrt{6}}{10}h, \frac{4+\sqrt{6}}{10}h\right] \end{cases}$$

from (6.3.12) it follows that

$$|\,r_i(x)\,| \le \frac{1}{2} \,\|\, D^3x \,\|_\infty \left\{ \int_0^{\frac{6-\sqrt{6}}{10}h} (20S^2 - 24hS + 6h^2)dS \right.$$

$$+ \int_{\frac{6-\sqrt{6}}{10}h}^{\frac{6+\sqrt{6}}{10}h} (-20S^2 + 24hS - 6h^2)dS + \int_{\frac{6+\sqrt{6}}{10}h}^{h} (20S^2 - 24hS + 6h^2)dS$$

$$+ \int_0^{\frac{4-\sqrt{6}}{10}h} (20\bar{S}^2 - 16h\bar{S} + 2h^2)d\bar{S} + \int_{\frac{4-\sqrt{6}}{10}h}^{\frac{4+\sqrt{6}}{10}h} (-20\bar{S}^2 + 16h\bar{S} - 2h^2)d\bar{S}$$

$$\left. + \int_{\frac{4+\sqrt{6}}{10}h}^{h} (20\bar{S}^2 - 16h\bar{S} + 2h^2)d\bar{S} \right\}$$

$$\text{(6.6.11)} \quad = 4\left(\frac{1}{6} + \frac{2\sqrt{6}}{25}\right) h^3 \,\|\, D^3x \,\|_\infty \;.$$

Now as in Lemma 6.4.4 we multiply both sides of (6.6.8) by the diagonal matrix $\square = [d_{ij}]$, where $d_{ii} = -1/(ah^2)$, $a \in \Re^+$, $1 \le i \le N$ to obtain

$$\text{(6.6.12)} \qquad \|\, e^2 \,\|_\infty \le \frac{1}{1 - \|\, A \,\|_\infty} \frac{1}{ah^2} \max_{1 \le i \le N} |\, r_i(x)\,|,$$

where $\square B = I + A$ with $\| A \|_\infty < 1$.

To find the smallest bound in (6.6.12), we need to maximize $(1 - \| A \|_\infty)a$ over $a \in \Re^+$. For this, from (6.6.3) we have

$$\| A \|_\infty = \left| \frac{6}{a} - 1 \right| + \frac{2}{a} = \begin{cases} \dfrac{8-a}{a}, & a \le 6 \\ \dfrac{a-4}{a}, & a \ge 6. \end{cases} \tag{6.6.13}$$

Thus, the condition $\| A \|_\infty < 1$ is equivalent to $\frac{8}{a} - 1 < 1$ $(1 - \frac{4}{a} < 1$ for $a \ge 6)$ which gives that $a > 4$. Hence, on using (6.6.13) we find

$$\max_{a>4} (1 - \| A \|_\infty)a = \max \left\{ \max_{4<a\le 6}(2a-8), \ \max_{a\ge 6}(4) \right\} = 4. \tag{6.6.14}$$

Using (6.6.11) and (6.6.14) in (6.6.12), we obtain

$$\| e^2 \|_\infty \le \left(\frac{1}{6} + \frac{2\sqrt{6}}{25} \right) h \| D^3 x \|_\infty . \quad \blacksquare$$

Theorem 6.6.4. Let $x(t) \in PC^{n,\infty}[a, b]$, $2 \le n \le 6$. Then,

$$\| D^k(x - S^{\Delta}_{3,3}x) \|_\infty \le \beta_{3,n,k} \, h^{n-k} \| D^n x \|_\infty , \quad 0 \le k \le n-1 \tag{6.6.15}$$

where the constants $\beta_{3,n,k}$ are given in the following table.

Table 6.6.1.

k \ n	2	3	4	5	6
0	0.24140625 + 0	0.17516953 − 1	0.19017601 − 2	0.27765625 − 3	0.65104166 − 4
1	0.80127342 + 0	0.56949061 − 1	0.62313767 − 2	0.89471171 − 3	0.20818149 − 3
2		0.57095918 + 0	0.57710354 − 1	0.83641667 − 2	0.19097222 − 2
3			0.70545305 + 0	0.10368000 + 0	0.25000000 − 1
4				0.74066173 + 0	0.18333333 + 0
5					0.66666667 + 0

Proof. Using Lemma 5.3.4, Theorem 5.3.8 and Lemma 6.6.3 in (6.2.6) the inequalities (6.6.15) are immediate. $\blacksquare$

Remark 6.6.3. As in Remark 6.4.3 in the following tables we compute the actual values of $\| D^k(x - S^{\Delta}_{3,3}x) \|_\infty$ for the same functions considered in Tables 6.4.2 and 6.4.3 and compare these with the corresponding right side bounds in (6.6.15).

Table 6.6.2.

$$x(t) = t^5 \sin\frac{1}{t}, \quad t \in \left[-\frac{1}{4}, \frac{1}{4}\right], \quad n = 2$$

N	7	15	19
$\| x - S^{\Delta}_{3,3}x \|_\infty$	0.63851462 – 5	0.20424209 – 5	0.14921598 – 5
Bound	0.26358701 – 3	0.65896752 – 4	0.42173922 – 4
$\| D(x - S^{\Delta}_{3,3}x) \|_\infty$	0.48483759 – 3	0.26367983 – 3	0.21308192 – 3
Bound	0.13998330 – 1	0.69991648 – 2	0.55993318 – 2

Table 6.6.3.

$$x(t) = t^6 \mid t \mid, \quad t \in [-1, 1], \quad n = 6$$

N	9	49	99
$\| x - S^{\Delta}_{3,3}x \|_\infty$	0.81178760 – 9	0.72885423 – 13	0.12029189 – 14
Bound	0.12817383 – 8	0.82031250 – 13	0.12817383 – 14
$\| D(x - S^{\Delta}_{3,3}x) \|_\infty$	0.51352716 – 7	0.22856072 – 10	0.75354019 – 12
Bound	0.81971464 – 7	0.26230868 – 10	0.81971464 – 12
$\| D^2(x - S^{\Delta}_{3,3}x) \|_\infty$	0.68897281 – 5	0.15584850 – 7	0.10226489 – 8
Bound	0.15039063 – 4	0.24062500 – 7	0.15039063 – 8
$\| D^3(x - S^{\Delta}_{3,3}x) \|_\infty$	0.16124777 – 2	0.14459242 – 4	0.18317840 – 5
Bound	0.39375000 – 2	0.31500000 – 4	0.39375000 – 5
$\| D^4(x - S^{\Delta}_{3,3}x) \|_\infty$	0.35699643 + 0	0.15719394 – 1	0.39748519 – 2
Bound	0.57750000 + 0	0.23100000 – 1	0.57750000 – 2
$\| D^5(x - S^{\Delta}_{3,3}x) \|_\infty$	0.32999822 + 2	0.70558485 + 1	0.35564254 + 1
Bound	0.42000000 + 2	0.84000000 + 1	0.42000000 + 1

∎

6.7 APPROXIMATED QUINTIC SPLINES : $\tau = 3$

Following Section 6.5, for a given function $x(t) \in PC^{6,\infty}[a,b]$ and a fixed partition Δ we shall construct approximates for the spline interpolate $S^{\Delta}_{3,3}x(t)$ when

1. the values of x'_i, $0 \le i \le N+1$ are unknown;
2. the values of x''_i, $i = 0, N+1$ are unknown;
3. the values of x'_i, $0 \le i \le N+1$, x''_i, $i = 0, N+1$ are unknown; and
4. the values of x_i, x'_i, $0 \le i \le N+1$, x''_i, $i = 0, N+1$ are unknown.

Case 1 As in Section 6.5 we approximate x'_i, $0 \le i \le N+1$ by the following relations : x'_0 by (6.5.3),

$$(6.7.1)\ x'_1 = -\frac{1}{5h}x_0 - \frac{13}{12h}x_1 + \frac{2}{h}x_2 - \frac{1}{h}x_3 + \frac{1}{3h}x_4 - \frac{1}{20h}x_5,$$

$$(6.7.2)\ x'_i = \frac{1}{20h}x_{i-2} - \frac{1}{2h}x_{i-1} - \frac{1}{3h}x_i + \frac{1}{h}x_{i+1} - \frac{1}{4h}x_{i+2} + \frac{1}{30h}x_{i+3},$$

$$2 \le i \le N-2$$

$$(6.7.3)\ x'_{N-1} = -\frac{1}{30h}x_{N-4} + \frac{1}{4h}x_{N-3} - \frac{1}{h}x_{N-2} + \frac{1}{3h}x_{N-1} + \frac{1}{2h}x_N - \frac{1}{20h}x_{N+1},$$

$$(6.7.4)\ x'_N = \frac{1}{20h}x_{N-4} - \frac{1}{3h}x_{N-3} + \frac{1}{h}x_{N-2} - \frac{2}{h}x_{N-1} + \frac{13}{12h}x_N + \frac{1}{5h}x_{N+1},$$

and x'_{N+1} by (6.5.4). It is easy to see that each of these relations has $O(h^6)$ truncation error.

Definition 6.7.1. We say $\bar{S}_1x(t)$ is an *approximate* for $S^{\Delta}_{3,3}x(t)$ if $\bar{S}_1x(t) \in S_{3,3}(\Delta)$ with $\bar{S}_1x(t_i) = x_i$, $D\bar{S}_1x(t_i) = \tilde{x}'_i$, $0 \le i \le N+1$ and $D^2\bar{S}_1x(t_i) = x''_i$, $i = 0, N+1$ where $\tilde{x}'_i$, $0 \le i \le N+1$ are the right sides of (6.5.3), (6.7.1) - (6.7.4), and (6.5.4) respectively.

We use (6.5.3), (6.7.1) - (6.7.4), (6.5.4) to replace x'_i, $0 \le i \le N+1$ in the system (6.6.1) and observe that the resulting unknowns $\bar{c}''_{1i}$, say, can be obtained uniquely in terms of x_i, $0 \le i \le N+1$, and x''_0 and x''_{N+1}. Further, by Remark 6.6.1, $\bar{S}_1x(t)$ can be explicitly expressed as

$$(6.7.5)\qquad \bar{S}_1 x(t) = \sum_{i=0}^{N+1} [h_{3,i,0}(t)x_i + h_{3,i,1}(t)\tilde{x}'_i] + h_{3,0,2}(t)x''_0$$

$$+\ h_{3,N+1,2}(t)x''_{N+1} + \sum_{i=1}^{N} h_{3,i,2}(t)\bar{c}''_{1i},$$

where $\tilde{x}'_i$, $0 \le i \le N+1$ are the approximates of x'_i, $0 \le i \le N+1$ given by the above relations.

Now as in Section 6.5, we begin with the inequality

$$(6.7.6)\qquad \| D^k(x - \bar{S}_1 x) \|_\infty \le \| D^k(x - S^{\Delta}_{3,3}x) \|_\infty + \| D^k(S^{\Delta}_{3,3}x - \bar{S}_1 x) \|_\infty, \quad 0 \le k \le 5$$

in which the first term of the right side can be estimated from Theorem 6.6.4, whereas for the second term we proceed as follows : from Remark 6.6.1 and (6.7.5), we have

$$(6.7.7)\qquad (\bar{S}_1 x - S^{\Delta}_{3,3}x)(t) = \sum_{i=0}^{N+1} h_{3,i,1}(t)\ \theta'_i + \sum_{i=1}^{N} h_{3,i,2}(t)\ \theta''_i,$$

where $\theta'_i = \tilde{x}'_i - x'_i$ and $\theta''_i = \bar{c}''_{1i} - c''_i$. Thus, as earlier it follows that

$$(6.7.8)\qquad \| D^k(\bar{S}_1 x - S^{\Delta}_{3,3}x) \|_\infty$$

$$\le \| \theta^1 \|_\infty \max_{0 \le i \le N} \max_{t_i \le t \le t_{i+1}} \left[| h^{(k)}_{3,i,1}(t) | + | h^{(k)}_{3,i+1,1}(t) | \right]$$

$$+ \| \theta^2 \|_\infty \max_{0 \le i \le N} \max_{t_i \le t \le t_{i+1}} \left[| h^{(k)}_{3,i,2}(t) | + | h^{(k)}_{3,i+1,2}(t) | \right], \quad 0 \le k \le 5$$

where θ^1 and θ^2 are the vectors $[\theta'_i]$ and $[\theta''_i]$.

Lemma 6.7.1. If $x(t) \in PC^{6,\infty}[a, b]$, then

$$(6.7.9)\qquad \| \theta^1 \|_\infty \le \max\left\{ \frac{1}{6};\ \frac{1}{30};\ \frac{1}{60},\ 2 \le i \le N-1;\ \frac{1}{30};\ \frac{1}{6} \right\} h^5 \| D^6 x \|_\infty$$

$$(6.7.10)\qquad = \frac{1}{6}\ h^5 \| D^6 x \|_\infty\ .$$

Proof. In Lemma 6.5.1 we have shown that $| \theta'_i | \le \frac{1}{6}\ h^5 \| D^6 x \|_\infty$ for $i = 0$ and $N+1$. Here, we shall prove only that $| \theta'_i | \le \frac{1}{60}\ h^5 \| D^6 x \|_\infty$ for $2 \le i \le N-2$, whereas the proof of this inequality for $i = N-1$, as well as of $|\theta_i'| \le \frac{1}{30} h^5 \| D^6 x \|_\infty$ for $i = 1$ and N is similar. For this, we introduce $r_i(x) = \theta_i'$, $2 \le i \le N-2$ so that

$$r_i(x) = \frac{1}{h}\left(\frac{1}{20}\, x_{i-2} - \frac{1}{2}\, x_{i-1} - \frac{1}{3}\, x_i + x_{i+1} - \frac{1}{4}\, x_{i+2} + \frac{1}{30}\, x_{i+3}\right) - x_i',$$

which in view of Theorem 5.2.1 is the same as

$$r_i(x) = \frac{1}{5!}\int_{t_{i-2}}^{t_{i+3}} (r_i)_t(t-s)_+^5\, D^6 x(s)\, ds, \tag{6.7.11}$$

where

$$\begin{aligned}(r_i)_t(t-s)_+^5 = \frac{1}{h}\Big[\frac{1}{20}(t_{i-2}-s)_+^5 - \frac{1}{2}(t_{i-1}-s)_+^5 - \frac{1}{3}(t_i-s)_+^5 + (t_{i+1}-s)_+^5 \\ -\frac{1}{4}(t_{i+2}-s)_+^5 + \frac{1}{30}(t_{i+3}-s)_+^5 - 5h(t_i-s)_+^4\Big]\end{aligned}$$

$$= \frac{1}{60h}\begin{cases} 3(h-S)^5 = \alpha_1(S), \ S = t_{i-1}-s, \ s \in [t_{i-2}, t_{i-1}], \ S \in [0,h] \\[2mm] 3(9S^5 - 40hS^4 + 60h^2S^3 - 20h^3S^2 - 30h^4S + 22h^5) = \alpha_2(S), \\ \qquad S = t_i - s, \ \ s \in [t_{i-1}, t_i], \ S \in [0,h] \\[2mm] 60S^5 - 15(h+S)^5 + 2(2h+S)^5 = \alpha_3(S), \ S = t_{i+1} - s, \\ \qquad s \in [t_i, t_{i+1}], \ S \in [0,h] \\[2mm] -15S^5 + 2(h+S)^5 = \alpha_4(S), \ S = t_{i+2} - s, \\ \qquad s \in [t_{i+1}, t_{i+2}], \ S \in [0,h] \\[2mm] 2S^5 = \alpha_5(S), \ S = t_{i+3} - s, \ s \in [t_{i+2}, t_{i+3}], \ S \in [0,h]. \end{cases}$$

Since $\alpha_i(S) \ge 0$, $1 \le i \le 5$, $S \in [0,h]$ from (6.7.11), we obtain

$$\begin{aligned} |r_i(x)| &\le \frac{1}{5!}\int_{t_{i-2}}^{t_{i+3}} |(r_i)_t(t-s)_+^5|\, |D^6x(s)|\, ds \\ &= \frac{1}{5!}\| D^6 x \|_\infty \frac{1}{60h}\sum_{i=1}^{5}\int_0^h \alpha_i(S)dS \\ &= \frac{1}{60}\, h^5 \| D^6 x\|_\infty . \qquad \blacksquare \end{aligned}$$

Lemma 6.7.2. If $x(t) \in PC^{6,\infty}[a,b]$, then

$$\| \theta^2 \|_\infty \le \frac{3}{10}\, h^4 \| D^6 x \|_\infty . \tag{6.7.12}$$

Proof. Following as in Lemma 6.5.2 it is easy to see that

$$\| \theta^2 \|_\infty \le \frac{1}{1- \| A \|_\infty} \frac{1}{ah^2} 8h \max_{1\le i\le N} | \theta'_{i+1} - \theta'_{i-1} | . \tag{6.7.13}$$

Further, as in Lemma 6.7.1 we find that

$$| \theta'_2 - \theta'_0 | \le \frac{3}{20} h^5 \| D^6x \|_\infty , \quad | \theta'_3 - \theta'_1 | \le \frac{1}{20} h^5 \| D^6x \|_\infty ,$$

$$| \theta'_{i+1} - \theta'_{i-1} | \le (0.30792196 - 1)h^5 \| D^6x \|_\infty , \quad 3 \le i \le N-3$$

$$| \theta'_{N-1} - \theta'_{N-3} | \le \frac{1}{30} h^5 \| D^6x \|_\infty , \quad | \theta'_N - \theta'_{N-2} | \le \frac{1}{60} h^5 \| D^6x \|_\infty ,$$

$$| \theta'_{N+1} - \theta'_{N-1} | \le \frac{3}{20} h^5 \| D^6x \|_\infty .$$

Using these estimates and the relation $\max_{a>4}(1- \| A \|_\infty)a = 4$, established in Lemma 6.6.3, in (6.7.13) we obtain

$$\| \theta^2 \|_\infty \le \frac{1}{4h^2} 8h \frac{3}{20} h^5 \| D^6x \|_\infty = \frac{3}{10} h^4 \| D^6x \|_\infty . \quad \blacksquare$$

Theorem 6.7.3. If $x(t) \in PC^{6,\infty}[a,b]$, then

$$\| D^k(x - \bar{S}_1x) \|_\infty \le \beta^1_{3,6,k} h^{6-k} \| D^6x \|_\infty , \quad 0 \le k \le 5 \tag{6.7.14}$$

where $\beta^1_{3,6,0} = 0.61523437 - 1$, $\beta^1_{3,6,1} = 0.19574236 + 0$, $\beta^1_{3,6,2} = 0.12641602 + 1$, $\beta^1_{3,6,3} = 0.13625000 + 2$, $\beta^1_{3,6,4} = 0.78183333 + 2$, and $\beta^1_{3,6,5} = 0.15666667 + 3$.

Proof. We use Lemmas 5.3.4, 6.7.1 and 6.7.2 in (6.7.8) to obtain an upper estimate for $\| D^k(S^\Delta_{3,3}x - \bar{S}_1x) \|_\infty$, this estimate together with Theorem 6.6.4 in (6.7.6) then gives (6.7.14). $\blacksquare$

Case 2

Definition 6.7.2. We say $\bar{S}_2x(t)$ is an *approximate* for $S^\Delta_{3,3}x(t)$ if $\bar{S}_2x(t) \in S_{3,3}(\Delta)$ with $D^k\bar{S}_2x(t_i) = x_i^{(k)}$, $0 \le i \le N+1$, $k = 0,1$ and $D^2\bar{S}_2x(t_i) = \tilde{x}''_i$, $i = 0, N+1$ where $\tilde{x}''_0$ and $\tilde{x}''_{N+1}$ are the right sides of (6.5.20) and (6.5.21), respectively.

Theorem 6.7.4. If $x(t) \in PC^{6,\infty}[a,b]$, then

$$\| D^k(x - \bar{S}_2x) \|_\infty \le \beta^2_{3,6,k} h^{6-k} \| D^6x \|_\infty , \quad 0 \le k \le 5 \tag{6.7.15}$$

where $\beta^2_{3,6,0} = 0.32315285 - 1$, $\beta^2_{3,6,1} = 0.12170569 + 0$, $\beta^2_{3,6,2} = 0.17144097 + 1$, $\beta^2_{3,6,3} = 0.16008333 + 2$, $\beta^2_{3,6,4} = 0.66400000 + 2$, and $\beta^2_{3,6,5} = 0.11483333 + 3$.

Proof. The proof is similar to that of Theorem 6.5.5. ∎

Case 3

Definition 6.7.3. We say $\bar{S}_3x(t)$ is an *approximate* for $S^{\Delta}_{3,3}x(t)$ if $\bar{S}_3x(t) \in S_{3,3}(\Delta)$ with $\bar{S}_3x(t_i) = x_i$, $D\bar{S}_3x(t_i) = \tilde{x}'_i$, $0 \le i \le N+1$ and $D^2\bar{S}_3x(t_i) = \tilde{x}''_i$, $i = 0, N+1$ where $\tilde{x}'_i$, $0 \le i \le N+1$ are the right sides of (6.5.3), (6.7.1) - (6.7.4), (6.5.4) and $\tilde{x}''_0$ and $\tilde{x}''_{N+1}$ are the right sides of (6.5.20) and (6.5.21), respectively.

Theorem 6.7.5. If $x(t) \in PC^{6,\infty}[a,b]$, then

$$\| D^k(x - \bar{S}_3x) \|_\infty \le \beta^3_{3,6,k}\, h^{6-k} \| D^6x \|_\infty, \quad 0 \le k \le 5 \tag{6.7.16}$$

where $\beta^3_{3,6,0} = 0.86644565 - 1$, $\beta^3_{3,6,1} = 0.28068201 + 0$, $\beta^3_{3,6,2} = 0.25967397 + 1$, $\beta^3_{3,6,3} = 0.25049288 + 2$, $\beta^3_{3,6,4} = 0.12160477 + 3$, and $\beta^3_{3,6,5} = 0.22524288 + 3$.

Proof. The proof is similar to that of Theorem 6.7.3. ∎

Case 4

Definition 6.5.4. We say $\bar{S}_4x(t)$ is an *approximate* for $S^{\Delta}_{3,3}x(t)$ if $\bar{S}_4x(t) \in S_{3,3}(\Delta)$ with $S_4x(t_i) = g_i$, $0 \le i \le N+1$, where the given g_i, $0 \le i \le N+1$ are such that

$$\max_{0 \le i \le N+1} | x_i - g_i | = \xi,$$

$D\bar{S}_4x(t_i) = \tilde{x}'_i$, $0 \le i \le N+1$ and $D^2\bar{S}_4x(t_i) = \tilde{x}''_i$, $i = 0, N+1$ where $\tilde{x}'_i$, $0 \le i \le N+1$ are the right sides of (6.5.3), (6.7.1) - (6.7.4), (6.5.4) and $\tilde{x}''_0$ and $\tilde{x}''_{N+1}$ are the right sides of (6.5.20) and (6.5.21) respectively, with x_i being replaced by g_i.

Theorem 6.7.6. If $x(t) \in PC^{6,\infty}[a,b]$, then

$$\| D^k(x - \bar{S}_4x) \|_\infty \le \beta^3_{3,6,k}\, h^{6-k} \| D^6x \|_\infty + \beta^4_{3,6,k}\, h^{-k}\, \xi, \quad 0 \le k \le 5 \tag{6.7.17}$$

where $\beta^4_{3,6,0} = 0.88015333 + 1$, $\beta^4_{3,6,1} = 0.29971861 + 2$, $\beta^4_{3,6,2} = 0.23674812 + 3$, $\beta^4_{3,6,3} = 0.23440000 + 4$, $\beta^4_{3,6,4} = 0.11904000 + 5$, and $\beta^4_{3,6,5} = 0.22528000 + 5$.

Proof. The proof is similar to that of Theorem 6.5.7. ∎

Remark 6.7.1. As in Remark 6.5.2 in the following tables we compute the actual values of $\| D^k(x - \bar{S}_i x) \|_\infty$; $1 \le i \le 3$, $0 \le k \le 5$ for the function $x(t) = t^6 \mid t \mid$ in the interval $[-1, 1]$ and compare these with the corresponding right side bounds.

Table 6.7.1.

N	9	49	99
$\| x - \bar{S}_1 x \|_\infty$	0.42287179 − 6	0.43434733 − 10	0.75143623 − 12
Bound	0.12112427 − 5	0.77519531 − 10	0.12112412 − 11
$\| D(x - \bar{S}_1 x) \|_\infty$	0.37500000 − 4	0.19200000 − 7	0.63264643 − 9
Bound	0.77073555 − 4	0.24663538 − 7	0.77073474 − 9
$\| D^2(x - \bar{S}_1 x) \|_\infty$	0.25895658 − 2	0.66108736 − 5	0.45853139 − 6
Bound	0.99552613 − 2	0.15928418 − 4	0.99552530 − 6
$\| D^3(x - \bar{S}_1 x) \|_\infty$	0.41624656 + 0	0.53086811 − 2	0.83564691 − 3
Bound	0.21459375 + 1	0.17167500 − 1	0.21459362 − 2
$\| D^4(x - \bar{S}_1 x) \|_\infty$	0.39068449 + 2	0.24911769 + 1	0.75442579 + 0
Bound	0.24627750 + 3	0.98511000 + 1	0.24627740 + 1
$\| D^5(x - \bar{S}_1 x) \|_\infty$	0.13433725 + 4	0.42451223 + 3	0.25276938 + 3
Bound	0.98700000 + 4	0.19740000 + 4	0.98699979 + 3

Table 6.7.2.

N	9	49	99
$\| x - \bar{S}_2 x \|_\infty$	0.16872030 − 6	0.16968446 − 10	0.27718343 − 12
Bound	0.63620717 − 6	0.40717259 − 10	0.63620717 − 12
$\| D(x - \bar{S}_2 x) \|_\infty$	0.12543052 − 4	0.63054209 − 8	0.20599614 − 9
Bound	0.47921617 − 4	0.15334918 − 7	0.47921617 − 9
$\| D^2(x - \bar{S}_2 x) \|_\infty$	0.35000000 − 2	0.87920000 − 5	0.57443751 − 6
Bound	0.13500977 − 1	0.21601563 − 4	0.13500977 − 5
$\| D^3(x - \bar{S}_2 x) \|_\infty$	0.59558215 + 0	0.74747186 − 2	0.97668455 − 3
Bound	0.25213125 + 1	0.20170500 − 1	0.25213125 − 2
$\| D^4(x - \bar{S}_2 x) \|_\infty$	0.44992144 + 2	0.28188069 + 1	0.73654928 + 0
Bound	0.20916000 + 3	0.83664000 + 1	0.20916000 + 1
$\| D^5(x - \bar{S}_2 x) \|_\infty$	0.14247572 + 4	0.44406773 + 3	0.23197864 + 3
Bound	0.72345000 + 4	0.14469000 + 4	0.72345000 + 3

Table 6.7.3.

N	9	49	99
$\Vert x - \bar{S}_3x \Vert_\infty$	0.25975801 − 6	0.26999766 − 10	0.44213857 − 12
Bound	0.17058149 − 5	0.10917215 − 9	0.17058149 − 11
$\Vert D(x - \bar{S}_3x) \Vert_\infty$	0.37500000 − 4	0.19200000 − 7	0.62812500 − 9
Bound	0.11051854 − 3	0.35365934 − 7	0.11051854 − 8
$\Vert D^2(x - \bar{S}_3x) \Vert_\infty$	0.35000000 − 2	0.87920000 − 5	0.57443751 − 6
Bound	0.20449325 − 1	0.32718921 − 4	0.20449325 − 5
$\Vert D^3(x - \bar{S}_3x) \Vert_\infty$	0.17930999 + 0	0.21818824 − 2	0.28441412 − 3
Bound	0.39452629 + 1	0.31562103 − 1	0.39452629 − 2
$\Vert D^4(x - \bar{S}_3x) \Vert_\infty$	0.58205986 + 1	0.62067597 + 0	0.17148605 + 0
Bound	0.38305504 + 3	0.15322201 + 2	0.38305504 + 1
$\Vert D^5(x - \bar{S}_3x) \Vert_\infty$	0.21801811 + 3	0.12723455 + 3	0.70498048 + 2
Bound	0.14190301 + 5	0.28380603 + 4	0.14190301 + 4

∎

6.8 CUBIC LIDSTONE - SPLINE INTERPOLATION

We need the following results.

Lemma 6.8.1. Let $1 \le i \le N$ but fixed, and $p(t)$, $q(t)$ be two cubic polynomials in $[t_{i-1}, t_i]$ and $[t_i, t_{i+1}]$, respectively. Suppose $D^{2k}p(t_i) = D^{2k}q(t_i) = x_i^{(2k)}$; $k = 0, 1$ then $Dp(t_i) = Dq(t_i)$ if and only if

$$\frac{h^2}{6}[D^2p(t_{i-1}) + 4\ x_i'' + D^2q(t_{i+1})] = p(t_{i-1}) - 2\ x_i + q(t_{i+1}). \tag{6.8.1}$$

Proof. The proof is similar to that of Lemma 6.3.1. ∎

Lemma 6.8.2. Let $h(t) \in L_2(\Delta)$ be a function for which, say, $c_i = h(t_i)$, $c_i'' = D^2 h(t_i)$, $0 \le i \le N+1$ exist. Then, $h(t) \in S_{2,2}(\Delta)$ if and only if c_i'', $1 \le i \le N$ satisfy the following relations

$$(6.8.2) \qquad c_{i-1}'' + 4\ c_i'' + c_{i+1}'' = \frac{6}{h^2}(c_{i-1} - 2\ c_i + c_{i+1}), \quad 1 \le i \le N.$$

Moreover, from the system (6.8.2) the unknowns c_i'', $1 \le i \le N$ can be obtained uniquely in terms of c_i, $0 \le i \le N+1$, c_0'' and c_{N+1}''.

Proof. The proof is similar to that of Lemma 6.3.2. ∎

Lemma 6.8.3. For a given $x(t) \in C^{(2)}[a,b]$, $LS_{2,2}^{\Delta}x(t)$ exists and is unique.

Proof. The proof is similar to that of Lemma 6.3.3. ∎

Remark 6.8.1. From Lemma 6.8.3 and (6.2.8) it is clear that $LS_{2,2}^{\Delta}x(t)$ can be expressed as

$$(6.8.3) \qquad LS_{2,2}^{\Delta}x(t) = \sum_{i=0}^{N+1} r_{2,i,0}(t)x_i + r_{2,0,1}(t)x_0'' + r_{2,N+1,1}(t)x_{N+1}''$$

$$+ \sum_{i=1}^{N} r_{2,i,1}(t)c_i'',$$

where c_i'', $1 \le i \le N$ satisfy (6.8.2). ∎

Lemma 6.8.4. If $x(t) \in PC^{n,\infty}[a,b]$, $2 \le n \le 4$ then

$$(6.8.4) \qquad \| d^2 \|_\infty \le \alpha_n\ h^{n-2} \| D^n x \|_\infty\ ,$$

where $\alpha_2 = 6$, $\alpha_3 = \frac{4\sqrt{3}}{9}$ and $\alpha_4 = \frac{1}{4}$.

Proof. The proof is similar to that of Lemma 6.3.5. ∎

Theorem 6.8.5. Let $x(t) \in PC^{n,\infty}[a,b]$, $2 \le n \le 4$. Then,

$$(6.8.5) \qquad \| D^k(x - LS_{2,2}^{\Delta}x) \|_\infty \le \gamma_{2,n,k}\ h^{n-k} \| D^n x \|_\infty\ , \quad 0 \le k \le n-1$$

where the constants $\gamma_{2,n,k}$ are given in the following table.

Table 6.8.1.

$k \backslash n$	2	3	4
0	1	$\frac{3+4\sqrt{3}}{72}$	$\frac{17}{384}$
1	4	$\frac{8\sqrt{3}}{27}$	$\frac{1}{6}$
2		$\frac{9+8\sqrt{3}}{18}$	$\frac{3}{8}$
3			1

Proof. Using Lemma 5.4.1, Theorems 5.4.2 - 5.4.4 and Lemma 6.8.4 in (6.2.11) the inequalities (6.8.5) are immediate. ■

6.9 QUINTIC LIDSTONE - SPLINE INTERPOLATION

We need the following results.

Lemma 6.9.1. Let $1 \le i \le N$ but fixed, and $p(t)$, $q(t)$ be two quintic polynomials in $[t_{i-1}, t_i]$ and $[t_i, t_{i+1}]$, respectively. Suppose $D^{2k}p(t_i) = D^{2k}q(t_i) = x_i^{(2k)}$, $0 \le k \le 2$ then $Dp(t_i) = Dq(t_i)$ if and only if

$$(6.9.1)\quad \frac{h^2}{6}[D^2p(t_{i-1}) + 4\ x_i'' + D^2q(t_{i+1})] = p(t_{i-1}) - 2\ x_i + q(t_{i+1}) + \frac{h^4}{360}[7D^4p(t_{i-1}) + 16\ x_i^{(4)} + 7D^4q(t_{i+1})],$$

and $D^3p(t_i) = D^3q(t_i)$ if and only if

$$(6.9.2)\quad D^2p(t_{i-1}) - 2\ x_i'' + D^2q(t_{i+1}) = \frac{h^2}{6}[D^4p(t_{i-1}) + 4\ x_i^{(4)} + D^4q(t_{i+1})].$$

Proof. The proof is similar to that of Lemma 6.4.1. ■

Lemma 6.9.2. Let $h(t) \in L_3(\Delta)$ be a function for which, say, $c_i = h(t_i)$, $c_i'' = D^2h(t_i)$ and $c_i^{(4)} = D^4h(t_i)$, $0 \le i \le N+1$ exist. Then, $h(t) \in S_{3,4}(\Delta)$ if and only if c_i'', $c_i^{(4)}$, $1 \le i \le N$ satisfy the following relations

$$(6.9.3)\quad 65\ c_1'' + 26\ c_2'' + c_3'' = -8\ c_0'' + \frac{2}{3}\ h^2\ c_0^{(4)} + \frac{20}{h^2}(4\ c_0 - 7\ c_1 + 2\ c_2 + c_3)$$

$$(6.9.4)\quad c_{i-2}'' + 26\ c_{i-1}'' + 66\ c_i'' + 26\ c_{i+1}'' + c_{i+2}''$$
$$= \frac{20}{h^2}(c_{i-2} + 2\ c_{i-1} - 6\ c_i + 2\ c_{i+1} + c_{i+2}),\quad 2 \le i \le N-1$$

$$(6.9.5)\quad c_{N-2}'' + 26\ c_{N-1}'' + 65\ c_N''$$
$$= -8\ c_{N+1}'' + \frac{2}{3}\ h^2\ c_{N+1}^{(4)} + \frac{20}{h^2}(c_{N-2} + 2\ c_{N-1} - 7\ c_N + 4\ c_{N+1})$$

$$(6.9.6)\quad 65\ c_1^{(4)} + 26\ c_2^{(4)} + c_3^{(4)}$$
$$= \frac{120}{h^2}\ c_0'' - 18\ c_0^{(4)} + \frac{120}{h^4}(-2\ c_0 + 5\ c_1 - 4\ c_2 + c_3)$$

$$(6.9.7)\quad c_{i-2}^{(4)} + 26\ c_{i-1}^{(4)} + 66\ c_i^{(4)} + 26\ c_{i+1}^{(4)} + c_{i+2}^{(4)}$$
$$= \frac{120}{h^4}(c_{i-2} - 4\ c_{i-1} + 6\ c_i - 4\ c_{i+1} + c_{i+2}),\quad 2 \le i \le N-1$$

$$(6.9.8)\quad c_{N-2}^{(4)} + 26\ c_{N-1}^{(4)} + 65\ c_N^{(4)}$$
$$= \frac{120}{h^2}\ c_{N+1}'' - 18\ c_{N+1}^{(4)} + \frac{120}{h^4}(c_{N-2} - 4\ c_{N-1} + 5\ c_N - 2\ c_{N+1}).$$

Moreover, from the systems (6.9.3) - (6.9.5) and (6.9.6) - (6.9.8) the unknowns c_i'' and $c_i^{(4)}$, $1 \le i \le N$ can be obtained uniquely in terms of c_i, $0 \le i \le N+1$, c_0'', c_{N+1}'', $c_0^{(4)}$, and $c_{N+1}^{(4)}$.

Proof. The proof is similar to that of Lemma 6.4.2. ∎

Lemma 6.9.3. For a given $x(t) \in C^{(4)}[a,b]$, $LS_{3,4}^{\Delta}x(t)$ exists and is unique.

Proof. The proof is similar to that of Lemma 6.3.3. ∎

Remark 6.9.1. From Lemma 6.9.3 and (6.2.8) it is clear that $LS_{3,4}^{\Delta}x(t)$ can be expressed as

$$(6.9.9)\quad LS_{3,4}^{\Delta}x(t) = \sum_{i=0}^{N+1} r_{3,i,0}(t)x_i + r_{3,0,1}(t)x_0'' + r_{3,N+1,1}(t)x_{N+1}''$$

$$+\sum_{i=1}^{N} r_{3,i,1}(t)c_i'' + r_{3,0,2}(t)x_0^{(4)} + r_{3,N+1,2}(t)x_{N+1}^{(4)} + \sum_{i=1}^{N} r_{3,i,2}(t)c_i^{(4)},$$

where c_i'' and $c_i^{(4)}$, $1 \le i \le N$ satisfy (6.9.3) - (6.9.8). ∎

Lemma 6.9.4. If $x(t) \in PC^{n,\infty}[a,b]$, $4 \le n \le 6$ then

(6.9.10) $$\| d^2 \|_\infty \le \bar{\alpha}_n \, h^{n-2} \, \| D^n x \|_\infty \,,$$

where $\bar{\alpha}_4 = 0.38179896 + 0$, $\bar{\alpha}_5 = 0.67806574 - 1$ and $\bar{\alpha}_6 = 0.18518519 - 1$.

Proof. The proof is similar to that of Lemma 6.4.4. ∎

Lemma 6.9.5. If $x(t) \in PC^{n,\infty}[a,b]$, $4 \le n \le 6$ then

(6.9.11) $$\| d^4 \|_\infty \le \beta_n \, h^{n-4} \, \| D^n x \|_\infty \,,$$

where $\beta_4 = 20$, $\beta_5 = 0.25465679 + 1$ and $\beta_6 = 0.83333333 + 0$.

Proof. The proof is similar to that of Lemma 6.4.4. ∎

Theorem 6.9.6. Let $x(t) \in PC^{n,\infty}[a,b]$, $4 \le n \le 6$. Then,

(6.9.12) $$\| D^k(x - LS_{3,4}^{\Delta}x) \|_\infty \le \gamma_{3,n,k} \, h^{n-k} \, \| D^n x \|_\infty \,, \quad 0 \le k \le n-1$$

where the constants $\gamma_{3,n,k}$ are given in the following table.

Table 6.9.1.

k \ n	4	5	6
0	0.33418320 + 0	0.45800925 − 1	0.14489293 − 1
1	0.11075662 + 1	0.15305465 + 0	0.48148148 − 1
2	0.31317990 + 1	0.42779423 + 0	0.13570602 + 0
3	0.11763598 + 2	0.15371972 + 1	0.49537037 + 0
4		0.30465679 + 1	0.95833333 + 0
5			0.21666667 + 1

Proof. Using Lemma 5.4.1, Theorems 5.4.2 - 5.4.4 and Lemmas 6.9.4 and 6.9.5 in (6.2.11) the inequalities (6.9.12) are immediate. ∎

Remark 6.9.2. The sharpness of the inequalities (6.9.12) also remains undecided. However, in Tables 6.9.2 and 6.9.3 we compute the actual values of $\| D^k(x - LS^{\Delta}_{3,4}x) \|_\infty$ for some simple functions and compare these with the corresponding right side bounds in (6.9.12).

Table 6.9.2.

$$x(t) = (1 - t^2)e^t,\ t \in [-1, 1],\ n = 6$$

N	9	24	49
$\| x - LS^{\Delta}_{3,4}x \|_\infty$	0.29912439 − 5	0.12513880 − 7	0.19666904 − 9
Bound	0.10586952 − 3	0.43364155 − 6	0.67756492 − 8
$\| D(x - LS^{\Delta}_{3,4}x) \|_\infty$	0.52868693 − 4	0.55591581 − 6	0.17508675 − 7
Bound	0.17590303 − 2	0.18012471 − 4	0.56288971 − 6
$\| D^2(x - LS^{\Delta}_{3,4}x) \|_\infty$	0.85958137 − 3	0.22728105 − 4	0.14347215 − 5
Bound	0.24789220 − 1	0.63460404 − 3	0.39662753 − 4
$\| D^3(x - LS^{\Delta}_{3,4}x) \|_\infty$	0.17228632 − 1	0.11434800 − 2	0.14459550 − 3
Bound	0.45244291 + 0	0.28956346 − 1	0.36195433 − 2
$\| D^4(x - LS^{\Delta}_{3,4}x) \|_\infty$	0.42994935 + 0	0.77150602 − 1	0.20047908 − 1
Bound	0.43764337 + 1	0.70022940 + 0	0.17505735 + 0
$\| D^5(x - LS^{\Delta}_{3,4}x) \|_\infty$	0.11720187 + 2	0.52185504 + 1	0.27049135 + 1
Bound	0.49472730 + 2	0.19789092 + 2	0.98945460 + 1

Table 6.9.3.

$x(t) = t(1-t)e^t,\ t \in [0,1],\ n = 6$

N	9	24	49
$\parallel x - LS^{\Delta}_{3,4}x \parallel_\infty$	0.40769351 − 7	0.16855348 − 9	0.26408407 − 11
Bound	0.14178953 − 5	0.58076993 − 8	0.90745302 − 10
$\parallel D(x - LS^{\Delta}_{3,4}x) \parallel_\infty$	0.14474043 − 5	0.15005063 − 7	0.47068971 − 9
Bound	0.47116884 − 4	0.48247689 − 6	0.15077403 − 7
$\parallel D^2(x - LS^{\Delta}_{3,4}x) \parallel_\infty$	0.47287676 − 4	0.12295117 − 5	0.77223926 − 7
Bound	0.13279939 − 2	0.33996645 − 4	0.21247903 − 5
$\parallel D^3(x - LS^{\Delta}_{3,4}x) \parallel_\infty$	0.19017616 − 2	0.12390957 − 3	0.15578538 − 4
Bound	0.48476026 − 1	0.31024657 − 2	0.38780821 − 3
$\parallel D^4(x - LS^{\Delta}_{3,4}x) \parallel_\infty$	0.10115788 + 0	0.17169979 − 1	0.43783821 − 2
Bound	0.93780723 + 0	0.15004916 + 0	0.37512289 − 1
$\parallel D^5(x - LS^{\Delta}_{3,4}x) \parallel_\infty$	0.54815309 + 1	0.23167478 + 1	0.11799139 + 1
Bound	0.21202599 + 2	0.84810394 + 1	0.42405197 + 1

∎

6.10 L_2 - ERROR BOUNDS FOR SPLINE INTERPOLATION

To obtain error bounds for the derivatives of the interpolation error $x(t) - S^{\Delta}_{m,\tau}x(t)$ in terms of the derivatives of $x(t)$ in L_2 - norm we begin with the following variational characterization of $S^{\Delta}_{m,\tau}x(t)$.

Theorem 6.10.1. Let Δ and $x_i^{(k)}$, $1 \le i \le N$, $0 \le k \le 2m-2-\tau$; $i = 0, N+1$, $0 \le k \le m-1$ be given and

$$V \equiv \Big\{\omega(t) \in PC^{m,2}[a,b] \ : \ \omega^{(k)}(t_i) = x_i^{(k)},$$
$$1 \le i \le N,\ 0 \le k \le 2m-2-\tau;\ i = 0, N+1,\ 0 \le k \le m-1\Big\}.$$

The variational problem of finding the function $p(t) \in V$ which minimizes $\| D^m\omega \|_2^2$ over all $\omega(t) \in V$ has the unique solution $S_{m,\tau}^{\Delta}x(t)$.

Proof. The proof is similar to that of Theorem 5.3.9. ∎

As in Section 5.3 the above theorem leads to the 'First Integral Relation'.

Corollary 6.10.2. If $x(t) \in PC^{m,2}[a,b]$, then

$$\| D^m S_{m,\tau}^{\Delta}x \|_2^2 + \| D^m S_{m,\tau}^{\Delta}x - D^m x \|_2^2 = \| D^m x \|_2^2 . \tag{6.10.1}$$ ∎

Once again by using the same type of integration by parts argument that we used in the proof of Theorem 5.3.9, we obtain the following result.

Theorem 6.10.3. If $y(t) \in PC^{2m,2}[a,b]$, $y^{(k)}(t_i) = x^{(k)}(t_i)$, $1 \le i \le N$, $0 \le k \le 2m-2-\tau$; $i = 0, N+1$, $0 \le k \le m-1$ then

$$\| D^m(y - S_{m,\tau}^{\Delta}x) \|_2^2 = (-1)^m(y - S_{m,\tau}^{\Delta}x, D^{2m}y). \tag{6.10.2}$$ ∎

As a particular case of (6.10.2) we obtain the 'Second Integral Relation'.

Corollary 6.10.4. If $x(t) \in PC^{2m,2}[a,b]$, then

$$\| D^m(x - S_{m,\tau}^{\Delta}x) \|_2^2 = (-1)^m(x - S_{m,\tau}^{\Delta}x, D^{2m}x). \tag{6.10.3}$$ ∎

A result which includes Theorem 5.3.16 is the following :

Theorem 6.10.5. Let $x(t) \in PC^{m+j,2}[a,b]$, $0 \le j \le m$. Then,

$$\| D^k(x - S_{m,\tau}^{\Delta}x) \|_2 \le \xi_{m,m+j,k}\, h^{m+j-k} \| D^{m+j}x \|_2 , \quad 0 \le k \le m \tag{6.10.4}$$

where

(i) $\xi_{m,m,k}$

$$= \lambda_{2m-1-\tau,k}^{-2m+1+\tau+k}\ \lambda_{2m-1-\tau,3m-2-2\tau}^{m-\tau-1}\ ;\quad 0 \le k \le 2m-1-\tau,\ \tau \le \left[\tfrac{3m-2}{2}\right]$$

$$= \pi^{-2\tau+3m-2}\ \lambda_{2m-1-\tau,k}^{-2m+1+\tau+k}\ \lambda_{2m-1-\tau,0}^{-2m+1+\tau} \times$$

$$\left\{\begin{array}{l} \displaystyle\prod_{s=1}^{((2\tau-3m+2)/2)} (2m-\tau+2s-1)^2,\ (2\tau-3m+2) \text{ is even} \\ \displaystyle(2m-\tau+1) \prod_{s=1}^{((2\tau-3m+1)/2)} (2m-\tau+2s)^2, \\ \hfill (2\tau-3m+2) \text{ is odd} \end{array}\right\}$$

$$0 \le k \le 2m-1-\tau,\ \left[\tfrac{3m-2}{2}\right] < \tau$$

$$= \pi^{-m+k}\left\{\begin{array}{l} \displaystyle\prod_{s=1}^{((m-k)/2)} (k-2m+\tau+2s+1)^2,\ (m-k) \text{ is even} \\ \displaystyle(k-2m+\tau+3) \prod_{s=1}^{((m-k-1)/2)} (k-2m+\tau+2s+2)^2, \\ \hfill (m-k) \text{ is odd} \end{array}\right\}$$

$$2m-\tau \le k \le m-1$$

$$= 1,\quad k = m$$

(ii) $\xi_{m,2m,k} = \xi_{m,m,k}\ \xi_{m,m,0}\ ,\quad 0 \le k \le m$

(iii) $\xi_{m,m+j,k} = \lambda_{m+j,k}^{-m-j+k} + 2^{m-j}(\tilde{k}_{m+j-1,m-j})^{1/2}\ \xi_{m,2m,k}\ ;$

$$1 \le j \le m-1,\ 0 \le k \le m$$

and the constants $\lambda_{n,k}$ and $\tilde{k}_{n,k}$ are defined in Theorem 5.2.4 and Remark 5.2.6, respectively.

Proof. (i) Since $D^\mu(x(t_i) - S_{m,\tau}^{\Delta}x(t_i)) = 0;\ 0 \le i \le N+1,\ 0 \le \mu \le 2m-2-\tau$ we can use Corollary 5.2.5 with $n = 2m-1-\tau$ and j such that $2n - j = m$, i.e., $j = 3m-2-2\tau$ (the existence of such a j in view of $\tau \le \left[\frac{3m-2}{2}\right]$ is guaranteed), to obtain

(6.10.5) $\displaystyle\int_{t_i}^{t_{i+1}} [D^k(x(t) - S_{m,\tau}^{\Delta}x(t))]^2\, dt$

$$\leq \left(\frac{t_{i+1}-t_i}{\lambda_{2m-1-\tau,k}}\right)^{4m-2-2\tau-2k} \left(\frac{t_{i+1}-t_i}{\lambda_{2m-1-\tau,3m-2-2\tau}}\right)^{2\tau+2-2m} \times$$

$$\int_{t_i}^{t_{i+1}} [D^m(x(t) - S^{\Delta}_{m,\tau}x(t))]^2\, dt \ ; \quad 0 \leq i \leq N,$$

$$0 \leq k \leq 2m-1-\tau,\ \tau \leq \left[\tfrac{3m-2}{2}\right].$$

Summing both sides of the inequality (6.10.5) with respect to i from 0 to N, and taking the square root of both sides of the resulting inequality, we obtain

$$\text{(6.10.6)} \qquad \| D^k(x - S^{\Delta}_{m,\tau}x) \|_2 \leq \ \xi_{m,m,k}\, h^{m-k} \| D^m(x - S^{\Delta}_{m,\tau}x) \|_2 \ ;$$

$$0 \leq k \leq 2m-1-\tau,\ \tau \leq \left[\tfrac{3m-2}{2}\right].$$

Now bounding the right side of (6.10.6) by the first integral relation (6.10.1) gives (6.10.4) for $j = 0$, $0 \leq k \leq 2m-1-\tau$, $\tau \leq \left[\frac{3m-2}{2}\right]$.

If $\left[\frac{3m-2}{2}\right] < \tau$, then once again in view of $D^{\mu}(x(t_i) - S^{\Delta}_{m,\tau}x(t_i)) = 0$; $0 \leq i \leq N+1$, $0 \leq \mu \leq 2m-2-\tau$ we can use Corollary 5.2.5 with $n = 2m-1-\tau$ and $j = 0$, to obtain

$$\text{(6.10.7)} \ \| D^k(x - S^{\Delta}_{m,\tau}x) \|_2 \leq \lambda_{2m-1-\tau,k}^{-2m+1+\tau+k}\ \lambda_{2m-1-\tau,0}^{-2m+1+\tau}\ h^{4m-2-2\tau-k} \times$$

$$\| D^{4m-2-2\tau}(x - S^{\Delta}_{m,\tau}x) \|_2\ , \quad 0 \leq k \leq 2m-1-\tau.$$

Next, for $m-1 \leq \tau \leq 2m-2$ we note that $D^{2m-2-\tau}(x(t) - S^{\Delta}_{m,\tau}x(t)) \in C^{(\tau-m+1)}[a,b]$ and vanishes at every t_i, $0 \leq i \leq N+1$. Thus, by Rolle's theorem for each $0 \leq \nu \leq \tau - m + 1$ there exist points $\left\{s_i^{(\nu)}\right\}_{i=0}^{N+1-\nu}$ in $[a,b]$ such that $s_i^{(0)} = t_i$, $0 \leq i \leq N+1$, and

(1) $D^{2m-2-\tau+\nu}(x(s_i^{(\nu)}) - S^{\Delta}_{m,\tau}x(s_i^{(\nu)})) = 0$; $0 \leq \nu \leq \tau-m+1$, $0 \leq i \leq N+1-\nu$

(2) $a = s_0^{(\nu)} < s_1^{(\nu)} < \cdots < s_{N+1-\nu}^{(\nu)} = b$, $0 \leq \nu \leq \tau - m + 1$

(3) $s_i^{(\nu)} \leq s_i^{(\nu+1)} < s_{i+1}^{(\nu)}$; $0 \leq \nu \leq \tau - m$, $0 \leq i \leq N - \nu$

(4) $| s_{i+1}^{(\nu)} - s_i^{(\nu)} | \leq (\nu+1)h$; $0 \leq \nu \leq \tau - m + 1$, $0 \leq i \leq N - \nu$.

Hence, in view of Corollary 5.2.3, we have

$$\int_{s_i^{(\nu)}}^{s_{i+1}^{(\nu)}} [D^{2m-2-\tau+\nu}(x(t) - S^{\Delta}_{m,\tau}x(t))]^2\, dt$$

$$\leq \left(\frac{(\nu+1)h}{\pi}\right)^4 \int_{s_i^{(\nu)}}^{s_{i+1}^{(\nu)}} [D^{2m-\tau+\nu}(x(t) - S_{m,\tau}^{\Delta}x(t))]^2\, dt \,;$$

$$0 \leq i \leq N-\nu,\ 0 \leq \nu \leq \tau - m$$

which on summing with respect to i from 0 to $N-\nu$, and taking the square root of both sides of the resulting inequality gives

$$(6.10.8) \quad \| D^{2m-2-\tau+\nu}(x - S_{m,\tau}^{\Delta}x) \|_2$$

$$\leq \left(\frac{(\nu+1)h}{\pi}\right)^2 \| D^{2m-\tau+\nu}(x - S_{m,\tau}^{\Delta}x) \|_2 \,, \quad 0 \leq \nu \leq \tau - m.$$

In a similar manner an application of Theorem 5.2.2 leads to

$$(6.10.9) \quad \| D^{2m-2-\tau+\nu}(x - S_{m,\tau}^{\Delta}x) \|_2$$

$$\leq \frac{(\nu+1)h}{\pi} \| D^{2m-\tau+\nu-1}(x - S_{m,\tau}^{\Delta}x) \|_2 \,, \quad 0 \leq \nu \leq \tau - m + 1.$$

Now using (6.10.8) repeatedly and (6.10.9) once, we obtain

$$(6.10.10) \quad \| D^{2m-\tau+\nu}(x - S_{m,\tau}^{\Delta}x) \|_2$$

$$\leq \left\{ \begin{array}{ll} \displaystyle\prod_{s=1}^{((\tau-m-\nu)/2)} (\nu+2s+1)^2, & (\tau-m-\nu) \text{ is even} \\ (\nu+3) \displaystyle\prod_{s=1}^{((\tau-m-\nu-1)/2)} (\nu+2s+2)^2, & (\tau-m-\nu) \text{ is odd} \end{array} \right\} \times$$

$$\left(\frac{h}{\pi}\right)^{\tau-m-\nu} \| D^m(x - S_{m,\tau}^{\Delta}x) \|_2 \,, \quad 0 \leq \nu \leq \tau - m - 1.$$

Bounding the right side of (6.10.7) first by (6.10.10) for $\nu = 2m - \tau - 2$ and then by (6.10.1) leads to (6.10.4) for $j = 0,\ 0 \leq k \leq 2m - 1 - \tau,\ \left[\frac{3m-2}{2}\right] < \tau$.

In (6.10.10) letting $\nu = k + \tau - 2m$ and bounding the right side by (6.10.1) gives (6.10.4) for $j = 0,\ 2m - \tau \leq k \leq m - 1$.

In (6.10.4), $\xi_{m,m,m} = 1$ obviously follows from (6.10.1).

(ii) From the proof of (i) it is clear that the inequality (6.10.6) indeed holds for all $0 \leq k \leq m$ and $m - 1 \leq \tau \leq 2m - 2$. Using the Cauchy - Schwarz inequality to the second integral relation (6.10.3), we get

$$\| D^m(x - S^{\Delta}_{m,\tau}x) \|_2^2 \le \| x - S^{\Delta}_{m,\tau}x \|_2 \ \| D^{2m}x \|_2 .$$

Combining this inequality with (6.10.6) for $k = 0$ gives that

$$\| D^m(x - S^{\Delta}_{m,\tau}x) \|_2 \le \xi_{m,m,0} \, h^m \ \| D^{2m}x \|_2 . \tag{6.10.11}$$

Now applying (6.10.11) to bound the right side of (6.10.6), we obtain

$$\| D^k(x - S^{\Delta}_{m,\tau}x) \|_2 \le \xi_{m,m,k} \, \xi_{m,m,0} \, h^{2m-k} \ \| D^{2m}x \|_2 , \quad 0 \le k \le m. \tag{6.10.12}$$

(iii) The proof is similar to that of Theorem 5.3.15 for the case $n = 4$, $0 \le k \le 3$ except that now we need to consider $q(t)$ the unique polynomial of degree $(2m + 2j - 1)$ in $[t_i, t_{i+1}]$, $0 \le i \le N$ but fixed, satisfying $D^{\mu}q(t_\ell) = D^{\mu}x(t_\ell)$; $0 \le \mu \le m + j - 1$, $\ell = i, i + 1$. For this polynomial it is clear that $S^{\Delta}_{m,\tau}x(t) = S^{\Delta}_{m,\tau}q(t)$, $t \in [t_i, t_{i+1}]$. ∎

Remark 6.10.1. For the case $\tau = m - 1$ the constants $\xi_{m,m+j,k}$ are the same as $\bar{c}_{m,m+j,k}$ appearing in (5.3.70). These constants for $m = 1, 2$ and 3 have been computed in Tables 5.3.10, 5.3.11 and 5.3.12, respectively. For the cases considered in Sections 6.3, 6.4 and 6.6 the constants $\xi_{m,m+j,k}$ are computed in the following tables.

Table 6.10.1.

$m = 2, \ \tau = 2$

k \ j	0	1	2
0	$\lambda_{1,0}^{-2} = \frac{1}{\pi^2}$	$\lambda_{3,0}^{-3} + 2\sqrt{15}\lambda_{1,0}^{-4} = 0.835514 - 1$	$\lambda_{1,0}^{-4} = \frac{1}{\pi^4}$
1	$\lambda_{1,0}^{-1} = \frac{1}{\pi}$	$\lambda_{3,1}^{-2} + 2\sqrt{15}\lambda_{1,0}^{-3} = 0.266034 + 0$	$\lambda_{1,0}^{-3} = \frac{1}{\pi^3}$
2	1	$\lambda_{3,2}^{-1} + 2\sqrt{15}\lambda_{1,0}^{-2} = 0.896105 + 0$	$\lambda_{1,0}^{-2} = \frac{1}{\pi^2}$

Table 6.10.2.

$m = 3, \ \tau = 4$

$k \backslash j$	0	1	2	3
0	$3\pi^{-1}\lambda_{1,0}^{-2} = \frac{3}{\pi^3}$	$\lambda_{4,0}^{-4} + 36\pi^{-2}\sqrt{\bar{k}_{3,2}}\lambda_{1,0}^{-4} = 0.858259 + 0$	$\lambda_{5,0}^{-5} + 18\pi^{-2}\sqrt{k_4}\lambda_{1,0}^{-4} = 0.182559 + 0$	$9\pi^{-2}\lambda_{1,0}^{-4} = \frac{9}{\pi^6}$
1	$3\pi^{-1}\lambda_{1,0}^{-1} = \frac{3}{\pi^2}$	$\lambda_{4,1}^{-3} + 36\pi^{-2}\sqrt{\bar{k}_{3,2}}\lambda_{1,0}^{-3} = 0.269665 + 1$	$\lambda_{5,1}^{-4} + 18\pi^{-2}\sqrt{k_4}\lambda_{1,0}^{-3} = 0.573550 + 0$	$9\pi^{-2}\lambda_{1,0}^{-3} = \frac{9}{\pi^5}$
2	$\frac{3}{\pi}$	$\lambda_{4,2}^{-2} + 36\pi^{-2}\sqrt{\bar{k}_{3,2}}\lambda_{1,0}^{-2} = 0.847705 + 1$	$\lambda_{5,2}^{-3} + 18\pi^{-2}\sqrt{k_4}\lambda_{1,0}^{-2} = 0.180221 + 1$	$9\pi^{-2}\lambda_{1,0}^{-2} = \frac{9}{\pi^4}$
3	1	$\lambda_{4,3}^{-1} + 12\pi^{-1}\sqrt{\bar{k}_{3,2}}\lambda_{1,0}^{-2} = 0.895446 + 1$	$\lambda_{5,3}^{-2} + 6\pi^{-1}\sqrt{k_4}\lambda_{1,0}^{-2} = 0.189254 + 1$	$3\pi^{-1}\lambda_{1,0}^{-2} = \frac{3}{\pi^3}$

Table 6.10.3.

$m = 3, \ \tau = 3$

$k \backslash j$	0	1	2	3
0	$\lambda_{2,0}^{-2}\lambda_{2,1}^{-1} = 0.711362 - 2$	$\lambda_{4,0}^{-4} + 4\sqrt{\bar{k}_{3,2}}\lambda_{2,0}^{-4}\lambda_{2,1}^{-2} = 0.490547 - 2$	$\lambda_{5,0}^{-5} + 2\sqrt{k_4}\lambda_{2,0}^{-4}\lambda_{2,1}^{-2} = 0.100079 - 2$	$\lambda_{2,0}^{-4}\lambda_{2,1}^{-2} = 0.506035 - 4$
1	$\lambda_{2,1}^{-2} = \frac{1}{4\pi^2}$	$\lambda_{4,1}^{-3} + 4\sqrt{\bar{k}_{3,2}}\lambda_{2,0}^{-2}\lambda_{2,1}^{-3} = 0.177083 - 1$	$\lambda_{5,1}^{-4} + 2\sqrt{k_4}\lambda_{2,0}^{-2}\lambda_{2,1}^{-3} = 0.358204 - 2$	$\lambda_{2,0}^{-2}\lambda_{2,1}^{-3} = 0.180190 - 3$
2	$\lambda_{2,1}^{-1} = \frac{1}{2\pi}$	$\lambda_{4,2}^{-2} + 4\sqrt{\bar{k}_{3,2}}\lambda_{2,0}^{-2}\lambda_{2,1}^{-2} = 0.112775 + 0$	$\lambda_{5,2}^{-3} + 2\sqrt{k_4}\lambda_{2,0}^{-2}\lambda_{2,1}^{-2} = 0.226397 - 1$	$\lambda_{2,0}^{-2}\lambda_{2,1}^{-2} = 0.113217 - 2$
3	1	$\lambda_{4,3}^{-1} + 4\sqrt{\bar{k}_{3,2}}\lambda_{2,0}^{-2}\lambda_{2,1}^{-1} = 0.738727 + 0$	$\lambda_{5,3}^{-2} + 2\sqrt{k_4}\lambda_{2,0}^{-2}\lambda_{2,1}^{-1} = 0.144576 + 0$	$\lambda_{2,0}^{-2}\lambda_{2,1}^{-1} = 0.711362 - 2$

∎

Our final result here generalizes Theorem 5.3.17.

Theorem 6.10.6. Let $p, q \geq 1$, $r = \max\{p, q, 2\}$, and $x(t) \in PC^{m+j,r}[a,b]$, $0 \leq j \leq m$. Then, for $1 \leq p \leq 2 \leq q$

$$\text{(6.10.13)} \quad \| D^k(x - S^{\Delta}_{m,\tau}x) \|_p \leq \xi_{m,m+j,k}\, h^{m+j-k}(b-a)^{p^{-1}-q^{-1}} \| D^{m+j}x \|_q\,, \qquad 0 \leq k \leq m$$

and for $p \geq 2$

$$\text{(6.10.14)} \quad \| D^k(x - S^{\Delta}_{m,\tau}x) \|_p \leq \frac{1}{2}\, \xi_{m,m+j,k+1} \begin{Bmatrix} 1, & 0 \leq k \leq 2m-2-\tau \\ (k-2m+3+\tau)^{2^{-1}+p^{-1}}, & 2m-1-\tau \leq k \leq m-1 \end{Bmatrix} \times h^{m+j-k-2^{-1}+p^{-1}} \| D^{m+j}x \|_2\,.$$

Proof. The proof is similar to that of Theorem 5.3.17. ∎

6.11 TWO VARIABLE SPLINE INTERPOLATION

For a fixed ρ and $m-1 \leq \tau \leq 2m-2$, we define the set $S_{m,\tau}(\rho)$ as follows :

$$\begin{aligned} S_{m,\tau}(\rho) &= S_{m,\tau}(\Delta) \oplus S_{m,\tau}(\Delta') \quad \text{(the tensor product)} \\ &= \Big\{ s(t,u) \in C^{(\tau,\tau)}([a,b] \times [c,d]) : s(t,u) \text{ is a two - dimensional} \\ &\qquad \text{polynomial of degree at most } (2m-1) \text{ in each variable} \\ &\qquad \text{and in each subrectangle } [t_i, t_{i+1}] \times [u_j, u_{j+1}]; \\ &\qquad\qquad 0 \leq i \leq N,\ 0 \leq j \leq M \Big\}. \end{aligned}$$

Since $S_{m,\tau}(\rho)$ is the tensor product of $S_{m,\tau}(\Delta)$ and $S_{m,\tau}(\Delta')$ which are of dimensions $[(2m-1-\tau)N + 2m]$ and $[(2m-1-\tau)M + 2m]$ respectively, $S_{m,\tau}(\rho)$ is of dimension $[(2m-1-\tau)N + 2m][(2m-1-\tau)M + 2m]$.

Definition 6.11.1. For a given $f(t,u) \in C^{(m-1,m-1)}([a,b] \times [c,d])$, we say $S^{\rho}_{m,\tau}f(t,u)$ is the $S_{m,\tau}(\rho)$ - *interpolate of* $f(t,u)$, also known as *spline interpolate of* $f(t,u)$, if $S^{\rho}_{m,\tau}f(t,u) \in S_{m,\tau}(\rho)$ with $D^{\mu}_t D^{\nu}_u S^{\rho}_{m,\tau} f(t_i, u_j) = f^{(\mu,\nu)}_{i,j}$, where μ, ν, i and j satisfy the following :

(1) if $0 \le \mu \le 2m-2-\tau$, $0 \le \nu \le 2m-2-\tau$ then $0 \le i \le N+1$, $0 \le j \le M+1$;
(2) if $2m-1-\tau \le \mu \le m-1$, $0 \le \nu \le 2m-2-\tau$ then $i = 0, N+1$, $0 \le j \le M+1$;
(3) if $0 \le \mu \le 2m-2-\tau$, $2m-1-\tau \le \nu \le m-1$ then $0 \le i \le N+1$, $j = 0,\ M+1$; and
(4) if $2m-1-\tau \le \mu \le m-1$, $2m-1-\tau \le \nu \le m-1$ then $(i,j) = (0,0),\ (0,M+1),\ (N+1,0),\ (N+1,M+1)$.

It is clear that for $\tau = m-1$, $S_{m,m-1}(\rho) = H_m(\rho)$ and $S^{\rho}_{m,m-1}f(t,u) \equiv H^{\rho}_m f(t,u)$, which has been discussed in Section 5.5. For $m-1 < \tau \le 2m-2$ we have the inclusion relation $S_{m,\tau}(\rho) \subset H_m(\rho)$. Therefore, in view of (5.5.1), $S^{\rho}_{m,\tau}f(t,u)$ can be written as

$$(6.11.1)\quad S^{\rho}_{m,\tau}f(t,u) = \sum_{i=0}^{N+1}\sum_{\mu=0}^{m-1}\sum_{j=0}^{M+1}\sum_{\nu=0}^{m-1} h_{m,i,\mu}(t)h_{m,j,\nu}(u)D^{\mu}_t D^{\nu}_u S^{\rho}_{m,\tau}f(t_i,u_j).$$

In (6.11.1), $D^{\mu}_t D^{\nu}_u S^{\rho}_{m,\tau}f(t_i,u_j)$ where μ, ν, i and j do not fulfill Definition 6.11.1 exist uniquely. In fact, for $m = 2$ and 3 we shall show that these unknown constants are the solutions of diagonally dominant systems of algebraic equations, and hence can be computed explicitly in terms of known quantities.

The following result provides a characterization of $S^{\rho}_{m,\tau}f(t,u)$ in terms of one - dimensional interpolation schemes.

Lemma 6.11.1. If $f(t,u) \in C^{(m-1,m-1)}([a,b]\times[c,d])$, then

$$(6.11.2)\qquad S^{\rho}_{m,\tau}f(t,u) = S^{\Delta'}_{m,\tau}S^{\Delta}_{m,\tau}f(t,u) = S^{\Delta}_{m,\tau}S^{\Delta'}_{m,\tau}f(t,u).$$

Proof. The proof is similar to that of Lemma 5.5.1. ∎

Now let $f(t,u) \in C^{(m-1,m-1)}([a,b]\times[c,d])$ be an arbitrary function. From Lemma 6.11.1 we have

$$\begin{aligned} & f - S^{\rho}_{m,\tau}f \\ (6.11.3)\qquad & = (f - S^{\Delta}_{m,\tau}f) + S^{\Delta}_{m,\tau}(f - S^{\Delta'}_{m,\tau}f) \\ (6.11.4)\qquad & = (f - S^{\Delta}_{m,\tau}f) + \left[S^{\Delta}_{m,\tau}(f - S^{\Delta'}_{m,\tau}f) - (f - S^{\Delta'}_{m,\tau}f)\right] + (f - S^{\Delta'}_{m,\tau}f) \end{aligned}$$

$$(6.11.5) \qquad = (f - S^{\Delta}_{m,\tau}f) + \left[S^{\Delta'}_{m,\tau}(f - S^{\Delta}_{m,\tau}f) - (f - S^{\Delta}_{m,\tau}f)\right] + (f - S^{\Delta'}_{m,\tau}f).$$

The following result follows from Lemma 6.3.2.

Lemma 6.11.2. For a given $h(t,u) \in H_2(\rho)$, we define $c^{\mu,\nu}_{i,j} = D^{\mu}_t D^{\nu}_u h(t_i, u_j)$, $0 \le \mu, \nu \le 1$, $0 \le i \le N+1$, $0 \le j \le M+1$. The function $h(t,u) \in S_{2,2}(\rho)$ if and only if $c^{\mu,\nu}_{i,j}$, where μ, ν, i and j are such that (1) if $\mu = 1$, $\nu = 0$, then $1 \le i \le N$, $0 \le j \le M+1$; (2) if $\mu = 0$, $\nu = 1$, then $0 \le i \le N+1$, $1 \le j \le M$; and (3) if $\mu = 1$, $\nu = 1$, then $1 \le i \le N$, $j = 0, M+1$ and $0 \le i \le N+1$, $1 \le j \le M$, satisfy the following relations :

$$(6.11.6) \qquad h(c^{1,\nu}_{i-1,j} + 4\ c^{1,\nu}_{i,j} + c^{1,\nu}_{i+1,j}) = -3(c^{0,\nu}_{i-1,j} - c^{0,\nu}_{i+1,j}),$$

where ν, i, j in (6.11.6) are such that if $\nu = 0$, then $1 \le i \le N$, $0 \le j \le M+1$, and if $\nu = 1$, then $1 \le i \le N$, $j = 0, M+1$; and

$$(6.11.7) \qquad \ell(c^{\mu,1}_{i,j-1} + 4\ c^{\mu,1}_{i,j} + c^{\mu,1}_{i,j+1}) = -3(c^{\mu,0}_{i,j-1} - c^{\mu,0}_{i,j+1}),$$

where μ, i, j in (6.11.7) are such that $\mu = 0, 1$, $0 \le i \le N+1$, $1 \le j \le M$.

Moreover, from (6.11.6) and (6.11.7) the unknowns $c^{\mu,\nu}_{i,j}$ where μ, ν, i and j satisfy the conditions of Lemma 6.11.2 can be obtained uniquely in terms of $c^{\mu,\nu}_{i,j}$ where μ, ν, i and j fulfill Definition 6.11.1 for $m = 2$, $\tau = 2$. ∎

Lemma 6.11.3. For a given $f(t,u) \in C^{(1,1)}([a,b] \times [c,d])$, $S^{\rho}_{2,2}f(t,u)$ exists and is unique.

Proof. For a given $g(t,u) \in C^{(1,1)}([a,b] \times [c,d])$, $H^{\rho}_2 g(t,u)$ exists and is unique. Further, by Lemma 6.11.2 for the given set of numbers $c^{\mu,\nu}_{i,j} = f^{(\mu,\nu)}_{i,j}$ where μ, ν, i and j fulfill Definition 6.11.1 for $m = 2$, $\tau = 2$, there exist unique $c^{\mu,\nu}_{i,j}$ where μ, ν, i and j satisfy the conditions of Lemma 6.11.2. Now, let $g(t,u) \in C^{(1,1)}([a,b] \times [c,d])$ be such that $g^{(\mu,\nu)}_{i,j} = c^{\mu,\nu}_{i,j}$, $0 \le \mu, \nu \le 1$, $0 \le i \le N+1$, $0 \le j \le M+1$. Then, again by Lemma 6.11.2, $H^{\rho}_2 g(t,u) \in S_{2,2}(\rho)$. However, from the definitions, this $H^{\rho}_2 g(t,u)$ is the same as $S^{\rho}_{2,2}f(t,u)$. ∎

Remark 6.11.1. In view of Remark 6.3.2, $S^{\rho}_{2,2}f(t,u)$ in terms of cardinal splines $\{s_i(t)\}_{i=0}^{N+3}$ and $\{s_j(u)\}_{j=0}^{M+3}$ can be explicitly expressed as

$$(6.11.8)\quad S_{2,2}^{\rho}f(t,u) = \sum_{i=0}^{N+1}\sum_{j=0}^{M+1} f_{i,j}s_i(t)s_j(u)$$

$$+\sum_{j=0}^{M+1}\left[f_{0,j}^{(1,0)}s_{N+2}(t) + f_{N+1,j}^{(1,0)}s_{N+3}(t)\right]s_j(u)$$

$$+\sum_{i=0}^{N+1}\left[f_{i,0}^{(0,1)}s_{M+2}(u) + f_{i,M+1}^{(0,1)}s_{M+3}(u)\right]s_i(t)$$

$$+\ f_{0,0}^{(1,1)}s_{N+2}(t)s_{M+2}(u) + f_{0,M+1}^{(1,1)}s_{N+2}(t)s_{M+3}(u)$$

$$+\ f_{N+1,0}^{(1,1)}s_{N+3}(t)s_{M+2}(u) + f_{N+1,M+1}^{(1,1)}s_{N+3}(t)s_{M+3}(u). \quad \blacksquare$$

Theorem 6.11.4. Let $f(t,u) \in PC^{4,\infty}([a,b]\times[c,d])$. Then,

$$(6.11.9)\qquad \| f - S_{2,2}^{\rho}f \|_\infty \le \alpha_{2,4,0}\ h^4 \ \| D_t^4 f \|_\infty + \alpha_{2,2,0}^2\ h^2\ell^2 \ \| D_u^2 D_t^2 f \|_\infty + \alpha_{2,4,0}\ \ell^4 \ \| D_u^4 f \|_\infty$$

and

$$(6.11.10)\ \| D_t(f - S_{2,2}^{\rho}f) \|_\infty \le \alpha_{2,4,1}\ h^3 \ \| D_t^4 f \|_\infty + \alpha_{2,2,0}\alpha_{2,2,1}\ h\ell^2 \ \| D_u^2 D_t^2 f \|_\infty + \alpha_{2,3,0}\ \ell^3 \ \| D_u^3 D_t f \|_\infty ,$$

where the constants $\alpha_{2,n,k}$ are defined in Table 6.3.1.

Proof. The proof uses the relation (6.11.4) and Theorem 6.3.6 and is similar to that of Theorem 5.5.3. ∎

Remark 6.11.2. In Theorem 6.11.4 the corresponding bound for $\| D_u(f - S_{2,2}^{\rho}f) \|_\infty$ can be obtained by replacing t with u and h with ℓ. ∎

The following result is an immediate consequence of Lemma 6.4.2.

Lemma 6.11.5. For a given $h(t,u) \in H_3(\rho)$, we define $c_{i,j}^{\mu,\nu} = D_t^{\mu} D_u^{\nu} h(t_i,u_j)$, $0 \le \mu,\nu \le 2,\ 0 \le i \le N+1,\ 0 \le j \le M+1$. The function $h(t,u) \in S_{3,4}(\rho)$ if and only if $c_{i,j}^{\mu,\nu}$, where μ, ν, i and j are such that (1) if $\mu = 1,2,\ \nu = 0$, then $1 \le i \le N,\ 0 \le j \le M+1$; (2) if $\mu = 0,\ \nu = 1,2$, then $0 \le i \le N+1,\ 1 \le j \le M$; and (3) if $\mu = 1,2,\ \nu = 1,2$, then $1 \le i \le N,\ j = 0, M+1$ and $0 \le i \le N+1,\ 1 \le j \le M$, satisfy the following relations :

$$(6.11.11)\quad h(227c_{1,j}^{1,\nu}+79c_{2,j}^{1,\nu}+3c_{3,j}^{1,\nu}) = -111hc_{0,j}^{1,\nu}-16h^2c_{0,j}^{2,\nu}-235c_{0,j}^{0,\nu}+65c_{1,j}^{0,\nu}+155c_{2,j}^{0,\nu}+15c_{3,j}^{0,\nu}$$

$$(6.11.12)\quad h(c_{i-2,j}^{1,\nu}+26c_{i-1,j}^{1,\nu}+66c_{i,j}^{1,\nu}+26c_{i+1,j}^{1,\nu}+c_{i+2,j}^{1,\nu}) = -5c_{i-2,j}^{0,\nu}-50c_{i-1,j}^{0,\nu}+50c_{i+1,j}^{0,\nu}+5c_{i+2,j}^{0,\nu}$$

$$(6.11.13)\quad h(3c_{N-2,j}^{1,\nu}+79c_{N-1,j}^{1,\nu}+227c_{N,j}^{1,\nu}) = -111hc_{N+1,j}^{1,\nu}+16h^2c_{N+1,j}^{2,\nu}-15c_{N-2,j}^{0,\nu}-155c_{N-1,j}^{0,\nu}-65c_{N,j}^{0,\nu}+235c_{N+1,j}^{0,\nu}$$

$$(6.11.14)\quad \frac{h^2}{20}(354c_{1,j}^{2,\nu}+201c_{2,j}^{2,\nu}+8c_{3,j}^{2,\nu}) = 12hc_{0,j}^{1,\nu}+\frac{23}{20}h^2c_{0,j}^{2,\nu}+37c_{0,j}^{0,\nu}-54c_{1,j}^{0,\nu}+9c_{2,j}^{0,\nu}+8c_{3,j}^{0,\nu}$$

$$(6.11.15)\quad \frac{h^2}{20}(c_{i-2,j}^{2,\nu}+26c_{i-1,j}^{2,\nu}+66c_{i,j}^{2,\nu}+26c_{i+1,j}^{2,\nu}+c_{i+2,j}^{2,\nu}) = c_{i-2,j}^{0,\nu}+2c_{i-1,j}^{0,\nu}-6c_{i,j}^{0,\nu}+2c_{i+1,j}^{0,\nu}+c_{i+2,j}^{0,\nu}$$

$$(6.11.16)\quad \frac{h^2}{20}(8c_{N-2,j}^{2,\nu}+201c_{N-1,j}^{2,\nu}+354c_{N,j}^{2,\nu}) = -12hc_{N+1,j}^{1,\nu}+\frac{23}{20}h^2c_{N+1,j}^{2,\nu}+8c_{N-2,j}^{0,\nu}+9c_{N-1,j}^{0,\nu}-54c_{N,j}^{0,\nu}+37c_{N+1,j}^{0,\nu}.$$

In the system (6.11.11) - (6.11.16) if $\nu = 0$, then in (6.11.12) and (6.11.15), $2 \le i \le N-1$, $0 \le j \le M+1$, and in (6.11.11), (6.11.13), (6.11.14) and (6.11.16), $0 \le j \le M+1$; also if $\nu = 1,2$ then in (6.11.12) and (6.11.15), $2 \le i \le N-1$, $j = 0, M+1$, and in (6.11.11), (6.11.13), (6.11.14) and (6.11.16), $j = 0, M+1$.

Further, we have

$$(6.11.17)\quad \ell(227c_{i,1}^{\mu,1}+79c_{i,2}^{\mu,1}+3c_{i,3}^{\mu,1}) = -111\ell c_{i,0}^{\mu,1}-16\ell^2c_{i,0}^{\mu,2}-235c_{i,0}^{\mu,0}+65c_{i,1}^{\mu,0}+155c_{i,2}^{\mu,0}+15c_{i,3}^{\mu,0}$$

$$(6.11.18)\quad \ell(c_{i,j-2}^{\mu,1}+26c_{i,j-1}^{\mu,1}+66c_{i,j}^{\mu,1}+26c_{i,j+1}^{\mu,1}+c_{i,j+2}^{\mu,1}) = -5c_{i,j-2}^{\mu,0}-50c_{i,j-1}^{\mu,0}+50c_{i,j+1}^{\mu,0}+5c_{i,j+2}^{\mu,0}$$

$$(6.11.19)\quad \ell(3c_{i,M-2}^{\mu,1}+79c_{i,M-1}^{\mu,1}+227c_{i,M}^{\mu,1}) = -111\ell c_{i,M+1}^{\mu,1}+16\ell^2c_{i,M+1}^{\mu,2}-15c_{i,M-2}^{\mu,0}-155c_{i,M-1}^{\mu,0}-65c_{i,M}^{\mu,0}+235c_{i,M+1}^{\mu,0}$$

$$(6.11.20)\quad \frac{\ell^2}{20}(354c_{i,1}^{\mu,2}+201c_{i,2}^{\mu,2}+8c_{i,3}^{\mu,2})$$

$$= 12\ell c_{i,0}^{\mu,1}+\frac{23}{20}\ell^2 c_{i,0}^{\mu,2}+37c_{i,0}^{\mu,0}-54c_{i,1}^{\mu,0}+9c_{i,2}^{\mu,0}+8c_{i,3}^{\mu,0}$$

$$(6.11.21)\quad \frac{\ell^2}{20}(c_{i,j-2}^{\mu,2}+26c_{i,j-1}^{\mu,2}+66c_{i,j}^{\mu,2}+26c_{i,j+1}^{\mu,2}+c_{i,j+2}^{\mu,2})$$

$$= c_{i,j-2}^{\mu,0}+2c_{i,j-1}^{\mu,0}-6c_{i,j}^{\mu,0}+2c_{i,j+1}^{\mu,0}+c_{i,j+2}^{\mu,0}$$

$$(6.11.22)\quad \frac{\ell^2}{20}(8c_{i,M-2}^{\mu,2}+201c_{i,M-1}^{\mu,2}+354c_{i,M}^{\mu,2})$$

$$= -12\ell c_{i,M+1}^{\mu,1}+\frac{23}{20}\ell^2 c_{i,M+1}^{\mu,2}+8c_{i,M-2}^{\mu,0}+9c_{i,M-1}^{\mu,0}-54c_{i,M}^{\mu,0}+37c_{i,M+1}^{\mu,0}.$$

For $0 \le \mu \le 2$ in equations (6.11.18) and (6.11.21), $2 \le j \le M-1$, $0 \le i \le N+1$, and in (6.11.17), (6.11.19), (6.11.20) and (6.11.22), $0 \le i \le N+1$.

Moreover, from the systems (6.11.11) - (6.11.13), (6.11.14) - (6.11.16), (6.11.17) - (6.11.19) and (6.11.20) - (6.11.22) the unknowns $c_{i,j}^{\mu,\nu}$ where μ, ν, i and j satisfy the conditions of Lemma 6.11.5 can be obtained uniquely in terms of $c_{i,j}^{\mu,\nu}$ where μ, ν, i and j fulfill Definition 6.11.1 for $m = 3$, $\tau = 4$. ∎

Lemma 6.11.6. For a given $f(t,u) \in C^{(2,2)}([a,b]\times[c,d])$, $S_{3,4}^{\varrho}f(t,u)$ exists and is unique.

Proof. The proof is similar to that of Lemma 6.11.3. ∎

Remark 6.11.3. In view of Remark 6.4.2, $S_{3,4}^{\varrho}f(t,u)$ in terms of cardinal splines $\{s_i(t)\}_{i=0}^{N+5}$ and $\{s_j(u)\}_{j=0}^{M+5}$ can be explicitly expressed as

$$(6.11.23)\quad S_{3,4}^{\varrho}f(t,u)$$

$$= \sum_{i=0}^{N+1}\sum_{j=0}^{M+1} f_{i,j}s_i(t)s_j(u)+\sum_{j=0}^{M+1}\Big[f_{0,j}^{(1,0)}s_{N+2}(t)+f_{N+1,j}^{(1,0)}s_{N+3}(t)$$

$$+ f_{0,j}^{(2,0)}s_{N+4}(t)+f_{N+1,j}^{(2,0)}s_{N+5}(t)\Big]s_j(u)+\sum_{i=0}^{N+1}\Big[f_{i,0}^{(0,1)}s_{M+2}(u)$$

$$+ f_{i,M+1}^{(0,1)}s_{M+3}(u)+f_{i,0}^{(0,2)}s_{M+4}(u)+f_{i,M+1}^{(0,2)}s_{M+5}(u)\Big]s_i(t)$$

$$+ f^{(1,1)}_{0,0} s_{N+2}(t)s_{M+2}(u) + f^{(1,1)}_{0,M+1} s_{N+2}(t)s_{M+3}(u) + f^{(1,1)}_{N+1,0} s_{N+3}(t)s_{M+2}(u)$$

$$+ f^{(1,1)}_{N+1,M+1} s_{N+3}(t)s_{M+3}(u) + f^{(2,1)}_{0,0} s_{N+4}(t)s_{M+2}(u) + f^{(2,1)}_{0,M+1} s_{N+4}(t)s_{M+3}(u)$$

$$+ f^{(2,1)}_{N+1,0} s_{N+5}(t)s_{M+2}(u) + f^{(2,1)}_{N+1,M+1} s_{N+5}(t)s_{M+3}(u) + f^{(1,2)}_{0,0} s_{N+2}(t)s_{M+4}(u)$$

$$+ f^{(1,2)}_{0,M+1} s_{N+2}(t)s_{M+5}(u) + f^{(1,2)}_{N+1,0} s_{N+3}(t)s_{M+4}(u) + f^{(1,2)}_{N+1,M+1} s_{N+3}(t)s_{M+5}(u)$$

$$+ f^{(2,2)}_{0,0} s_{N+4}(t)s_{M+4}(u) + f^{(2,2)}_{0,M+1} s_{N+4}(t)s_{M+5}(u) + f^{(2,2)}_{N+1,0} s_{N+5}(t)s_{M+4}(u)$$

$$+ f^{(2,2)}_{N+1,M+1} s_{N+5}(t)s_{M+5}(u). \qquad \blacksquare$$

Theorem 6.11.7. Let $f(t,u) \in PC^{4,\infty}([a,b] \times [c,d])$. Then,

$$\text{(6.11.24)} \quad \| f - S^{\rho}_{3,4} f \|_\infty \le \alpha_{3,4,0}\, h^4 \| D^4_t f \|_\infty + \alpha^2_{3,2,0}\, h^2 \ell^2 \| D^2_u D^2_t f \|_\infty + \alpha_{3,4,0}\, \ell^4 \| D^4_u f \|_\infty$$

and

$$\text{(6.11.25)} \quad \| D_t(f - S^{\rho}_{3,4} f) \|_\infty \le \alpha_{3,4,1}\, h^3 \| D^4_t f \|_\infty + \alpha_{3,2,0}\alpha_{3,2,1}\, h\ell^2 \| D^2_u D^2_t f \|_\infty + \alpha_{3,3,0}\, \ell^3 \| D^3_u D_t f \|_\infty ,$$

where the constants $\alpha_{3,n,k}$ are defined in Table 6.4.1.

Proof. Since as a function of t, $(f - S^{\Delta'}_{3,4} f) \in PC^{2,\infty}[a,b]$, an application of Theorem 6.4.6 in (6.11.4) gives

$$\text{(6.11.26)} \quad \| f - S^{\rho}_{3,4} f \|_\infty \le \alpha_{3,4,0}\, h^4 \| D^4_t f \|_\infty + \alpha_{3,2,0}\, h^2 \| D^2_t (f - S^{\Delta'}_{3,4} f) \|_\infty + \alpha_{3,4,0}\, \ell^4 \| D^4_u f \|_\infty .$$

Now since as a function of u, $D^2_t f \in PC^{2,\infty}[c,d]$, and $D^2_t S^{\Delta'}_{3,4} f = S^{\Delta'}_{3,4} D^2_t f$, Theorem 6.4.6 can be used again to obtain

$$\text{(6.11.27)} \qquad \| D^2_t (f - S^{\Delta'}_{3,4} f) \|_\infty \le \alpha_{3,2,0}\, \ell^2 \| D^2_u D^2_t f \|_\infty .$$

Using (6.11.27) in (6.11.26) we find (6.11.24). The proof of (6.11.25) is similar. $\blacksquare$

Theorem 6.11.8. Let $f(t,u) \in PC^{5,\infty}([a,b] \times [c,d])$. Then,

(6.11.28) $\| f - S^{\rho}_{3,4} f \|_\infty \le \alpha_{3,5,0}\, h^5 \| D_t^5 f \|_\infty + \alpha_{3,2,0}\alpha_{3,3,0}\, h^3 \ell^2 \| D_u^2 D_t^3 f \|_\infty$

$+ \alpha_{3,5,0}\, \ell^5 \| D_u^5 f \|_\infty$

(6.11.29) $\| f - S^{\rho}_{3,4} f \|_\infty \le \alpha_{3,5,0}\, h^5 \| D_t^5 f \|_\infty + \alpha_{3,2,0}\alpha_{3,3,0}\, h^2 \ell^3 \| D_u^3 D_t^2 f \|_\infty$

$+ \alpha_{3,5,0}\, \ell^5 \| D_u^5 f \|_\infty$

(6.11.30) $\| D_t(f - S^{\rho}_{3,4} f) \|_\infty \le \alpha_{3,5,1}\, h^4 \| D_t^5 f \|_\infty$

$+ \alpha_{3,2,0}\alpha_{3,3,1}\, h^2 \ell^2 \| D_u^2 D_t^3 f \|_\infty + \alpha_{3,4,0}\, \ell^4 \| D_u^4 D_t f \|_\infty$

(6.11.31) $\| D_t(f - S^{\rho}_{3,4} f) \|_\infty \le \alpha_{3,5,1}\, h^4 \| D_t^5 f \|_\infty$

$+ \alpha_{3,2,1}\alpha_{3,3,0}\, h \ell^3 \| D_u^3 D_t^2 f \|_\infty + \alpha_{3,4,0}\, \ell^4 \| D_u^4 D_t f \|_\infty$

and

(6.11.32) $\| D_t^2(f - S^{\rho}_{3,4} f) \|_\infty \le \alpha_{3,5,2}\, h^3 \| D_t^5 f \|_\infty$

$+ \alpha_{3,2,0}\alpha_{3,3,2}\, h \ell^2 \| D_u^2 D_t^3 f \|_\infty + \alpha_{3,3,0}\, \ell^3 \| D_u^3 D_t^2 f \|_\infty ,$

where the constants $\alpha_{3,n,k}$ are defined in Table 6.4.1.

Proof. The proof is similar to that of Theorem 6.11.7. ∎

Theorem 6.11.9. Let $f(t,u) \in PC^{6,\infty}([a,b] \times [c,d])$. Then,

(6.11.33) $\| f - S^{\rho}_{3,4} f \|_\infty \le \alpha_{3,6,0}\, h^6 \| D_t^6 f \|_\infty + \alpha_{3,2,0}\alpha_{3,4,0}\, h^4 \ell^2 \| D_u^2 D_t^4 f \|_\infty$

$+ \alpha_{3,6,0}\, \ell^6 \| D_u^6 f \|_\infty$

(6.11.34) $\| f - S^{\rho}_{3,4} f \|_\infty \le \alpha_{3,6,0}\, h^6 \| D_t^6 f \|_\infty + \alpha_{3,3,0}^2\, h^3 \ell^3 \| D_u^3 D_t^3 f \|_\infty$

$+ \alpha_{3,6,0}\, \ell^6 \| D_u^6 f \|_\infty$

(6.11.35) $\| f - S^{\rho}_{3,4} f \|_\infty \le \alpha_{3,6,0}\, h^6 \| D_t^6 f \|_\infty + \alpha_{3,2,0}\alpha_{3,4,0}\, h^2 \ell^4 \| D_u^4 D_t^2 f \|_\infty$

$+ \alpha_{3,6,0}\, \ell^6 \| D_u^6 f \|_\infty$

$$(6.11.36)\ \| D_t(f - S^{\rho}_{3,4}f) \|_\infty \le \alpha_{3,6,1}\, h^5 \| D_t^6 f \|_\infty + \alpha_{3,2,0}\alpha_{3,4,1}\, h^3\ell^2 \| D_u^2 D_t^4 f \|_\infty + \alpha_{3,5,0}\, \ell^5 \| D_u^5 D_t f \|_\infty$$

$$(6.11.37)\ \| D_t(f - S^{\rho}_{3,4}f) \|_\infty \le \alpha_{3,6,1}\, h^5 \| D_t^6 f \|_\infty + \alpha_{3,3,0}\alpha_{3,3,1}\, h^2\ell^3 \| D_u^3 D_t^3 f \|_\infty + \alpha_{3,5,0}\, \ell^5 \| D_u^5 D_t f \|_\infty$$

$$(6.11.38)\ \| D_t(f - S^{\rho}_{3,4}f) \|_\infty \le \alpha_{3,6,1}\, h^5 \| D_t^6 f \|_\infty + \alpha_{3,2,1}\alpha_{3,4,0}\, h\ell^4 \| D_u^4 D_t^2 f \|_\infty + \alpha_{3,5,0}\, \ell^5 \| D_u^5 D_t f \|_\infty$$

$$(6.11.39)\ \| D_t^2(f - S^{\rho}_{3,4}f) \|_\infty \le \alpha_{3,6,2}\, h^4 \| D_t^6 f \|_\infty + \alpha_{3,2,0}\alpha_{3,4,2}\, h^2\ell^2 \| D_u^2 D_t^4 f \|_\infty + \alpha_{3,4,0}\, \ell^4 \| D_u^4 D_t^2 f \|_\infty$$

$$(6.11.40)\ \| D_t^2(f - S^{\rho}_{3,4}f) \|_\infty \le \alpha_{3,6,2}\, h^4 \| D_t^6 f \|_\infty + \alpha_{3,3,0}\alpha_{3,3,2}\, h\ell^3 \| D_u^3 D_t^3 f \|_\infty + \alpha_{3,4,0}\, \ell^4 \| D_u^4 D_t^2 f \|_\infty$$

and

$$(6.11.41)\ \| D_t^3(f - S^{\rho}_{3,4}f) \|_\infty \le \alpha_{3,6,3}\, h^3 \| D_t^6 f \|_\infty + \alpha_{3,2,0}\alpha_{3,4,3}\, h\ell^2 \| D_u^2 D_t^4 f \|_\infty + \alpha_{3,3,0}\, \ell^3 \| D_u^3 D_t^3 f \|_\infty ,$$

where the constants $\alpha_{3,n,k}$ are defined in Table 6.4.1.

Proof. The proof is similar to that of Theorem 6.11.7. ∎

Remark 6.11.4. In Theorems 6.11.7 - 6.11.9 the corresponding bounds for $\| D_u^k(f - S^{\rho}_{3,4}f) \|_\infty$ can be obtained by replacing t with u and h with ℓ. With such a replacement a particular inequality (6.11.•) will be denoted by (6.11.•)′.

Remark 6.11.5. In Table 6.11.1 we compute the actual values of $\| D_t^k(f - S^{\rho}_{3,4}f) \|_\infty$ and $\| D_u^k(f - S^{\rho}_{3,4}f) \|_\infty$, $0 \le k \le 3$ for the function $f(t,u) = t(1 - e^{tu}) \in C^{(\infty,\infty)}([0,1] \times [0,1])$ and use Theorem 6.11.9 to compare these with the corresponding right side bounds.

Table 6.11.1.

$N = M$	7	9	15
$\| f - S^{\rho}_{3,4} f \|_{\infty}$	0.54018019 − 8	0.14792303 − 8	0.93839104 − 10
Bound (6.11.33)	0.63203026 − 3	0.16568294 − 3	0.98754729 − 5
Bound (6.11.34)	0.45752894 − 3	0.11993846 − 3	0.71488897 − 5
Bound (6.11.35)	0.26873306 − 3	0.70446759 − 4	0.41989540 − 5
$\| D_t(f - S^{\rho}_{3,4} f) \|_{\infty}$	0.12516071 − 6	0.42885312 − 7	0.43611023 − 8
Bound (6.11.36)	0.15959848 − 1	0.52297231 − 2	0.49874526 − 3
Bound (6.11.37)	0.11529662 − 1	0.37780399 − 2	0.36030197 − 3
Bound (6.11.38)	0.67825408 − 2	0.22225030 − 2	0.21195440 − 3
$\| D_u(f - S^{\rho}_{3,4} f) \|_{\infty}$	0.32055214 − 7	0.88441303 − 8	0.63126050 − 9
Bound (6.11.36)′	0.68348617 − 2	0.22396475 − 2	0.21358943 − 3
Bound (6.11.37)′	0.11570850 − 1	0.37915364 − 2	0.36158909 − 3
Bound (6.11.38)′	0.15974819 − 1	0.52346288 − 2	0.49921311 − 3
$\| D^2_t(f - S^{\rho}_{3,4} f) \|_{\infty}$	0.55136004 − 5	0.23584122 − 5	0.38352312 − 6
Bound (6.11.39)	0.97122799 + 0	0.39781499 + 0	0.60701749 − 1
Bound (6.11.40)	0.73259435 + 0	0.30007064 + 0	0.45787146 − 1
$\| D^2_u(f - S^{\rho}_{3,4} f) \|_{\infty}$	0.80901403 − 6	0.34415261 − 6	0.55515350 − 7
Bound (6.11.39)′	0.41733812 + 0	0.17094170 + 0	0.26083632 − 1
Bound (6.11.40)′	0.73552195 + 0	0.30126978 + 0	0.45970121 − 1
$\| D^3_t(f - S^{\rho}_{3,4} f) \|_{\infty}$	0.34484214 − 3	0.17660074 − 3	0.43120302 − 4
Bound (6.11.41)	0.87612562 + 2	0.44857632 + 2	0.10951570 + 2
$\| D^3_u(f - S^{\rho}_{3,4} f) \|_{\infty}$	0.49280003 − 4	0.25231112 − 4	0.61595014 − 5
Bound (6.11.41)′	0.37358944 + 2	0.19127779 + 2	0.46698679 + 1

■

The following result follows from Lemma 6.6.1.

Lemma 6.11.10. For a given $h(t,u) \in H_3(\rho)$, we define $c^{\mu,\nu}_{i,j} = D^{\mu}_t D^{\nu}_u h(t_i, u_j)$, $0 \le \mu, \nu \le 2$, $0 \le i \le N+1$, $0 \le j \le M+1$. The function $h(t,u) \in S_{3,3}(\rho)$ if and only if $c^{\mu,\nu}_{i,j}$, where μ, ν, i and j are such that (1) if $\mu = 0, 1$, $\nu = 2$, then $0 \le i \le N+1$, $1 \le j \le M$; (2) if $\mu = 2$, $\nu = 0, 1$, then $1 \le i \le N$, $0 \le j \le M+1$; and (3) if $\mu = 2$, $\nu = 2$, then $1 \le i \le N$, $j = 0, M+1$ and

$0 \le i \le N+1,\ 1 \le j \le M$, satisfy the following relations :

$$(6.11.42)\quad h^2(c_{i-1,j}^{2,\nu} - 6\ c_{i,j}^{2,\nu} + c_{i+1,j}^{2,\nu}) = -20(c_{i-1,j}^{0,\nu} - 2\ c_{i,j}^{0,\nu} + c_{i+1,j}^{0,\nu}) - 8h(c_{i-1,j}^{1,\nu} - c_{i+1,j}^{1,\nu}),$$

where ν, i, j in (6.11.42) are such that if $\nu = 0, 1$, then $1 \le i \le N$, $0 \le j \le M+1$, and if $\nu = 2$, then $1 \le i \le N$, $j = 0, M+1$; and

$$(6.11.43)\quad l^2(c_{i,j-1}^{\mu,2} - 6\ c_{i,j}^{\mu,2} + c_{i,j+1}^{\mu,2}) = -20(c_{i,j-1}^{\mu,0} - 2\ c_{i,j}^{\mu,0} + c_{i,j+1}^{\mu,0}) - 8\ell(c_{i,j-1}^{\mu,1} - c_{i,j+1}^{\mu,1}),$$

where μ, i, j in (6.11.43) are such that $0 \le \mu \le 2$, $0 \le i \le N+1$, $1 \le j \le M$.

Moreover, from (6.11.42) and (6.11.43) the unknowns $c_{i,j}^{\mu,\nu}$ where μ, ν, i and j satisfy the conditions of Lemma 6.11.10 can be obtained uniquely in terms of $c_{i,j}^{\mu,\nu}$ where μ, ν, i and j fulfill Definition 6.11.1 for $m = 3$, $\tau = 3$. ∎

Lemma 6.11.11. For a given $f(t,u) \in C^{(2,2)}([a,b] \times [c,d])$, $S_{3,3}^{\rho} f(t,u)$ exists and is unique.

Proof. The proof is similar to that of Lemma 6.11.3. ∎

Remark 6.11.6. In view of Remark 6.6.2, $S_{3,3}^{\rho} f(t,u)$ in terms of cardinal splines $\{s_i(t)\}_{i=0}^{2N+5}$ and $\{s_j(u)\}_{j=0}^{2M+5}$ can be explicitly expressed as

$$(6.11.44)\quad S_{3,3}^{\rho} f(t,u) = \sum_{i=0}^{N+1} \sum_{j=0}^{M+1} \Big[f_{i,j} s_i(t) s_j(u) + f_{i,j}^{(1,1)} s_{i+N+2}(t) s_{j+M+2}(u) + f_{i,j}^{(0,1)} s_i(t) s_{j+M+2}(u) + f_{i,j}^{(1,0)} s_{i+N+2}(t) s_j(u) \Big]$$
$$+ \sum_{j=0}^{M+1} \Big\{ \Big[f_{0,j}^{(2,0)} s_{2N+4}(t) + f_{N+1,j}^{(2,0)} s_{2N+5}(t) \Big] s_j(u) + \Big[f_{0,j}^{(2,1)} s_{2N+4}(t) + f_{N+1,j}^{(2,1)} s_{2N+5}(t) \Big] s_{j+M+2}(u) \Big\}$$
$$+ \sum_{i=0}^{N+1} \Big\{ \Big[f_{i,0}^{(0,2)} s_{2M+4}(u) + f_{i,M+1}^{(0,2)} s_{2M+5}(u) \Big] s_i(t) + \Big[f_{i,0}^{(1,2)} s_{2M+4}(u) + f_{i,M+1}^{(1,2)} s_{2M+5}(u) \Big] s_{i+N+2}(t) \Big\}$$

$$+ f^{(2,2)}_{0,0} s_{2N+4}(t)s_{2M+4}(u) + f^{(2,2)}_{0,M+1} s_{2N+4}(t)s_{2M+5}(u)$$

$$+ f^{(2,2)}_{N+1,0} s_{2N+5}(t)s_{2M+4}(u) + f^{(2,2)}_{N+1,M+1} s_{2N+5}(t)s_{2M+5}(u). \quad \blacksquare$$

Remark 6.11.7. In Theorems 6.11.7 - 6.11.9 inequalities (6.11.24) - (6.11.41) hold if we replace $S^{\rho}_{3,4}f(t,u)$ by $S^{\rho}_{3,3}f(t,u)$ and $\alpha_{3,n,k}$ by $\beta_{3,n,k}$, where $\beta_{3,n,k}$ are defined in Table 6.6.1. An inequality (6.11.•) with such a replacement will be denoted as (6.11.•)*. Further, as in Remark 6.11.4 the corresponding bounds for $\| D^k_u(f - S^{\rho}_{3,3}f) \|_{\infty}$ can be obtained by replacing t with u and h with ℓ in the inequalities (6.11.•)*. With such a replacement a particular inequality (6.11.•)* will be denoted by (6.11.•)′*. ■

Remark 6.11.8. In Table 6.11.2 we compute the actual values of $\| D^k_t(f - S^{\rho}_{3,3}f) \|_{\infty}$ and $\| D^k_u(f - S^{\rho}_{3,3}f) \|_{\infty}$, $0 \le k \le 3$ for the same function $f(t,u) = t(1 - e^{tu}) \in C^{(\infty,\infty)}([0,1] \times [0,1])$ and compare these with the corresponding right side bounds.

Our last result here provides error bounds in L_2 - norm.

Theorem 6.11.12. Let $f(t,u) \in PC^{2m,2}([a,b] \times [c,d])$. Then,

$$(6.11.45) \quad \| D^k_t(f - S^{\rho}_{m,\tau}f) \|_2$$

$$\le \xi_{m,2m,k}\, h^{2m-k} \| D^{2m}_t f \|_2 + \xi_{m,m,0}\xi_{m,m,k}\, h^{m-k}\ell^m \| D^m_u D^m_t f \|_2$$

$$+ \xi_{m,2m-k,0}\, \ell^{2m-k} \| D^{2m-k}_u D^k_t f \|_2, \quad 0 \le k \le m-1$$

and

$$(6.11.46) \quad \| D^m_t(f - S^{\rho}_{m,\tau}f) \|_2 \le \xi_{m,2m,m}\, h^m \| D^{2m}_t f \|_2$$

$$+ 2\, \xi_{m,m,0}\, \ell^m \| D^m_u D^m_t f \|_2,$$

where the constants $\xi_{m,n,k}$ are defined in Theorem 6.10.5.

Proof. The proof is similar to that of Theorem 5.5.7. ■

Table 6.11.2.

$N = M$	7	9	15
$\Vert f - S^{\rho}_{3,3} f \Vert_\infty$	0.35190999 − 8	0.96441899 − 9	0.61482375 − 10
Bound (6.11.33)*	0.35292222 − 6	0.92516444 − 7	0.55144098 − 8
Bound (6.11.34)*	0.23767153 − 6	0.62304167 − 7	0.37136177 − 8
Bound (6.11.35)*	0.15297836 − 6	0.40102358 − 7	0.23902868 − 8
$\Vert D_t(f - S^{\rho}_{3,3} f) \Vert_\infty$	0.88434697 − 7	0.30338230 − 7	0.31011702 − 8
Bound (6.11.36)*	0.93917330 − 5	0.30774831 − 5	0.29349166 − 6
Bound (6.11.37)*	0.63231718 − 5	0.20719769 − 5	0.19759912 − 6
Bound (6.11.38)*	0.42008262 − 5	0.13765267 − 5	0.13127582 − 6
$\Vert D_u(f - S^{\rho}_{3,3} f) \Vert_\infty$	0.20852825 − 7	0.57554557 − 8	0.49168181 − 9
Bound (6.11.36)′*	0.45997610 − 5	0.15072497 − 5	0.14374253 − 6
Bound (6.11.37)′*	0.67723471 − 5	0.22191627 − 5	0.21163585 − 6
Bound (6.11.38)′*	0.99592163 − 5	0.32634360 − 5	0.31122551 − 6
$\Vert D_t^2(f - S^{\rho}_{3,3} f) \Vert_\infty$	0.46969770 − 5	0.20339370 − 5	0.33825538 − 6
Bound (6.11.39)*	0.72292781 − 3	0.29611123 − 3	0.45182988 − 4
Bound (6.11.40)*	0.53252683 − 3	0.21812299 − 3	0.33282927 − 4
$\Vert D_u^2(f - S^{\rho}_{3,3} f) \Vert_\infty$	0.69507418 − 6	0.29905670 − 6	0.49211798 − 7
Bound (6.11.39)′*	0.38001464 − 3	0.15565400 − 3	0.23750915 − 4
Bound (6.11.40)′*	0.57793036 − 3	0.23672028 − 3	0.36120648 − 4
$\Vert D_t^3(f - S^{\rho}_{3,3} f) \Vert_\infty$	0.39981576 − 3	0.20838908 − 3	0.52274663 − 4
Bound (6.11.41)*	0.73721155 − 1	0.37745231 − 1	0.92151444 − 2
$\Vert D_u^3(f - S^{\rho}_{3,3} f) \Vert_\infty$	0.57759921 − 4	0.30040286 − 4	0.75107188 − 5
Bound (6.11.41)′*	0.34950424 − 1	0.17894617 − 1	0.43688010 − 2

∎

6.12 TWO VARIABLE LIDSTONE - SPLINE INTERPOLATION

We begin with the following :

Definition 6.12.1. For a given $f(t,u) \in C^{(2m-2,2m-2)}([a,b] \times [c,d])$, we say

$LS^{\rho}_{m,2m-2}f(t,u)$ is the *Lidstone* $S_{m,2m-2}(\rho)$ *- interpolate of* $f(t,u)$, also known as *Lidstone - spline interpolate of* $f(t,u)$, if $LS^{\rho}_{m,2m-2}f(t,u) \in S_{m,2m-2}(\rho)$ with $D_t^{2\mu}D_u^{2\nu}LS^{\rho}_{m,2m-2}f(t_i,u_j) = f_{i,j}^{(2\mu,2\nu)}$, where μ, ν, i and j satisfy the following :
(1) if $\mu = 0,\ \nu = 0$ then $0 \le i \le N+1,\ 0 \le j \le M+1$;
(2) if $1 \le \mu \le m-1,\ \nu = 0$ then $i = 0, N+1,\ 0 \le j \le M+1$;
(3) if $\mu = 0,\ 1 \le \nu \le m-1$ then $0 \le i \le N+1,\ j = 0, M+1$; and
(4) if $1 \le \mu \le m-1,\ 1 \le \nu \le m-1$ then $(i,j) = (0,0),\ (0,M+1),\ (N+1,0),\ (N+1,M+1)$.

In view of (5.6.1) it is clear that $LS^{\rho}_{m,2m-2}f(t,u)$ can be written as

$$(6.12.1)\quad LS^{\rho}_{m,2m-2}f(t,u) = \sum_{i=0}^{N+1}\sum_{\mu=0}^{m-1}\sum_{j=0}^{M+1}\sum_{\nu=0}^{m-1} r_{m,i,\mu}(t)r_{m,j,\nu}(u)D_t^{2\mu}D_u^{2\nu}LS^{\rho}_{m,2m-2}f(t_i,u_j).$$

Like (6.11.1) in (6.12.1) also, $D_t^{2\mu}D_u^{2\nu}LS^{\rho}_{m,2m-2}f(t_i,u_j)$ where μ, ν, i and j do not fulfill Definition 6.12.1 exist uniquely. In fact, for $m = 2$ and 3 we shall show that these unknown constants are the solutions of diagonally dominant systems of algebraic equations.

The following result provides a characterization of $LS^{\rho}_{m,2m-2}f(t,u)$ in terms of one - dimensional interpolation schemes.

Lemma 6.12.1. If $f(t,u) \in C^{(2m-2,2m-2)}([a,b]\times[c,d])$, then

$$(6.12.2)\quad LS^{\rho}_{m,2m-2}f(t,u) = LS^{\Delta'}_{m,2m-2}LS^{\Delta}_{m,2m-2}f(t,u) = LS^{\Delta}_{m,2m-2}LS^{\Delta'}_{m,2m-2}f(t,u).$$

Proof. The proof is similar to that of Lemma 5.5.1. ∎

Now let $f(t,u) \in C^{(2m-2,2m-2)}([a,b]\times[c,d])$ be an arbitrary function. From Lemma 6.12.1 we have

$$f - LS^{\rho}_{m,2m-2}f$$

$$(6.12.3)\quad = (f - LS^{\Delta}_{m,2m-2}f) + LS^{\Delta}_{m,2m-2}(f - LS^{\Delta'}_{m,2m-2}f)$$

$$(6.12.4)\quad = (f - LS^{\Delta}_{m,2m-2}f) + \Big[LS^{\Delta}_{m,2m-2}(f - LS^{\Delta'}_{m,2m-2}f)$$

$$-(f - LS^{\Delta'}_{m,2m-2}f)\big] + (f - LS^{\Delta'}_{m,2m-2}f)$$

$$(6.12.5) \qquad = (f - LS^{\Delta}_{m,2m-2}f) + \big[LS^{\Delta'}_{m,2m-2}(f - LS^{\Delta}_{m,2m-2}f) - (f - LS^{\Delta}_{m,2m-2}f)\big] + (f - LS^{\Delta'}_{m,2m-2}f).$$

The following result follows from Lemma 6.8.2.

Lemma 6.12.2. Let $h(t,u) \in L_2(\rho)$, be a function for which, say, $c_{i,j}^{2\mu,2\nu} = D_t^{2\mu} D_u^{2\nu} h(t_i,u_j)$; $0 \le \mu,\nu \le 1$, $0 \le i \le N+1$, $0 \le j \le M+1$ exist. Then, $h(t,u) \in S_{2,2}(\rho)$ if and only if $c_{i,j}^{2\mu,2\nu}$, where μ, ν, i and j are such that (1) if $\mu = 1$, $\nu = 0$, then $1 \le i \le N$, $0 \le j \le M+1$; (2) if $\mu = 0$, $\nu = 1$, then $0 \le i \le N+1$, $1 \le j \le M$; and (3) if $\mu = 1$, $\nu = 1$, then $1 \le i \le N$, $j = 0, M+1$ and $0 \le i \le N+1$, $1 \le j \le M$, satisfy the following relations :

$$(6.12.6) \qquad h^2(c_{i-1,j}^{2,2\nu} + 4\, c_{i,j}^{2,2\nu} + c_{i+1,j}^{2,2\nu}) = 6(c_{i-1,j}^{0,2\nu} - 2\, c_{i,j}^{0,2\nu} + c_{i+1,j}^{0,2\nu}),$$

where ν, i, j in (6.12.6) are such that if $\nu = 0$, then $1 \le i \le N$, $0 \le j \le M+1$, and if $\nu = 1$, then $1 \le i \le N$, $j = 0, M+1$; and

$$(6.12.7) \qquad \ell^2(c_{i,j-1}^{2\mu,2} + 4\, c_{i,j}^{2\mu,2} + c_{i,j+1}^{2\mu,2}) = 6(c_{i,j-1}^{2\mu,0} - 2\, c_{i,j}^{2\mu,0} + c_{i,j+1}^{2\mu,0}),$$

where μ, i, j in (6.12.7) are such that $\mu = 0, 1$, $0 \le i \le N+1$, $1 \le j \le M$.

Moreover, from (6.12.6) and (6.12.7) the unknowns $c_{i,j}^{2\mu,2\nu}$ where μ, ν, i and j satisfy the conditions of Lemma 6.12.2 can be obtained uniquely in terms of $c_{i,j}^{2\mu,2\nu}$ where μ, ν, i and j fulfill Definition 6.12.1 for $m = 2$. ∎

Lemma 6.12.3. For a given $f(t,u) \in C^{(2,2)}([a,b] \times [c,d])$, $LS^{\rho}_{2,2}f(t,u)$ exists and is unique.

Proof. The proof is similar to that of Lemma 6.11.3. ∎

Theorem 6.12.4. Let $f(t,u) \in PC^{4,\infty}([a,b] \times [c,d])$. Then,

$$(6.12.8) \qquad \| f - LS^{\rho}_{2,2}f \|_\infty \le \gamma_{2,4,0}\, h^4 \| D_t^4 f \|_\infty + \gamma_{2,2,0}^2\, h^2\ell^2 \| D_u^2 D_t^2 f \|_\infty + \gamma_{2,4,0}\, \ell^4 \| D_u^4 f \|_\infty$$

and

$$(6.12.9)\quad \| D_t(f - LS^{\rho}_{2,2}f) \|_{\infty} \le \gamma_{2,4,1}\, h^3 \| D_t^4 f \|_{\infty}$$

$$+ \gamma_{2,2,0}\gamma_{2,2,1}\, h\ell^2 \| D_u^2 D_t^2 f \|_{\infty} + \gamma_{2,3,0}\, \ell^3 \| D_u^3 D_t f \|_{\infty},$$

where the constants $\gamma_{2,n,k}$ are defined in Table 6.8.1.

Proof. The proof uses the relation (6.12.4) and Theorem 6.8.5 and is similar to that of Theorem 5.5.3. ∎

Remark 6.12.1. In Theorem 6.12.4 the corresponding bound for $\| D_u(f - LS^{\rho}_{2,2}f) \|_{\infty}$ can be obtained by replacing t with u and h with ℓ. ∎

In view of the above considerations and Lemma 6.9.2 it is clear that results analogous to Lemmas 6.11.5 and 6.11.6 for the case $m = 3$ (which corresponds to the $LS^{\rho}_{3,4}f(t,u)$ interpolate) can be stated rather easily.

Theorem 6.12.5. Let $f(t,u) \in PC^{n,n,\infty}([a,b] \times [c,d])$, $4 \le n \le 6$. Then,

$$(6.12.10)\qquad \| D_t^k(f - LS^{\rho}_{3,4}f) \|_{\infty}$$

$$\le \gamma_{3,n,k}\, h^{n-k} \| D_t^n f \|_{\infty} + \gamma_{3,n,k}\gamma_{3,n,0}\, h^{n-k}\ell^n \| D_u^n D_t^n f \|_{\infty}$$

$$+ \gamma_{3,n,0}\, \ell^n \| D_u^n D_t^k f \|_{\infty}, \quad 0 \le k \le n-1$$

where the constants $\gamma_{3,n,k}$ are defined in Table 6.9.1.

Proof. The proof is similar to that of Theorem 5.6.2. ∎

Remark 6.12.2. In Theorem 6.12.5 the corresponding bounds for $\| D_u^k(f - LS^{\rho}_{3,4}f) \|_{\infty}$ can be obtained by replacing t with u and h with ℓ. ∎

Remark 6.12.3. As in Remarks 6.11.5 and 6.11.8 in Table 6.12.1 for $n = 6$ we compute the actual values of $\| D_t^k(f - LS^{\rho}_{3,4}f) \|_{\infty}$ and $\| D_u^k(f - LS^{\rho}_{3,4}f) \|_{\infty}$, $0 \le k \le 5$ for the same function $f(t,u) = t(1 - e^{tu}) \in C^{(\infty,\infty)}([0,1] \times [0,1])$ and use Theorem 6.12.5 to compare these with the corresponding right side bounds.

Table 6.12.1.

$N = M$	7	9	15
$\parallel f - LS^{\rho}_{3,4}f \parallel_\infty$	0.30053037 − 7	0.79267291 − 8	0.47702953 − 9
Bound	0.12022773 − 5	0.31510933 − 6	0.18780776 − 7
$\parallel D_t(f - LS^{\rho}_{3,4}f) \parallel_\infty$	0.84897714 − 6	0.28014313 − 6	0.26965530 − 7
Bound	0.29169307 − 4	0.94774178 − 5	0.89250457 − 6
$\parallel D_u(f - LS^{\rho}_{3,4}f) \parallel_\infty$	0.16362769 − 6	0.51461613 − 7	0.43901072 − 8
Bound	0.10463016 − 4	0.30031132 − 5	0.22576743 − 6
$\parallel D^2_t(f - LS^{\rho}_{3,4}f) \parallel_\infty$	0.22226736 − 4	0.91750474 − 5	0.14149493 − 5
Bound	0.63917381 − 3	0.26048616 − 3	0.39535400 − 4
$\parallel D^2_u(f - LS^{\rho}_{3,4}f) \parallel_\infty$	0.31517736 − 5	0.13044340 − 5	0.20175806 − 6
Bound	0.12465393 − 3	0.45928225 − 4	0.61665526 − 5
$\parallel D^3_t(f - LS^{\rho}_{3,4}f) \parallel_\infty$	0.71469950 − 3	0.36895889 − 3	0.91138015 − 4
Bound	0.18469208 − 1	0.94407284 − 2	0.23020948 − 2
$\parallel D^3_u(f - LS^{\rho}_{3,4}f) \parallel_\infty$	0.10240725 − 3	0.52814012 − 4	0.13031980 − 4
Bound	0.27924699 − 2	0.13884489 − 2	0.33121300 − 3
$\parallel D^4_t(f - LS^{\rho}_{3,4}f) \parallel_\infty$	0.29006293 − 1	0.18961716 − 1	0.76511253 − 2
Bound	0.28530337 + 0	0.18244285 + 0	0.71235952 − 1
$\parallel D^4_u(f - LS^{\rho}_{3,4}f) \parallel_\infty$	0.41865442 − 2	0.27311926 − 2	0.10986838 − 2
Bound	0.41396749 − 1	0.26223958 − 1	0.10185700 − 1
$\parallel D^5_t(f - LS^{\rho}_{3,4}f) \parallel_\infty$	0.12960879 + 1	0.10582995 + 1	0.68218151 + 0
Bound	0.51563344 + 1	0.41234140 + 1	0.25767384 + 1
$\parallel D^5_u(f - LS^{\rho}_{3,4}f) \parallel_\infty$	0.18694901 + 0	0.15235634 + 0	0.97926395 − 1
Bound	0.73973501 + 0	0.58980712 + 0	0.36814392 + 0

∎

6.13 SOME APPLICATIONS

Applications of spline interpolates abound in literature, to cite only a few, these have been used for the numerical differentiation and integration; for the construction of approximate solutions of ordinary, partial differential and integral equations; least squares problems; eigenvalue problems, and optimal control problems, e.g., [2,4-6,8,14,18,19]. For these problems the inequalities obtained in this and the previous chapters can be used to obtain a priori as well as posteriori error bounds between the exact and approximate solutions. As an example, here we shall provide the details only for the Fredholm type linear integral equations.

Linear Integral Equations

'Everyone knows' that the method of degenerate kernels for linear Fredholm integral equations of the second kind leads to an algebraic system of equations, and thus it provides approximations to the solutions of integral equations rather easily. Here, we shall use biquintic splines to degenerate the kernels, and as an application of the results established in Section 6.11 provide explicit a priori as well as posteriori error bounds between the exact and approximate solutions.

We shall consider the linear Fredholm integral equation of the second kind

$$(6.13.1) \qquad \phi(t) = \int_c^d f(t,u)\,\phi(u)\,du + \psi(t), \quad t \in [a,b]$$

where $f(t,u) \in C^{(2,2)}([a,b] \times [c,d])$ and $\psi(t) \in C[a,b]$.

In general the kernel $f(t,u)$ in (6.13.1) is not degenerate, i.e., it cannot be expressed as

$$f(t,u) = \sum_{\mu=1}^{\xi} \sum_{\nu=1}^{\eta} \beta_{\mu\nu} A_\mu(t) B_\nu(u),$$

however, we can approximate $f(t,u)$ by its biquintic spline interpolate, $S_{3,4}^{\rho} f(t, u)$, which in view of (6.11.23) is in degenerate form. With this approximation the resulting integral equation appears as

$$(6.13.2) \qquad \bar{\phi}(t) = \int_c^d S_{3,4}^{\rho} f(t,u)\,\bar{\phi}(u)\,du + \psi(t), \quad t \in [a,b]$$

which determines an approximate solution $\bar{\phi}(t)$.

For the simplicity of notation, we shall assume that the biquintic spline interpolate of $f(t,u)$ can be written as

$$S^{\rho}_{3,4}f(t,u) = \sum_{\mu=1}^{\xi}\sum_{\nu=1}^{\eta} \beta_{\mu\nu}A_\mu(t)B_\nu(u),$$

so that (6.13.2) takes the form

$$\bar{\phi}(t) = \sum_{\mu=1}^{\xi} A_\mu(t)\alpha_\mu + \psi(t), \quad t \in [a,b] \tag{6.13.3}$$

where

$$\alpha_\mu = \int_c^d \sum_{\nu=1}^{\eta} \beta_{\mu\nu}B_\nu(u)\bar{\phi}(u)\,du, \quad 1 \le \mu \le \xi. \tag{6.13.4}$$

From (6.13.3) the approximate solution $\bar{\phi}(t)$ can be obtained if we can determine the $\xi \times 1$ vector $\alpha = [\alpha_\mu]$. For this, we substitute (6.13.3) in (6.13.4), to obtain

$$\alpha_\mu = \sum_{\nu=1}^{\eta}\int_c^d \beta_{\mu\nu}B_\nu(u)\sum_{\mu=1}^{\xi}\alpha_\mu A_\mu(u)\,du + \sum_{\nu=1}^{\eta}\int_c^d \beta_{\mu\nu}B_\nu(u)\psi(u)\,du, \quad 1 \le \mu \le \xi \tag{6.13.5}$$

which in system form can be written as $\alpha = P\alpha + q$, or

$$(I-P)\alpha = q, \tag{6.13.6}$$

where $P = [p_{ij}]$ is an $\xi \times \xi$ matrix

$$p_{ij} = \sum_{\nu=1}^{\eta}\int_c^d \beta_{i\nu}B_\nu(u)A_j(u)\,du, \quad 1 \le i,j \le \xi \tag{6.13.7}$$

and $q = [q_i]$ is an $\xi \times 1$ vector

$$q_i = \sum_{\nu=1}^{\eta}\int_c^d \beta_{i\nu}B_\nu(u)\psi(u)\,du, \quad 1 \le i \le \xi. \tag{6.13.8}$$

It is clear from (6.13.3) and (6.13.6) that a unique $\bar{\phi}(t)$ exists if and only if the matrix $(I-P)$ is nonsingular. To provide sufficient conditions for the existence of a unique $\bar{\phi}(t)$, we introduce the operators R and S on $C[a,b]$ as follows

$$R[\phi] = \int_c^d f(t,u)\,\phi(u)\,du$$

and

$$S[\phi] = \int_c^d S^{\rho}_{3,4}f(t,u)\ \phi(u)\ du$$

so that (6.13.1) and (6.13.2) in operator form can be written as

$$(I - R)[\phi] = \psi \tag{6.13.9}$$

and

$$(I - S)[\bar{\phi}] = \psi, \tag{6.13.10}$$

respectively.

Definition 6.13.1. Let $T : C[a,b] \to C[a,b]$ be an operator defined by $T[\phi] = \psi$. We say that the operator T is *invertible* if $T[\phi] = \psi$ has a unique solution $\phi \in C[a,b]$ for each $\psi \in C[a,b]$.

Lemma 6.13.1. If $(I - R)$ is invertible and

$$\tau = (d - c) \parallel f - S^{\rho}_{3,4}f \parallel_\infty \parallel (I - R)^{-1} \parallel_\infty < 1 \tag{6.13.11}$$

then $(I - S)$ is invertible, i.e., (6.13.2) has a unique solution $\bar{\phi}(t)$.

Proof. We have to show that for all $\psi \in C[a,b]$, $(I - S)[\bar{\phi}] = \psi$ has a unique solution $\bar{\phi} \in C[a,b]$. For an arbitrary $\psi \in C[a,b]$, since $(I - R)$ is invertible, we let $g \in C[a,b]$ be the unique solution of $(I - R)[g] = \psi$. Clearly, we need only to consider

$$(I - R)^{-1}(I - S)[\bar{\phi}] = g \tag{6.13.12}$$

and to show that this has a unique solution $\bar{\phi} \in C[a,b]$.

For this, we define $W = (I - R)^{-1}(S - R)$ and observe that

$$\begin{aligned}(I - R)^{-1}(I - S) &= I - (I - R)^{-1}[(I - R) - (I - S)] \\ &= I - (I - R)^{-1}(S - R) = I - W.\end{aligned} \tag{6.13.13}$$

Further, since

$$(R - S)[\phi] = \int_c^d (f - S^{\rho}_{3,4}f)(t,u)\ \phi(u)\ du$$

we have

$$\| (R-S)[\phi] \|_\infty \le \| f - S^{\rho}_{3,4} f \|_\infty \| \phi \|_\infty (d-c).$$

Hence, $\| (R-S) \|_\infty \le (d-c) \| f - S^{\rho}_{3,4} f \|_\infty$, and so

$$\begin{aligned} \text{(6.13.14)} \quad \| W \|_\infty &= \| (I-R)^{-1}(S-R) \|_\infty \le \| (I-R)^{-1} \|_\infty \| S-R \|_\infty \\ &\le \| (I-R)^{-1} \|_\infty (d-c) \| f - S^{\rho}_{3,4} f \|_\infty \\ &= \tau < 1. \end{aligned}$$

Using (6.13.13) in (6.13.12), we find $(I-W)[\bar{\phi}] = g$, which is the same as

$$\bar{\phi} = W[\bar{\phi}] + g = V[\bar{\phi}]. \tag{6.13.15}$$

The operator V maps $C[a,b]$ into itself and for all $\bar{\phi}$, $\bar{\varphi} \in C[a,b]$ it follows from (6.13.15) and (6.3.14) that

$$\begin{aligned} \| V[\bar{\phi}] - V[\bar{\varphi}] \|_\infty &= \| W[\bar{\phi} - \bar{\varphi}] \|_\infty \le \| W \|_\infty \| \bar{\phi} - \bar{\varphi} \|_\infty \\ &\le \tau \| \bar{\phi} - \bar{\varphi} \|_\infty , \end{aligned}$$

where $\tau < 1$. Hence, V is a contraction mapping on $C[a,b]$. Thus, in conclusion (6.13.15) and consequently (6.13.12) has a unique solution $\bar{\phi} \in C[a,b]$. ∎

Remark 6.13.1. From Lemma 6.13.1 and an extension of Lemma 6.3.4 for bounded linear operators in Banach spaces, we can obtain a bound for $\| \bar{\phi} \|_\infty$ as follows

$$\begin{aligned} \| \bar{\phi} \|_\infty &\le \| (I-W)^{-1} \|_\infty \| g \|_\infty \\ &\le \| (I-W)^{-1} \|_\infty \| (I-R)^{-1} \|_\infty \| \psi \|_\infty \\ &\le \frac{1}{1 - \| W \|_\infty} \| (I-R)^{-1} \|_\infty \| \psi \|_\infty \\ &\le \frac{1}{1-\tau} \| (I-R)^{-1} \|_\infty \| \psi \|_\infty . \end{aligned}$$ ∎

Theorem 6.13.2. If $(I-R)$ is invertible and (6.13.11) holds, then

$$\| \phi - \bar{\phi} \|_\infty \le \frac{\tau}{1-\tau} \| \phi \|_\infty \tag{6.13.16}$$

and

$$\|\phi - \bar{\phi}\|_\infty \le \tau \|\bar{\phi}\|_\infty \quad . \tag{6.13.17}$$

(Inequality (6.13.16) gives a priori error bound, whereas (6.13.17) provides a posteriori error bound.)

Proof. To prove (6.13.16), we write

$$\begin{aligned}\bar{\phi} &= (I-S)^{-1}[\psi] = (I-S)^{-1}(I-R)[\phi] \\ &= (I-S)^{-1}[(I-S)+(S-R)][\phi] \\ &= [I+(I-S)^{-1}(S-R)][\phi] \\ &= \phi + (I-S)^{-1}(S-R)[\phi],\end{aligned}$$

which implies that

$$\|\phi - \bar{\phi}\|_\infty \le \|(I-S)^{-1}(S-R)\|_\infty \|\phi\|_\infty \quad . \tag{6.13.18}$$

Next since

$$(I-R)^{-1}(I-S) = I - (I-R)^{-1}(S-R)$$

or

$$(I-R)^{-1} = (I-S)^{-1} - (I-R)^{-1}(S-R)(I-S)^{-1}$$

or

$$(I-R)^{-1}(S-R) = (I-S)^{-1}(S-R) - (I-R)^{-1}(S-R)(I-S)^{-1}(S-R)$$

or

$$W = Z - WZ,$$

where $Z = (I-S)^{-1}(S-R)$, it follows that

$$\begin{aligned}\|W\|_\infty &\ge \|Z\|_\infty - \|WZ\|_\infty \\ &\ge \|Z\|_\infty - \|W\|_\infty \|Z\|_\infty = \|Z\|_\infty (1 - \|W\|_\infty).\end{aligned}$$

However, from (6.13.14), $\|W\|_\infty \le \tau < 1$ and hence we find that

$$\|Z\|_\infty = \|(I-S)^{-1}(S-R)\|_\infty \le \frac{\|W\|_\infty}{1-\|W\|_\infty} \le \frac{\tau}{1-\tau}. \tag{6.13.19}$$

Using (6.13.19) in (6.13.18) gives (6.13.16).

To show (6.13.17), we write

$$\begin{aligned}\phi &= (I-R)^{-1}[\psi] = (I-R)^{-1}(I-S)[\bar{\phi}] \\ &= (I-R)^{-1}[(I-R)+(R-S)][\bar{\phi}] \\ &= \bar{\phi} + (I-R)^{-1}(R-S)[\bar{\phi}] = \bar{\phi} - W[\bar{\phi}],\end{aligned}$$

which implies that

$$\| \phi - \bar{\phi} \|_\infty \le \| W \|_\infty \| \bar{\phi} \|_\infty \le \tau \| \bar{\phi} \|_\infty . \quad \blacksquare$$

From Theorems 6.11.7 - 6.11.9 the following corollaries are immediate.

Corollary 6.13.3. If $(I-R)$ is invertible, $f(t,u) \in PC^{4,\infty}([a,b]\times[c,d])$, and ρ is such that

$$\tau = (\alpha_{3,4,0}\, h^4 \| D_t^4 f \|_\infty + \alpha_{3,2,0}^2\, h^2\ell^2 \| D_u^2 D_t^2 f \|_\infty + \alpha_{3,4,0}\, \ell^4 \| D_u^4 f \|_\infty) \times$$
$$(d-c) \| (I-R)^{-1} \|_\infty < 1,$$

then $(I-S)$ is invertible. Moreover, (6.13.16) and (6.13.17) hold. ∎

Corollary 6.13.4. If $(I-R)$ is invertible, $f(t,u) \in PC^{5,\infty}([a,b]\times[c,d])$, and ρ is such that

$$\tau_1 = (\alpha_{3,5,0}\, h^5 \| D_t^5 f \|_\infty + \alpha_{3,2,0}\alpha_{3,3,0}\, h^3\ell^2 \| D_u^2 D_t^3 f \|_\infty$$
$$+ \alpha_{3,5,0}\, \ell^5 \| D_u^5 f \|_\infty)(d-c) \| (I-R)^{-1} \|_\infty < 1,$$

or

$$\tau_2 = (\alpha_{3,5,0}\, h^5 \| D_t^5 f \|_\infty + \alpha_{3,2,0}\alpha_{3,3,0}\, h^2\ell^3 \| D_u^3 D_t^2 f \|_\infty$$
$$+ \alpha_{3,5,0}\, \ell^5 \| D_u^5 f \|_\infty)(d-c) \| (I-R)^{-1} \|_\infty < 1,$$

then $(I-S)$ is invertible. Moreover, (6.13.16) and (6.13.17) with τ replaced by appropriate τ_1 or τ_2 hold. ∎

Corollary 6.13.5. If $(I-R)$ is invertible, $f(t,u) \in PC^{6,\infty}([a,b]\times[c,d])$, and ρ is such that

$$\tau_1 = (\alpha_{3,6,0}\, h^6 \parallel D_t^6 f \parallel_\infty + \alpha_{3,2,0}\alpha_{3,4,0}\, h^4\ell^2 \parallel D_u^2 D_t^4 f \parallel_\infty + \alpha_{3,6,0}\, \ell^6 \parallel D_u^6 f \parallel_\infty)(d-c) \parallel (I-R)^{-1} \parallel_\infty < 1,$$

or

$$\tau_2 = (\alpha_{3,6,0}\, h^6 \parallel D_t^6 f \parallel_\infty + \alpha_{3,3,0}^2\, h^3\ell^3 \parallel D_u^3 D_t^3 f \parallel_\infty + \alpha_{3,6,0}\, \ell^6 \parallel D_u^6 f \parallel_\infty) \times (d-c) \parallel (I-R)^{-1} \parallel_\infty < 1,$$

or

$$\tau_3 = (\alpha_{3,6,0}\, h^6 \parallel D_t^6 f \parallel_\infty + \alpha_{3,2,0}\alpha_{3,4,0}\, h^2\ell^4 \parallel D_u^4 D_t^2 f \parallel_\infty + \alpha_{3,6,0}\, \ell^6 \parallel D_u^6 f \parallel_\infty)(d-c) \parallel (I-R)^{-1} \parallel_\infty < 1,$$

then $(I-S)$ is invertible. Moreover, (6.13.16) and (6.13.17) with τ replaced by appropriate τ_1 or τ_2 or τ_3 hold. ∎

Example 6.13.1. Consider the integral equation

$$\phi(t) = \int_0^1 t(1-e^{tu})\, \phi(u)\, du + e^t - t, \quad t \in [0,1] \tag{6.13.20}$$

whose exact solution is known to be $\phi(t) \equiv 1$. In this equation the operator R is given by

$$R[\phi] = \int_0^1 t(1-e^{tu})\, \phi(u)\, du,$$

and hence

$$\begin{aligned} \parallel R[\phi] \parallel_\infty &\le \max_{0\le t\le 1} \int_0^1 \mid t(1-e^{tu}) \mid du \parallel \phi \parallel_\infty \\ &= \max_{0\le t\le 1} (e^t - t - 1) \parallel \phi \parallel_\infty \\ &= (e-2) \parallel \phi \parallel_\infty , \end{aligned}$$

which gives that $\parallel R \parallel_\infty \le (e-2) < 1$.

From the extension of Lemma 6.3.4 for bounded linear operators in Banach spaces, the term $\parallel (I-R)^{-1} \parallel_\infty$ in τ_3 (Corollary 6.13.5) can be replaced by $(3-e)^{-1}$. With this modification the two error bounds (6.13.16) and (6.13.17) will be larger. We shall find $\bar{\phi}(t)$ and the actual value of $\parallel \phi - \bar{\phi} \parallel_\infty$, and compare this with the modified error bounds.

Choosing N and M such that in Corollary 6.13.5, $\tau_3 < 1$. We first obtain the biquintic spline interpolate of the kernel $f(t,u) = t(1-e^{tu}) \in C^{(\infty,\infty)}([0,1]\times[0,1])$. For this, in view of Remark 6.11.3 we need only to construct the cardinal splines $s_i(t),\ 0 \le i \le N+5$ and $s_j(u),\ 0 \le j \le M+5$. From (6.4.19) we note that only the values of c_i' and c_i'' need to be computed for each cardinal spline as the explicit expressions of the functions $h_{3,i,j}(t)$ are known. Thus, we need to solve the systems (6.4.3) - (6.4.5) and (6.4.6) - (6.4.8). To find $\bar{\phi}(t)$ we need to solve the system (6.13.6) to obtain the vector α, then from (6.13.3) it follows that

$$\bar{\phi}(t) = \sum_{\mu=0}^{N+5} s_\mu(t)\alpha_\mu + e^t - t, \quad t \in [0,1].$$

In Table 6.13.1 we present the actual value of $\| \phi - \bar{\phi} \|_\infty = \max_{0\le i\le N} \max_{t_i \le t \le t_{i+1}} | 1 - \bar{\phi}(t) |$, and the two modified error bounds.

Table 6.13.1.

$N = M$	7	9	15
$\| \phi - \bar{\phi} \|_\infty$	$0.74488406-9$	$0.20377618-9$	$0.12562705-10$
$\dfrac{\tau_3}{1-\tau_3} \| \phi \|_\infty$	$0.95481824-3$	$0.25012365-3$	$0.14905025-4$
$\tau_3 \| \bar{\phi} \|_\infty$	$0.95390990-3$	$0.25006100-3$	$0.14904783-4$

∎

Remark 6.13.2. Throughout, in the above we can replace $S^{\rho}_{3,4}f$ by $S^{\rho}_{3,3}f$. With such a change τ defined in (6.13.11) will be denoted as τ^*. Inequalities (6.13.16) and (6.13.17) then hold with τ replaced by τ^*. Further, the Corollaries 6.13.3 - 6.13.5 remain valid with τ changed to τ^* and $\alpha_{3,n,k}$ replaced by the corresponding $\beta_{3,n,k}$, where $\beta_{3,n,k}$ are given in Table 6.6.1. ∎

Remark 6.13.3. Once again we consider the integral equation (6.13.20) and choose N and M such that $\tau^* < 1$. As in Example 6.13.1 to obtain the biquintic spline interpolate $S^{\rho}_{3,3}f$ of the kernel $f(t,u) = t(1-e^{tu})$, in view of Remark 6.11.6, we need only to construct the cardinal splines $s_i(t)$, $0 \le i \le 2N+5$ and $s_j(u)$, $0 \le j \le 2M+5$. For this, from (6.6.5) we note that only the values of c''_i need to be computed for each cardinal spline as the explicit expressions of the functions $h_{3,i,j}(t)$ are known. Thus, we need to solve the system (6.6.1). To find $\bar{\phi}(t)$ we need to solve the system (6.13.6) to obtain the vector α, then from (6.13.3) it follows that

$$\bar{\phi}(t) = \sum_{\mu=0}^{2N+5} s_\mu(t)\alpha_\mu + e^t - t, \quad t \in [0,1].$$

Once again as in Example 6.13.1 in Table 6.13.2 we present the actual value of $\| \phi - \bar{\phi} \|_\infty = \max_{0\le i\le N} \max_{t_i\le t\le t_{i+1}} | 1 - \bar{\phi}(t) |$, and the two corresponding modified error bounds.

Table 6.13.2.

$N = M$	7	9	15
$\| \phi - \bar{\phi} \|_\infty$	0.47280278 – 9	0.13057944 – 9	0.85620677 – 11
$\frac{\tau_3^*}{1-\tau_3^*} \| \phi \|_\infty$	0.54301944 – 6	0.14234923 – 6	0.84846739 – 8
$\tau_3^* \| \bar{\phi} \|_\infty$	0.54301914 – 6	0.14234921 – 6	0.84846738 – 8

■

REFERENCES

1. **R.P.Agarwal** and **P.J.Y.Wong**, Explicit error bounds for the derivatives of spline interpolation in L_2 - norm, to appear.

2. **J.H.Ahlberg, E.N.Nilson** and **J.L.Walsh**, The Theory of Splines and their Applications, Academic Press, New York, 1967.

3. **G.Birkhoff** and **C.DeBoor,** Error bounds for spline interpolation, J. Math. Mech. 13(1964), 827-836.

4. **G.Birkhoff, M.H.Schultz** and **R.S.Varga,** Piecewise Hermite interpolation in one and two variables with applications to partial differential equations, Numerische Mathematik 11(1968), 232-256.

5. **P.G.Ciarlet, M.H.Schultz** and **R.S.Varga,** Numerical methods of high order accuracy for nonlinear boundary value problems, I. one dimensional problem, Numerische Mathematik 9(1967), 394-430.

6. **P.G.Ciarlet, M.H.Schultz** and **R.S.Varga,** Numerical methods of high order accuracy for nonlinear boundary value problems, II. nonlinear boundary conditions, Numerische Mathematik 11(1968), 331-345.

7. **C.A.Hall,** On error bounds for spline interpolation, J. Approximation Theory 1(1968), 209-218.

8. **M.Isa** and **R.A.Usmani,** Quintic spline solution of a boundary value problem, Intern. J. Computer Math. 11(1982), 169-184.

9. **T.R.Lucas,** A generalization of L - splines, Numerische Mathematik 15(1970), 359-370.

10. **I.J.Schoenberg,** Contributions to the problem of approximation of equidistant data by analytic functions, parts A and B. Quart. Appl. Math. 4(1946), 45-99, 112-141.

11. **M.H.Schultz** and **R.S.Varga,** L - splines, Numerische Mathematik 10(1967), 345-369.

12. **M.H.Schultz,** Error bounds for polynomial spline interpolation, Mathematics of Computation 24(1970), 507-515.

13. **M.H.Schultz,** Error bounds for a bivariate interpolation scheme, J. Approximation Theory 8(1973), 189-194.

14. **M.H.Schultz,** Spline Analysis, Prentice-Hall, Englewood Cliffs, N.J., 1973.

15. **L.Schumaker,** Spline Functions : Basic Theory, John Wiley, New York, 1981.

16. **B.K.Swartz,** $O(h^{2n+2-\ell})$ bounds on some spline interpolation errors, Bull. Amer. Math. Soc. 74(1968), 1072-1078.

17. **B.K.Swartz** and **R.S.Varga,** Error bounds for spline and L - spline interpolation, J. Approximation Theory 6(1972), 6-49.

18. **R.A.Usmani** and **S.A.Warsi,** Quintic spline solutions of boundary value problems, Computers Math. Applic. 6(1980), 197-203.

19. **R.A.Usmani** and **S.A.Warsi,** Smooth spline solutions for boundary value problems in plate deflection theory, Computers Math. Applic. 6(1980), 205-211.

20. **R.A.Usmani,** Applied Linear Algebra, Marcel Dekker, Inc. 1987.

21. **P.J.Y.Wong** and **R.P.Agarwal,** Explicit error estimates for quintic and biquintic spline interpolation, Computers Math. Applic. 18(1989), 701-722.

22. **P.J.Y.Wong** and **R.P.Agarwal**, Quintic spline solutions of Fredholm integral equations of the second kind, Intern. J. Computer Math. 33 (1990), 237-249.

23. **P.J.Y.Wong** and **R.P.Agarwal**, Explicit error estimates for quintic and biquintic spline interpolation II, to appear.

24. **P.J.Y.Wong** and **R.P.Agarwal**, Sharp error bounds for the derivatives of Lidstone - spline interpolation , to appear.

Name Index

Agarwal R P 21, 30, 59, 102, 105, 124, 168, 171, 183, 186, 189, 190, 191, 198, 216, 225, 278, 280, 360, 361, 362
Ahlberg J H 247, 278, 282, 283, 351, 360
Akrivis G 21, 59
Anon S 104, 168
Baldwin P 21, 59
Beckenbach E 220, 256, 278
Beesack P R 75, 168
Bellman R 220, 225, 256, 278
Berezin I S 62, 169
Bessmertnyh G A 105, 124, 167, 169
Birkhoff G 125, 169, 217, 278, 282, 351, 360
Boas R P 1, 2, 14, 59
Borg G 161, 169
Boutayeb A 21, 59, 61
Brink J 102, 105, 169, 278
Chawla M M 21, 60
Chow S N 149, 169
Ciarlet P G 115, 169, 217, 279, 351, 360
Cimmino G 221, 279
Coppel W A 75, 169, 186, 190
Courant R 163, 169
Das K M 98, 163, 169, 170, 222, 279
Davis P J 1, 60, 62, 170, 172, 190
DeBoor C 282, 360
Dörfler P 224, 279
Dunninger D R 149, 169
Ehme J 186, 190
Elias U 186, 190
Eloe P W 186, 190
Erbe L 198, 216
Fink A M 279
Forster P 222, 279
Fort T 9, 60
Gautschi W 119, 170
Goetgheluck P 225, 279
Gontscharoff V L 172, 190
Gupta R C 225, 278

Hall C A 282, 288, 311, 360
Hankerson D 186, 190
Hardy G H 220, 279
Hartman P 162, 170
Henderson J 186, 190
Hilbert D 163, 169
Hille E 225, 279
Hinton D B 163, 170
Howell G 1, 61, 203, 216
Hukuhara M 105, 124, 170
Isa M 351, 360
Jahn J R 21, 60
Jordan C 6, 9, 10, 13, 60
Katsifarakis K L 234, 280
Katti C P 21, 60
Kobelkov Y P 105, 170
Kobyakov I I 105, 170
Kuttler J R 149, 170
Lasota A 149, 169
Latour J 21, 60
Levin A Ju 75, 105, 124, 164, 167, 169, 170, 186, 190, 198, 216
Lidstone G J 1, 60
Littlewood J E 220, 279
Lucas T R 282, 360
Luke Y L 4, 8, 60
MacRobert T M 145, 170
Milne-Thomson L M 7, 60
Milovanovic G V 102, 170
Mitrinovič D S 104, 171, 279
Muldowney J 186, 190
Nehari Z 76, 171, 186, 191
Nilson E N 247, 278, 282, 283, 351, 360
Noor Muhammad Aslam 21, 60
Pécarić J E 102, 170, 279
Peterson A C 186, 191, 198, 216
Phillips G M 96, 171
Pokornyi Yu V 75, 171
Polya G 220, 279
Poritsky H 1, 60
Priver A 125, 169
Protter M H 149, 171
Schmidt E 223, 279
Schoenberg L J 1, 60, 282, 361
Schultz M H 115, 169, 217, 218, 278, 279, 280, 282, 288, 351, 360, 361

Schumaker L 282, 283, 361
Šeda V 150, 171
Sheng Qin 171, 186, 190
Spiegel E A 21, 60
Swartz B 282, 361
Szegö G 116, 171, 225, 279
Szidarovszky F 95, 171
Tamarkin J D 225, 279
Taylor P J 96, 171
Tirmizi S I 21, 60, 61
Toomre J 21, 60
Torchinsky A 104, 171
Tumura M 105, 124, 171
Twizell E H 21, 59, 61
Umamaheswaram S 186, 191
Usmani R A 21, 61, 186, 189, 288, 351, 360, 361
Varga R S 115, 169, 217, 278, 279, 282, 351, 360, 361
Varma A K 1, 61, 203, 216, 234, 280
Vatsala A S 98, 163, 169, 170, 222, 279
Venkata Rama M 186, 191
Walsh J L 247, 278, 282, 283, 351, 360
Warsi S A 351, 361
Weinberger H F 149, 171
Whittaker J M 1, 5, 61
Widder D V 1, 2, 11, 12, 61
Wong P J Y 21, 59, 168, 171, 186, 190, 191, 216, 278, 280, 360, 361, 362
Yakowitz S 95, 171
Zaidman S 104, 171
Zhidkov N P 62, 169

Other *Mathematics and Its Applications* titles of interest:

A.M. Samoilenko: *Elements of the Mathematical Theory of Multi-Frequency Oscillations*. 1991, 314 pp. ISBN 0-7923-1438-7

Yu.L. Dalecky and S.V. Fomin: *Measures and Differential Equations in Infinite-Dimensional Space*. 1991, 338 pp. ISBN 0-7923-1517-0

W. Mlak: *Hilbert Space and Operator Theory*. 1991, 296 pp. ISBN 0-7923-1042-X

N.J. Vilenkin and A.U. Klimyk: *Representations of Lie Groups and Special Functions. Volume 1: Simplest Lie Groups, Special Functions, and Integral Transforms*. 1991, 608 pp. ISBN 0-7923-1466-2

K. Gopalsamy: *Stability and Oscillations in Delay Differential Equations of Population Dynamics*. 1992, 502 pp. ISBN 0-7923-1594-4

N.M. Korobov: *Exponential Sums and their Applications*. 1992, 210 pp. ISBN 0-7923-1647-9

Chuang-Gan Hu and Chung-Chun Yang: *Vector-Valued Functions and their Applications*. 1991, 172 pp. ISBN 0-7923-1605-3

Z. Szmydt and B. Ziemian: *The Mellin Transformation and Fuchsian Type Partial Differential Equations*. 1992, 224 pp. ISBN 0-7923-1683-5

L.I. Ronkin: *Functions of Completely Regular Growth*. 1992, 394 pp. ISBN 0-7923-1677-0

R. Delanghe, F. Sommen and V. Soucek: *Clifford Algebra and Spinor-valued Functions. A Function Theory of the Dirac Operator*. 1992, 486 pp. ISBN 0-7923-0229-X

A. Tempelman: *Ergodic Theorems for Group Actions*. 1992, 400 pp. ISBN 0-7923-1717-3

D. Bainov and P. Simenov: *Integral Inequalities and Applications*. 1992, 426 pp. ISBN 0-7923-1714-9

I. Imai: *Applied Hyperfunction Theory*. 1992, 460 pp. ISBN 0-7923-1507-3

Yu.I. Neimark and P.S. Landa: *Stochastic and Chaotic Oscillations*. 1992, 502 pp. ISBN 0-7923-1530-8

H.M. Srivastava and R.G. Buschman: *Theory and Applications of Convolution Integral Equations*. 1992, 240 pp. ISBN 0-7923-1891-9

A. van der Burgh and J. Simonis (eds.): *Topics in Engineering Mathematics*. 1992, 266 pp. ISBN 0-7923-2005-3

F. Neuman: *Global Properties of Linear Ordinary Differential Equations*. 1992, 320 pp. ISBN 0-7923-1269-4

A. Dvurecenskij: *Gleason's Theorem and its Applications*. 1992, 334 pp. ISBN 0-7923-1990-7

Zeitfracht Medien GmbH
Ferdinand-Jühlke-Straße 7
99095 Erfurt, Deutschland
produktsicherheit@kolibri360.de